State-Space Methods for Time Series Analysis

Theory, Applications and Software

1. Stochastic Population Models in Ecology and Epidemiology *M.S. Barlett* (1960)
2. Queues *D.R. Cox and W.L. Smith* (1961)
3. Monte Carlo Methods *J.M. Hammersley and D.C. Handscomb* (1964)
4. The Statistical Analysis of Series of Events *D.R. Cox and P.A.W. Lewis* (1966)
5. Population Genetics *W.J. Ewens* (1969)
6. Probability, Statistics and Time *M.S. Barlett* (1975)
7. Statistical Inference *S.D. Silvey* (1975)
8. The Analysis of Contingency Tables *B.S. Everitt* (1977)
9. Multivariate Analysis in Behavioural Research *A.E. Maxwell* (1977)
10. Stochastic Abundance Models *S. Engen* (1978)
11. Some Basic Theory for Statistical Inference *E.J.G. Pitman* (1979)
12. Point Processes *D.R. Cox and V. Isham* (1980)
13. Identification of Outliers *D.M. Hawkins* (1980)
14. Optimal Design *S.D. Silvey* (1980)
15. Finite Mixture Distributions *B.S. Everitt and D.J. Hand* (1981)
16. Classification *A.D. Gordon* (1981)
17. Distribution-Free Statistical Methods, 2nd edition *J.S. Maritz* (1995)
18. Residuals and Influence in Regression *R.D. Cook and S. Weisberg* (1982)
19. Applications of Queueing Theory, 2nd edition *G.F. Newell* (1982)
20. Risk Theory, 3rd edition *R.E. Beard, T. Pentikäinen and E. Pesonen* (1984)
21. Analysis of Survival Data *D.R. Cox and D. Oakes* (1984)
22. An Introduction to Latent Variable Models *B.S. Everitt* (1984)
23. Bandit Problems *D.A. Berry and B. Fristedt* (1985)
24. Stochastic Modelling and Control *M.H.A. Davis and R. Vinter* (1985)
25. The Statistical Analysis of Composition Data *J. Aitchison* (1986)
26. Density Estimation for Statistics and Data Analysis *B.W. Silverman* (1986)
27. Regression Analysis with Applications *G.B. Wetherill* (1986)
28. Sequential Methods in Statistics, 3rd edition *G.B. Wetherill and K.D. Glazebrook* (1986)
29. Tensor Methods in Statistics *P. McCullagh* (1987)
30. Transformation and Weighting in Regression *R.J. Carroll and D. Ruppert* (1988)
31. Asymptotic Techniques for Use in Statistics *O.E. Bandorff-Nielsen and D.R. Cox* (1989)
32. Analysis of Binary Data, 2nd edition *D.R. Cox and E.J. Snell* (1989)
33. Analysis of Infectious Disease Data *N.G. Becker* (1989)
34. Design and Analysis of Cross-Over Trials *B. Jones and M.G. Kenward* (1989)
35. Empirical Bayes Methods, 2nd edition *J.S. Maritz and T. Lwin* (1989)
36. Symmetric Multivariate and Related Distributions *K.T. Fang, S. Kotz and K.W. Ng* (1990)
37. Generalized Linear Models, 2nd edition *P. McCullagh and J.A. Nelder* (1989)
38. Cyclic and Computer Generated Designs, 2nd edition *J.A. John and E.R. Williams* (1995)
39. Analog Estimation Methods in Econometrics *C.F. Manski* (1988)
40. Subset Selection in Regression *A.J. Miller* (1990)
41. Analysis of Repeated Measures *M.J. Crowder and D.J. Hand* (1990)
42. Statistical Reasoning with Imprecise Probabilities *P. Walley* (1991)
43. Generalized Additive Models *T.J. Hastie and R.J. Tibshirani* (1990)
44. Inspection Errors for Attributes in Quality Control *N.L. Johnson, S. Kotz and X. Wu* (1991)
45. The Analysis of Contingency Tables, 2nd edition *B.S. Everitt* (1992)
46. The Analysis of Quantal Response Data *B.J.T. Morgan* (1992)
47. Longitudinal Data with Serial Correlation—A State-Space Approach *R.H. Jones* (1993)

48. Differential Geometry and Statistics *M.K. Murray and J.W. Rice* (1993)
49. Markov Models and Optimization *M.H.A. Davis* (1993)
50. Networks and Chaos—Statistical and Probabilistic Aspects
O.E. Barndorff-Nielsen, J.L. Jensen and W.S. Kendall (1993)
51. Number-Theoretic Methods in Statistics *K.-T. Fang and Y. Wang* (1994)
52. Inference and Asymptotics *O.E. Barndorff-Nielsen and D.R. Cox* (1994)
53. Practical Risk Theory for Actuaries *C.D. Daykin, T. Pentikäinen and M. Pesonen* (1994)
54. Biplots *J.C. Gower and D.J. Hand* (1996)
55. Predictive Inference—An Introduction *S. Geisser* (1993)
56. Model-Free Curve Estimation *M.E. Tarter and M.D. Lock* (1993)
57. An Introduction to the Bootstrap *B. Efron and R.J. Tibshirani* (1993)
58. Nonparametric Regression and Generalized Linear Models *P.J. Green and B.W. Silverman* (1994)
59. Multidimensional Scaling *T.F. Cox and M.A.A. Cox* (1994)
60. Kernel Smoothing *M.P. Wand and M.C. Jones* (1995)
61. Statistics for Long Memory Processes *J. Beran* (1995)
62. Nonlinear Models for Repeated Measurement Data *M. Davidian and D.M. Giltinan* (1995)
63. Measurement Error in Nonlinear Models *R.J. Carroll, D. Rupert and L.A. Stefanski* (1995)
64. Analyzing and Modeling Rank Data *J.J. Marden* (1995)
65. Time Series Models—In Econometrics, Finance and Other Fields
D.R. Cox, D.V. Hinkley and O.E. Barndorff-Nielsen (1996)
66. Local Polynomial Modeling and its Applications *J. Fan and I. Gijbels* (1996)
67. Multivariate Dependencies—Models, Analysis and Interpretation *D.R. Cox and N. Wermuth* (1996)
68. Statistical Inference—Based on the Likelihood *A. Azzalini* (1996)
69. Bayes and Empirical Bayes Methods for Data Analysis *B.P. Carlin and T.A Louis* (1996)
70. Hidden Markov and Other Models for Discrete-Valued Time Series *I.L. MacDonald and W. Zucchini* (1997)
71. Statistical Evidence—A Likelihood Paradigm *R. Royall* (1997)
72. Analysis of Incomplete Multivariate Data *J.L. Schafer* (1997)
73. Multivariate Models and Dependence Concepts *H. Joe* (1997)
74. Theory of Sample Surveys *M.E. Thompson* (1997)
75. Retrial Queues *G. Falin and J.G.C. Templeton* (1997)
76. Theory of Dispersion Models *B. Jørgensen* (1997)
77. Mixed Poisson Processes *J. Grandell* (1997)
78. Variance Components Estimation—Mixed Models, Methodologies and Applications *P.S.R.S. Rao* (1997)
79. Bayesian Methods for Finite Population Sampling *G. Meeden and M. Ghosh* (1997)
80. Stochastic Geometry—Likelihood and computation
O.E. Barndorff-Nielsen, W.S. Kendall and M.N.M. van Lieshout (1998)
81. Computer-Assisted Analysis of Mixtures and Applications—Meta-Analysis, Disease Mapping and Others
D. Böhning (1999)
82. Classification, 2nd edition *A.D. Gordon* (1999)
83. Semimartingales and their Statistical Inference *B.L.S. Prakasa Rao* (1999)
84. Statistical Aspects of BSE and vCJD—Models for Epidemics *C.A. Donnelly and N.M. Ferguson* (1999)
85. Set-Indexed Martingales *G. Ivanoff and E. Merzbach* (2000)
86. The Theory of the Design of Experiments *D.R. Cox and N. Reid* (2000)
87. Complex Stochastic Systems *O.E. Barndorff-Nielsen, D.R. Cox and C. Klüppelberg* (2001)
88. Multidimensional Scaling, 2nd edition *T.F. Cox and M.A.A. Cox* (2001)
89. Algebraic Statistics—Computational Commutative Algebra in Statistics
G. Pistone, E. Riccomagno and H.P. Wynn (2001)
90. Analysis of Time Series Structure—SSA and Related Techniques
N. Golyandina, V. Nekrutkin and A.A. Zhigljavsky (2001)
91. Subjective Probability Models for Lifetimes *Fabio Spizzichino* (2001)
92. Empirical Likelihood *Art B. Owen* (2001)
93. Statistics in the 21st Century *Adrian E. Raftery, Martin A. Tanner, and Martin T. Wells* (2001)
94. Accelerated Life Models: Modeling and Statistical Analysis
Vilijandas Bagdonavicius and Mikhail Nikulin (2001)

95. Subset Selection in Regression, Second Edition *Alan Miller* (2002)
96. Topics in Modelling of Clustered Data *Marc Aerts, Helena Geys, Geert Molenberghs, and Louise M. Ryan* (2002)
97. Components of Variance *D.R. Cox and P.J. Solomon* (2002)
98. Design and Analysis of Cross-Over Trials, 2nd Edition *Byron Jones and Michael G. Kenward* (2003)
99. Extreme Values in Finance, Telecommunications, and the Environment *Bärbel Finkenstädt and Holger Rootzén* (2003)
100. Statistical Inference and Simulation for Spatial Point Processes *Jesper Møller and Rasmus Plenge Waagepetersen* (2004)
101. Hierarchical Modeling and Analysis for Spatial Data *Sudipto Banerjee, Bradley P. Carlin, and Alan E. Gelfand* (2004)
102. Diagnostic Checks in Time Series *Wai Keung Li* (2004)
103. Stereology for Statisticians *Adrian Baddeley and Eva B. Vedel Jensen* (2004)
104. Gaussian Markov Random Fields: Theory and Applications *Håvard Rue and Leonhard Held* (2005)
105. Measurement Error in Nonlinear Models: A Modern Perspective, Second Edition *Raymond J. Carroll, David Ruppert, Leonard A. Stefanski, and Ciprian M. Crainiceanu* (2006)
106. Generalized Linear Models with Random Effects: Unified Analysis via H-likelihood *Youngjo Lee, John A. Nelder, and Yudi Pawitan* (2006)
107. Statistical Methods for Spatio-Temporal Systems *Bärbel Finkenstädt, Leonhard Held, and Valerie Isham* (2007)
108. Nonlinear Time Series: Semiparametric and Nonparametric Methods *Jiti Gao* (2007)
109. Missing Data in Longitudinal Studies: Strategies for Bayesian Modeling and Sensitivity Analysis *Michael J. Daniels and Joseph W. Hogan* (2008)
110. Hidden Markov Models for Time Series: An Introduction Using R *Walter Zucchini and Iain L. MacDonald* (2009)
111. ROC Curves for Continuous Data *Wojtek J. Krzanowski and David J. Hand* (2009)
112. Antedependence Models for Longitudinal Data *Dale L. Zimmerman and Vicente A. Núñez-Antón* (2009)
113. Mixed Effects Models for Complex Data *Lang Wu* (2010)
114. Intoduction to Time Series Modeling *Genshiro Kitagawa* (2010)
115. Expansions and Asymptotics for Statistics *Christopher G. Small* (2010)
116. Statistical Inference: An Integrated Bayesian/Likelihood Approach *Murray Aitkin* (2010)
117. Circular and Linear Regression: Fitting Circles and Lines by Least Squares *Nikolai Chernov* (2010)
118. Simultaneous Inference in Regression *Wei Liu* (2010)
119. Robust Nonparametric Statistical Methods, Second Edition *Thomas P. Hettmansperger and Joseph W. McKean* (2011)
120. Statistical Inference: The Minimum Distance Approach *Ayanendranath Basu, Hiroyuki Shioya, and Chanseok Park* (2011)
121. Smoothing Splines: Methods and Applications *Yuedong Wang* (2011)
122. Extreme Value Methods with Applications to Finance *Serguei Y. Novak* (2012)
123. Dynamic Prediction in Clinical Survival Analysis *Hans C. van Houwelingen and Hein Putter* (2012)
124. Statistical Methods for Stochastic Differential Equations *Mathieu Kessler, Alexander Lindner, and Michael Sørensen* (2012)
125. Maximum Likelihood Estimation for Sample Surveys *R. L. Chambers, D. G. Steel, Suojin Wang, and A. H. Welsh* (2012)
126. Mean Field Simulation for Monte Carlo Integration *Pierre Del Moral* (2013)
127. Analysis of Variance for Functional Data *Jin-Ting Zhang* (2013)
128. Statistical Analysis of Spatial and Spatio-Temporal Point Patterns, Third Edition *Peter J. Diggle* (2013)
129. Constrained Principal Component Analysis and Related Techniques *Yoshio Takane* (2014)
130. Randomised Response-Adaptive Designs in Clinical Trials *Anthony C. Atkinson and Atanu Biswas* (2014)
131. Theory of Factorial Design: Single- and Multi-Stratum Experiments *Ching-Shui Cheng* (2014)
132. Quasi-Least Squares Regression *Justine Shults and Joseph M. Hilbe* (2014)
133. Data Analysis and Approximate Models: Model Choice, Location-Scale, Analysis of Variance, Nonparametric Regression and Image Analysis *Laurie Davies* (2014)
134. Dependence Modeling with Copulas *Harry Joe* (2014)
135. Hierarchical Modeling and Analysis for Spatial Data, Second Edition *Sudipto Banerjee, Bradley P. Carlin, and Alan E. Gelfand* (2014)

136. Sequential Analysis: Hypothesis Testing and Changepoint Detection *Alexander Tartakovsky, Igor Nikiforov, and Michèle Basseville* (2015)

137. Robust Cluster Analysis and Variable Selection *Gunter Ritter* (2015)

138. Design and Analysis of Cross-Over Trials, Third Edition *Byron Jones and Michael G. Kenward* (2015)

139. Introduction to High-Dimensional Statistics *Christophe Giraud* (2015)

140. Pareto Distributions: Second Edition *Barry C. Arnold* (2015)

141. Bayesian Inference for Partially Identified Models: Exploring the Limits of Limited Data *Paul Gustafson* (2015)

142. Models for Dependent Time Series *Granville Tunnicliffe Wilson, Marco Reale, John Haywood* (2015)

143. Statistical Learning with Sparsity: The Lasso and Generalizations *Trevor Hastie, Robert Tibshirani, and Martin Wainwright* (2015)

144. Measuring Statistical Evidence Using Relative Belief *Michael Evans* (2015)

145. Stochastic Analysis for Gaussian Random Processes and Fields: With Applications *Vidyadhar S. Mandrekar and Leszek Gawarecki* (2015)

146. Semialgebraic Statistics and Latent Tree Models *Piotr Zwiernik* (2015)

147. Inferential Models: Reasoning with Uncertainty *Ryan Martin and Chuanhai Liu* (2016)

148. Perfect Simulation *Mark L. Huber* (2016)

149. State-Space Methods for Time Series Analysis: Theory, Applications and Software *Jose Casals, Alfredo Garcia-Hiernaux, Miguel Jerez, Sonia Sotoca, and A. Alexandre Trindade* (2016)

Monographs on Statistics and Applied Probability 149

State-Space Methods for Time Series Analysis

Theory, Applications and Software

Jose Casals
Universidad Complutense de Madrid, Spain

Alfredo Garcia-Hiernaux
Universidad Complutense de Madrid, Spain

Miguel Jerez
Universidad Complutense de Madrid, Spain

Sonia Sotoca
Universidad Complutense de Madrid, Spain

A. Alexandre Trindade
Texas Tech University, Lubbock, U.S.A.

CRC Press
Taylor & Francis Group
Boca Raton London New York

CRC Press is an imprint of the
Taylor & Francis Group, an **informa** business
A CHAPMAN & HALL BOOK

CRC Press
Taylor & Francis Group
6000 Broken Sound Parkway NW, Suite 300
Boca Raton, FL 33487-2742

First issued in paperback 2020

Version Date: 20160211

ISBN 13: 978-0-3675-7058-3 (pbk)
ISBN 13: 978-1-4822-1959-3 (hbk)

Visit the Taylor & Francis Web site at
http://www.taylorandfrancis.com

and the CRC Press Web site at
http://www.crcpress.com

A wise man once said that books by young writers should be dedicated to their parents, while mature authors must bestow this honor on their children.
Sadly this book is dedicated, with love, to our children.

Contents

List of Figures xvii

List of Tables xix

Preface xxiii

About the Authors xxvii

1 Introduction 1

2 Linear State-Space models 5
- 2.1 The multiple error model 6
  - 2.1.1 Model formulation 6
  - 2.1.2 Similar transformations 8
  - 2.1.3 Properties of the State-Space model 8
- 2.2 Single error models 12
  - 2.2.1 Model formulation 12
  - 2.2.2 State estimation in the SEM 12
  - 2.2.3 Obtaining the SEM equivalent to a MEM 13
  - 2.2.4 Obtaining the SEM equivalent to a general linear process 15

3 Model transformations 19
- 3.1 Model decomposition 19
  - 3.1.1 Deterministic and stochastic subsystems 19
  - 3.1.2 Implied univariate models 21
  - 3.1.3 Block-diagonal forms 22
- 3.2 Model combination 24
  - 3.2.1 Endogeneization of stochastic inputs 24
  - 3.2.2 Colored errors 26
- 3.3 Change of variables in the output 28
  - 3.3.1 Observation errors 28
  - 3.3.2 Missing values and aggregated data 29
  - 3.3.3 Temporal aggregation 30
- 3.4 Uses of these transformations 31

4 Filtering and Smoothing 33
4.1 The conditional moments of a State-Space model 33
4.2 The Kalman Filter 34
4.3 Decomposition of the smoothed moments 36
4.4 Smoothing for a general State-Space model 37
4.5 Smoothing for fixed-coefficients and single-error models 39
4.6 Uncertainty of the smoothed estimates in a SEM 40
4.7 Examples 42
4.7.1 Recursive Least Squares 42
4.7.2 Cleaning the Wolf sunspot series 42
4.7.3 Extracting the Hodrick-Prescott trend 44

5 Likelihood computation for fixed-coefficients models 45
5.1 Maximum likelihood estimation 45
5.1.1 Problem statement 45
5.1.2 Prediction error decomposition 46
5.1.3 Initialization of the Kalman filter in the stationary case 47
5.2 The likelihood for a nonstationary model 48
5.2.1 Diffuse likelihood 49
5.2.2 Minimally conditioned likelihood 50
5.2.3 Likelihood computation for a fixed-parameter SEM 52
5.2.4 Initialization of the Kalman filter in the nonstationary case 52
5.3 The likelihood for a model with inputs 54
5.3.1 Models with deterministic inputs 54
5.3.2 Models with stochastic inputs 55
5.4 Examples 57
5.4.1 Models for the Airline Passenger series 57
5.4.2 Modeling the series of Housing Starts and Sales 60

6 The likelihood of models with varying parameters 65
6.1 Regression with time-varying parameters 66
6.1.1 SS formulation 66
6.1.2 Maximum likelihood estimation 67
6.2 Periodic models 68
6.2.1 All the seasons have the same dynamic structure 69
6.2.2 The *s* models do not have the same dynamic structure 70
6.2.3 Stationarity and invertibility 71
6.2.4 Maximum likelihood estimation 72
6.3 The likelihood of models with GARCH errors 74
6.4 Examples 76
6.4.1 A time-varying CAPM regression 76
6.4.2 A periodic model for West German Consumption 78
6.4.3 A model with vector-GARCH errors for two exchange rate series 79

7 Subspace methods **83**
7.1 Theoretical foundations 83
7.1.1 Subspace structure and notation 84
7.1.2 Assumptions, projections and model reduction 86
7.1.3 Estimating the system matrices 87
7.2 System order estimation 89
7.2.1 Preliminary data analysis methods 89
7.2.2 Model comparison methods 91
7.3 Constrained estimation 92
7.3.1 State sequence structure 92
7.3.2 Subspace-based likelihood 93
7.4 Multiplicative seasonal models 94
7.5 Examples 95
7.5.1 Univariate models 95
7.5.2 A multivariate model for the interest rates 99

8 Signal extraction **107**
8.1 Input and error-related components 108
8.1.1 The deterministic and stochastic subsystems 108
8.1.2 Enforcing minimality 109
8.2 Estimation of the deterministic components 110
8.2.1 Estimating the total effect of the inputs 111
8.2.2 Estimating the individual effect of each input 113
8.3 Decomposition of the stochastic component 114
8.3.1 Characterization of the structural components 114
8.3.2 Estimation of the structural components 116
8.4 Structure of the method 116
8.5 Examples 116
8.5.1 Comparing different methods with simulated data 116
8.5.2 Common features in wheat prices 120
8.5.3 The effect of advertising on sales 123

9 The VARMAX representation of a State-Space model **129**
9.1 Notation and previous results 130
9.2 Obtaining the VARMAX form of a State-Space model 131
9.2.1 From State-Space to standard VARMAX 132
9.2.2 From State-Space to canonical VARMAX 133
9.3 Practical applications and examples 135
9.3.1 The VARMAX form of some common State-Space models 135
9.3.2 Identifiability and conditioning of the estimates 135
9.3.3 Fitting an errors-in-variables model to Wolf's sunspot series 139
9.3.4 "Bottom-up" modeling of quarterly US GDP trend 140

10 Aggregation and disaggregation of time series **145**
10.1 The effect of aggregation on an SS model 146
10.1.1 The high-frequency model in stacked form 146
10.1.2 Aggregation relationships 148
10.1.3 Relationships between the models for high, low, and mixed-frequency data 149
10.1.4 The effect of aggregation on predictive accuracy 150
10.2 Observability in the aggregated model 150
10.2.1 Unobservable modes 150
10.2.2 Observability and fixed-interval smoothing 152
10.2.3 An algorithm to aggregate a linear model: theory and examples 153
10.3 Specification of the high-frequency model 154
10.3.1 Enforcing approximate consistency 154
10.3.2 "Bottom-up" determination of the quarterly model 158
10.4 Empirical example 158
10.4.1 Annual model 159
10.4.2 Decomposition of the quarterly indicator 159
10.4.3 Specification and estimation of the quarterly model 160
10.4.4 Diagnostics 160
10.4.5 Forecast accuracy and non-conformable samples 162
10.4.6 Comparison with alternative methods 163

11 Cross-sectional extension: longitudinal and panel data **167**
11.1 Model formulation 168
11.2 The Kalman Filter 169
11.2.1 Case of uncorrelated state and observational errors 170
11.2.2 Case of correlated state and observational errors 171
11.3 The linear mixed model in SS form 171
11.4 Maximum likelihood estimation 174
11.5 Missing data modifications 176
11.5.1 Missingness in responses only 176
11.5.2 Missingness in both responses and covariates: method 1 178
11.5.3 Missingness in both responses and covariates: method 2 180
11.6 Real data examples 182
11.6.1 A LMM for the mare ovarian follicles data 182
11.6.2 Smoothing and prediction of missing values for the beluga whales data 183

Appendices **191**

A Some results in numerical algebra and linear systems **193**
A.1 QR Decomposition 193
A.2 Schur decomposition 195
A.3 The Hessenberg Form 196
A.4 SVD Decomposition 197
A.5 Canonical Correlations 198
A.6 Algebraic Lyapunov and Sylvester equations 198
A.7 Numerical solution of a Sylvester equation 200
A.8 Block-diagonalization of a matrix 201
A.9 Reduced rank least squares 202
A.10 Riccati equations 203
A.10.1 Definition 204
A.10.2 Solving the ARE in the general case 205
A.10.3 Solving the ARE for GARCH models 207
A.11 Kalman filter 208

B Asymptotic properties of maximum likelihood estimates **211**
B.1 Preliminaries 211
B.2 Basic likelihood results for the State-Space model 214
B.2.1 The information matrix 214
B.2.2 Regularity conditions 216
B.2.3 Choice of estimation method 217
B.3 The State-Space model with cross-sectional extension 218

C Software (E^4) **223**
C.1 Models supported in E^4 224
C.1.1 State-Space model 224
C.1.2 The THD format 224
C.1.3 Basic models 226
C.1.3.1 Mathematical definition 226
C.1.3.2 Definition in THD format 227
C.1.4 Models with GARCH errors 228
C.1.4.1 Mathematical definition of the GARCH process 228
C.1.4.2 Defining a model with GARCH errors in THD format 229
C.1.5 Nested models 230
C.2 Overview of computational procedures 233
C.2.1 Standard procedures for time series analysis 233
C.2.2 Signal extraction methods 235
C.2.3 Likelihood and model estimation 238
C.3 Who can benefit from E^4? 241

D Downloading E^4 and the examples in this book **243**
D.1 The E^4 website 243
D.2 Downloading and installing E^4 243
D.3 Downloading the code for the examples in this book 245

Bibliography **247**

Author Index **261**

Subject Index **265**

List of Figures

4.1 Percent relative corrections on the original Wolf numbers: $(z_t^*/z_t - 1) \times 100$. 43
4.2 Changes in the HP trend. 44

5.1 International Airline Passengers (thousands) from 1949 to 1960. Source: see Box, Jenkins and Reinsel [30]. 58
5.2 Standardized residuals of the transfer function (5.67). 60
5.3 Standardized residuals of the transfer function (5.68). 61
5.4 Monthly U.S. single-family Housing Starts and Housing Sales, in thousands of units, from January 1965 through May 1975. 61
5.5 Additive seasonal component of (log) Housing Starts (thick line) and Housing Sales (thin line), from January 1965 through May 1975. 63
5.6 Seasonality and calendar-adjusted (log) Housing Starts and Housing Sales series, from January 1965 through May 1975. 64

6.1 Fixed-interval smoothed estimates for β_t (solid line) in the time-varying parameter CAPM regression with $\pm 2\sigma$ limits (dashed line). 78
6.2 Quarterly unadjusted West German Consumption. The data is in billions of Deutsch Marks. 78
6.3 Conditional correlation between Deutsch Mark and Japanese Yen returns. 81

7.1 Data in examples 7.5.1 and 7.5.2. Top: Airline passengers time series from Box, Jenkins and Reinsel [30]. Bottom: US short-term interest rates. 96

8.1 Comparison between the model components obtained from the STSM (left) and the equivalent SEM (right). 119
8.2 Trace of the smoothed covariance matrices corresponding to the STSM and the equivalent SEM. 120
8.3 Trace of the smoother covariance of Model (8.54). 123
8.4 Common features of (log) wheat prices. 124

8.5 Monthly series of sales and advertising of the Lydia Pinkham vegetable compound from January 1954 to June 1960. Source: Palda [175]. 125
8.6 Sales versus the exponential of the stochastic component for the Lydia Pinkham data. The distance between both series (gray area) is an estimate of the effect of advertising over sales. 126
8.7 Estimated value added by advertising, computed in each period as the estimate of sales generated by advertising minus the advertising investment (Lydia Pinkham data). 127
8.8 Multiplicative effect of advertising (thick line) versus multiplicative seasonality of sales (thin line) for the Lydia Pinkham data. The values between November 1957 and February 1958 are smoothed interpolations. 127

9.1 Parameters of the ARMA model (9.24) and smallest singular value (SV) of the Jacobian for several values of ϕ in the AR(1) plus error model. 139

10.1 Decomposition of the quarterly indicator. The adjusted indicator includes the trend, irregular, and level change components, and excludes seasonality and calendar affects. 161
10.2 Standardized plot of quarterly estimates of VAI (thick line) and adjusted indicator (thin line). The last values are forecasts computed from 2002.1 to 2002.4 quarters using model (10.27). 162

11.1 Observed and predicted time series of ovarian follicles for the 11th mare in dataset `Ovary` of Section 11.6.1. The 12-step ahead predictions (open circles) and corresponding 95% prediction bands (dashed lines), are obtained by running the Kalman recursions for the SS formulation of the LMM described for the series. 184
11.2 Kalman smoothing and prediction for beluga whale Casey in the `whale.dat` dataset of Subsection 11.6.2. Shown are smoothed values (solid lines) and 5-step ahead predictions (open circles) for nursing time (top panel), number of bouts (middle panel), and number of lockons (bottom panel), from period 100 onward. Corresponding 95% confidence bands for the smoothed and predicted values are shown as dotted and dashed lines, respectively. 187

List of Tables

2.1 Some common specifications for the structural time series components, see [118]. The cycle parameters are such that $0 \leq \rho < 1$ and $0 \leq \lambda_c < \pi$. The white noise errors η_t, ζ_t, κ_t, κ_t^*, and ω_t are often constrained to be independent. In the models for the seasonal component, the letter S denotes the seasonal period. 7

4.1 Estimates and standard errors for the AR(2)Autoregression(s) plus noise model fitted to the Wolf sunspot series. 43

5.1 Estimation results for the airline models (5.64) and (5.65). 59

6.1 Estimation results for model (6.77)–(6.79). 77

6.2 Estimation results for the model (6.80). The values on convergence of the (log) likelihood function and the information criteria are, respectively, $\ell^*(\hat{\theta})$=-313.766; AIC = -5.715, SBC = -5.515 79

7.1 Definition of the weight matrices $\boldsymbol{W_1}$ and $\boldsymbol{W_2}$ in equation (7.13). Note that when the CVA approach is used, then the singular values obtained from the SVD coincide with the canonical correlation coefficients between $\boldsymbol{Z}_f$ and $\boldsymbol{V}_p$. 87

7.2 Regular system order estimation for z_t. 97

7.3 Seasonal system order estimation for z_t with seasonal structure ($s = 12$) Block Hankel matrices. 97

7.4 Regular system order estimation for $\hat{a}_{2t}$. 98

7.5 Seasonal system order estimation for $\hat{a}_{2t}$ with the seasonal Block Hankel matrix and $s = 12$. 98

7.6 Datasets used to compare the SM, SUBML and ML methods in Tables 7.7, 7.8, and 7.9. 100

7.7 Univariate models. Comparison of models estimated for the data described in Table 7.6 - Part 1. Standard errors are in parenthesis. $\hat{\sigma}_a^2$ denotes the residual variance. SM is a subspace methods estimation with weights corresponding to the CVA algorithm (see Table 7.1) and using an extended observability matrix approach (see Section 7.1). 101

7.8 Univariate models. Comparison of models estimated for the data described in Table 7.6 - Part 2. Standard errors are in parenthesis. $\hat{\sigma}_a^2$ denotes the residual variance. SUBML values were obtained with the algorithm described in Section 7.3, by pruning the non-significant parameters found in Table 7.7. 102
7.9 Univariate models. Comparison of models estimated for the data described in Table 7.6 - Part 3. Standard errors are in parenthesis. $\hat{\sigma}_a^2$ denotes the residual variance. ML estimates were initiated with the parameters obtained by SUBML shown in Table 7.8. 103
7.10 System order estimation for $\boldsymbol{Y}_{1t}$. 104
7.11 Sequence of identifiedIdentifiability models for the process $\boldsymbol{Y}_{1t}$. The figures in parentheses are the standard errors of the parameters and the symbol "-" denotes that the corresponding parameter has been constrained to its current value, so it does not have a standard error. 105

8.1 Association between state variables and structural components according to the eigenvalues of the transition matrix and their spectral properties. 115
8.2 Definition of the data generating process in STSM and ARIMA forms. 117
8.3 Structure of the block-diagonal SEM representation of the data generating process and association of the state variables to structural components. 118
8.4 Comparison between the ARIMA representations of the STSM and the block-diagonal SEM form. 118
8.5 Structure of the block-diagonal SEM equivalent to model (8.54) and association of the state variables to structural components. 122

9.1 Structural and ARIMAX forms for some common univariate specifications. Both representations are observationally equivalent. 136
9.2 Structural and VARMAX forms for some common bivariate specifications. Both representations are observationally equivalent. 137
9.3 Characterization of the identifiability of an SS model using the rank of the Jacobian ($\boldsymbol{J}$) measuring how a change in the structural model parameters affects the parameters in the reduced form. 138
9.4 Step-by-step results from a bottom-up modeling of the US GDP series. 143

10.1 Aggregation of several univariate models. The columns labelled "states" show the number of dynamic components in the minimal SS representation of the corresponding model. Therefore, the difference between the number of states in the quarterly and annual models is the number of dynamic components that become unobservable after aggregation. The quarterly variables z_t, z_{1t} and z_{2t} are assumed to be flows, so their annual aggregate is the sum of the corresponding quarterly values. 155
10.2 Aggregation of several bivariate models. All the quarterly variables z_{1t} and z_{2t} are assumed to be flows, so their annual aggregate is the sum of the corresponding quarterly values. If a variable were to be aggregated as an average of the quarterly values, the model would require an appropriate re-scaling. 156
10.3 Original data and interpolations obtained with full and non-conformable samples. Underlined values correspond to interpolations, retropolations or forecasts. The column "Indicator" refers to the calendar and seasonally adjusted values computed with model (10.26). 164
10.4 Ranking of RMSEs for end-of year forecasts of annual VAI from 1997 to 2001. RMSEs are computed for forecasts of both, VAI levels and growth rates. The columns RMSE% show the corresponding RMSEs normalized so that the RMSE of Model (10.29) is 100%. 165

11.1 Final parameter estimates for the SS model fitted to the beluga whale data of Subsection 11.6.2. Estimates were obtained via the EM algorithm initialized with the values in square brackets. The final estimate of the initial state mean was $\widehat{\mu}^T = [3.296, 1.009, 1.390, -8.848 \times 10^{-3}, 6.653 \times 10^{-1}]$, starting from the zero vector. 186
11.2 Five-step ahead predictions and 95% confidence bounds for Casey in the beluga whale data of Subsection 11.6.2. 188

C.1 Standard statistics procedures to specify and simulate a time series model. 234
C.2 Signal extraction functions. These are the main procedures offered for time series disaggregation, structural decomposition, smoothing and forecasting. 236
C.3 Model estimation functions. These are the main procedures which can be applied to compute the Gaussian likelihood, its first-order derivatives, and information matrix. These functions can be fed to the standard E^4 optimizer (`e4min`) to compute ML estimates. Initial estimates can be computed using a specialized function (`e4preest`). 239

Preface

The story of how any book comes into being is typically complex. Even moreso in the case of this one...

Some of us started this project (at least in our minds) during our undergraduate studies, when we discovered the awesome Box and Jenkins [29, 30] book on time series analysis. This text, together with the influence of some excellent instructors, was key to our decision to pursue an academic career in Statistics/Econometrics and seeded an early ambition to write a book on this matter.

Some or many years later, depending on each particular case, we also discovered the elegance and clarity of the State-Space approach, which was a revelation comparable to the previous one.

In the 1980s, State-Space methods were applied almost exclusively to problems in engineering but some excellent books, such as those by Jazwinsky [130] or Anderson and Moore [6], sprouted spontaneously on the shelves of many Departments of Mathematics and Economics, frequently with some volumes of the dominant journals on systems engineering. This progressive input from the engineering field led to the increased interest of time series analysts to adopt these methods. As a consequence, some important books on the State-Space approach were written by authors with a mainstream time series background, such as Hannan and Deistler [116], Harvey [118] or West and Harrison [224, 225]. Also, it has become customary to include a chapter on State-Space models in time series textbooks such as those by Hamilton [115], Wei [221], or Lutkepohl [158].

State-Space methods also appear frequently in the foundations of some well-known statistics and econometrics packages. This is the case of the STATESPACE, UCM, KALDFS or RANTWO procedures in SAS/ETS and SAS/IML, see [193], or the default likelihood evaluation method for ARIMA models implemented in GRETL, see [13]. In R [186], the popular free software environment for statistical computing and graphics, the main State-Space modeling packages are dlm, dse, sspir, KFAS, and FKF, but there is a proliferation of many others; see the review articles by Petris and Petrone [182] and Tusell [213]. Last, an important software package specifically devoted to State-Space modeling is SsfPack, a library of routines written in C and linked to Ox, see Koopman, Shephard and Doornik [141] and Pelagatti [179].

Returning to our personal history, some of us discovered this literature in the 1980s and soon learned that the State-Space model was a simple and comprehensive formulation that could accommodate any of the standard linear processes that we used in our work.

This idea was awesome. Bear in mind that in those years, time series software was scarce and underpowered, so we had to work very hard at coding our own procedures. Moreover, the software that we developed for an ARIMA process required painful adaptations before it could be applied to other formulations, such as transfer functions or VARMAX. After doing so, the different versions of the basic algorithm would require their specific documentation, debugging, and updating cycles.

In comparison with this messy situation, the State-Space approach offered us a formal framework where any result or procedure developed for the basic model could be seamlessly applied to any standard formulation written in State-Space form. In short: you only needed a single version of each basic procedure, no matter if it was related to likelihood computation, interpolation, or extrapolation. Moreover, you could accommodate with a reasonable effort nonstandard situations such as observation errors, aggregation constraints or missing in-sample values, also taking advantage of the excellent "off-the-shelf" numerical procedures and code that our colleagues in the systems engineering field had been developing and sharing over the years.

These compelling advantages convinced us that the State-Space approach would be productive for our academic work. The first results exceeded our expectations and, after many articles and PhD Theses (including some of our own!) we want to synthesize and present these advantages to a wider audience by means of a book including theory, examples, and software.

Therefore, this monograph is the result of an extended process of maturation and discovery. It is addressed to postgraduate students, professional statisticians and academics. Our goal is to share with its readers the reasons why we are so enthusiastic about the State-Space approach. Because our potential audience is relatively wide, most of the concepts and methods presented are explained on an intuitive basis. Despite our effort to favor intuition we recognize that the subject discussed here is inherently difficult, so grasping 100% of the content will require an advanced level in linear algebra, dynamical systems, and time series analysis.

Concerning the scope of the book, we opted to concentrate on discussing the computational processes that can be applied to a specified linear model that is being entertained. This specific focus lead to omitting some substantive topics which would deserve specialized monographs in their own right. In particular, we do not cover the nonlinear/non-Gaussian case. Also, we barely discuss how one might arrive at an initial model specification, except in Chapter 7. Last, we do not provide a systematic coverage of the structural time series modeling approach, initially due to Harvey [118] and later developed by many impressive works, such as those of Koopman [139], Koopman and Shephard [140] or Harvey, Ruiz and Shephard [119] [1].

On the other hand, we do provide a broad presentation of what can be done after a tentative specification has been determined. In particular, we discuss two classical topics in the State-Space literature: model estimation and signal extraction. Besides, we pay special attention to the manipulation of the model and, accordingly, present many procedures to combine, decompose, aggregate and disaggregate a State-Space

[1] For a comprehensive overview of the foundations of this methodology, the reader should check the books by Commandeur and Koopman [55] and Durbin and Koopman [82].

form. Finally, we also cover the connection between mainstream time series models and the State-Space representation. We do so by showing how one can obtain the State-Space formulation of a general linear process (see Subsection 2.2.4) and, conversely, provide procedures to transform a State-Space model into an equivalent VARMAX representation (see Chapter 9).

An important feature of this book is that it comes with a substantial software package.

As we explained above, our original interest in State-Space methods had much to do with developing flexible and powerful software tools for time series analysis. Our first attempts were coded in standard high-level languages such as FORTRAN or Pascal. They were a huge advance in comparison with the previous situation, but software development and maintenance was hard and time-consuming. Also, it was difficult to share this code with other researchers. For these reasons, in the early 1990s we launched the E^4 project, which was specifically devised to solve these issues.

The name E^4 refers to the Spanish: "**E**stimación de modelos **E**conométricos en **E**spacio de los **E**stados" (State-Space Estimation of Econometric Models). Under this name, we started developing a MATLAB Toolbox which collects all the computational procedures resulting from our research, together with the administrative and analytical functions required to conform a fully operational time series analysis package. This toolbox is freely distributed under the terms of the GNU General Public License through the website `www.ucm.es/e-4/` which, besides source code, provides a complete user manual and other reference materials.

The connection between E^4 and this book is very clear: all the procedures presented here are implemented as E^4 functions, so the reader will be able to apply them to his/her own work. To facilitate this process, appendices C and D provide additional information about E^4 and all the information needed to download it and to replicate all the practical examples in this book.

Additional material is available at `www.crcpress.com/9781482219593` and the aforementioned E^4 website, `www.ucm.es/e-4/`.

We do not wish to end this Preface without thanking our editor, Rob Calver, for planting the initial "seed" in our minds by means of a timely e-mail in late 2012; a message which hung in the balance between "deletion" and "response" for several weeks before succumbing to the latter. We are especially grateful to his editorial assistants, Sarah Gelson, Sarfraz Khan, and Alexander Edwards, for their responsive and knowledgeable help in all the phases of this project. Their continual requests for progress reports in particular, helped us avoid delving into the classical vices of any author: unproductive procrastination and frolicking. Last, we are deeply indebted to Laurie Oknowsky, Michael W. Davidson and the Taylor & Francis production team for their excellent work in the final phases of this project. Shashi Kumar provided fast and flawless support to solve our LaTeX issues.

The first four authors gratefully acknowledge financial support from *Ministerio de Economia y Competitividad*, through Grant ECO2011-23972. The last author wishes to thank Texas Tech University for teaching-load reductions in the final two years leading up to publication. Finally, four anonymous reviewers provided constructive criticism and suggestions which improved this book in many ways.

Somosaguas, Madrid, Spain
Lubbock, Texas, USA
October 2015

About the Authors

Jose Casals is the Head of Global Risk Management at Bankia. He also works part-time as an Associate Professor of Econometrics at Universidad Complutense de Madrid.

Alfredo Garcia-Hiernaux is an Associate Professor of Econometrics at Universidad Complutense and freelance consultant.

Miguel Jerez is an Associate Professor of Econometrics at Universidad Complutense and a freelance consultant. For six years he was Executive Vice-President at Caja de Madrid.

Sonia Sotoca is an Associate Professor of Econometrics at Universidad Complutense.

These authors are engaged in a long-term research project to apply State-Space techniques to standard econometric problems. Their common interests include State-Space methods and time series econometrics. A list of their publications on these topics can be found at: `www.ucm.es/grupoee/publicaciones`

Many algorithms derived from their State-Space research are implemented in E^4 (`www.ucm.es/info/icae/e4`) a MATLAB toolbox freely offered under the terms of the GNU General Public License. They can be reached at the same address: Departamento de Fundamentos del Analisis Economico II, Universidad Complutense de Madrid, Campus de Somosaguas s/n, 28223 Madrid (SPAIN).

A. Alexandre (Alex) Trindade is Professor of Statistics in the Department of Mathematics & Statistics at Texas Tech University, and Adjunct Professor in the Graduate School of Biomedical Sciences, Texas Tech University Health Sciences Center. His research spans a broad swath of theoretical and computational statistics. Prior to Texas Tech, he held appointments at the University of Florida, Colorado State University, the University of New Mexico, and the University of Oklahoma.

Alex can be reached at the Department of Mathematics & Statistics, Texas Tech University, Broadway and Boston, Lubbock, TX 79409-1042, USA.

Chapter 1

Introduction

This book provides a comprehensive approach to the many computational procedures that can be applied to a previously specified linear model in State-Space (hereafter, SS) form. Our interest in this field is motivated by the important advantages of the SS approach. Some of them are:

1. Any linear time series model can be written in an equivalent SS form[1]. In particular, ARIMAX, VARMAX or transfer functions can be represented in this way, see Subsection 2.2.4. In other cases, such as that of structural time series models (see Harvey [118]) the model is directly defined in SS form, so no translation is required.
2. Conversely, any linear SS model can be written in an equivalent VARMAX form, see Chapter 9.
3. The flexibility of the SS representation allows one to accommodate nonstandard situations like time-varying parameters (Pagan [173] and Watson and Engle [219]), unobserved variables (Harvey [118]), observation errors (Terceiro [206]), aggregation constraints (Casals, Sotoca and Jerez [46]) or missing in-sample values (Ansley and Kohn [7], Kohn and Ansley [137] and Naranjo, Trindade and Casella [165]).
4. System engineering provides a wealth of excellent "off-the-shelf" procedures and ideas to solve many relevant problems such as the specification, estimation and optimal control of SS models.

These ideas have deep implications. Consider for example points 1 and 2 above. Taken together they imply that the specific form of a model is a matter of choice as: (a) one can work with a SS representation and finally translate it to a VARMAX form, if this is convenient, or (b) build on a VARMAX model and run chosen parts of the analysis using the equivalent SS model.

[1]Throughout this book, we will consider that two models are equivalent if they result in the same probability distribution of the sample. This concept is also known as "observational equivalence" in the literature.

The ability to translate a VARMAX model to the SS form and vice versa also implies that the flexibility and power of the SS procedures, highlighted in points 3 and 4 above, can be immediately put into practice in mainstream time series analysis. Specifically, any identification, estimation, decomposition or aggregation method devised for the linear SS model can be applied to mainstream time series specifications, such as VAR, VARMAX, structural econometric models or transfer functions.

Finally, a model is a mathematical representation of a system. A procedure is a tool to manipulate a model. Models and procedures are different things and, therefore, a desirable feature for any procedure is to be independent of particular model formulations. Accordingly, it makes sense to concentrate the development of computational procedures on the SS model, using the generality of this formulation as an "abstraction layer" to separate them from specific model families. In our experience, this brings flexibility, reliability and consistency to statistical software.

The structure of this book is as follows. Chapter 2 discusses the formulation of the SS model. To this end, we define the general form of the multiple-error linear SS model and describe how one can obtain equivalent formulations by applying a similar transformation to the state variables. We also present the single-error SS model, which is an alternative formulation with clear advantages when one wants to estimate the state variables. The last section explains how to obtain the SS model matrices corresponding to a generalized VARMAX process.

Chapter 3 illustrates the flexibility of the SS representation by presenting three classes of SS model transformations: decomposition, combination, and change of output variables. Decomposing an SS model consists in breaking it into several meaningful sub-models, each one with its own SS representation. Some examples are: (a) the deterministic-stochastic decomposition, which breaks the model into additive components affected either by the inputs or the errors of the system, (b) the decomposition of an SS model into the univariate models implied by its structure and (c) the block-diagonal SS form, which breaks the model structure into trend, seasonal, cyclical, and irregular components.

On the other hand, we use the term "model combination" or "model nesting" to describe the construction of a complex SS representation by combining several simpler SS models. The link between the different sub-models may be given either by their inputs ("combination in inputs" or "endogeneization") or their errors ("combination in errors"). Building on these procedures one can tackle nontrivial problems such as defining a complex model as the product of several multiplicative factors or linking the model for the system output with those of the inputs into a unique representation where all the variables (inputs and outputs) are endogenous.

Finally, the change of output variables in the SS model consists of replacing the endogenous variable with another one which takes into account nonstandard data features such as, e.g., observation errors, missing values, or aggregated data.

Chapter 4 covers the main state estimation algorithms considered in the SS literature: filtering and smoothing. Both focus on estimating the sequence of state variables with different information sets. In particular, filtering algorithms provide the expected value and error covariance of the states at time $t+1$, conditional on all the sample information available up to time t[2]. On the other hand, smoothing procedures provide the conditional first and second-order moments of the states with flexible information sets. Anderson and Moore [6] distinguish between fixed-point, fixed-lag and fixed-interval smoothing. For the purposes of this book we will concentrate on the latter, which provides the first and second-order moments of the states conditional on all the sample information.

Filtering and fixed-interval smoothing procedures are applied in this book with many purposes in mind. In particular, we use the Kalman filter to compute the innovations required to evaluate the likelihood for a SS model, see Terceiro [206], while fixed-interval smoothing methods are applied to estimate unobserved components in structural time series models, and to deal with samples containing missing values or aggregated data. The last Section in Chapter 4 includes some examples of these applications.

Chapters 5 and 6 describe in detail how one can compute the Gaussian likelihood for the unknown coefficients in the SS matrices of a given model. In particular, Chapter 5 concentrates on models with fixed coefficients, while Chapter 6 covers the case where some coefficients or covariances are time-varying, including periodic models, regression equations with time-varying coefficients or models with GARCH errors. Combining these algorithms with a standard iterative optimization procedure allows one to obtain maximum-likelihood (ML) estimates for any unknown parameter in the SS model matrices. Likelihood computation also provides a basis from which to approximate the information matrix, see Terceiro [206], and for inference based on the likelihood ratio, see Casella and Berger [48].

Chapter 7 provides an introduction to the so-called "subspace methods." This term refers to a dominant trend in modern system identification, see e.g. Ljung [152] or Qin [185]. In the context of this book, we focus on their application to compute rank-deficient Generalized Least Squares (hereafter, GLS) estimates for the model parameters, see Ljung [152]. The fast and stable estimates provided by these procedures can be used either as an alternative to ML, or as preliminary estimates, useful to initiate an iterative estimation procedure.

Chapter 8 deals with a classical subject in the SS literature, generally described as "signal extraction." This term refers to the estimation of signals buried in a vector of time series. The time series literature often reduces it to breaking a single variable into the sum of its "structural components," which are the trend, cycle, seasonal, and irregular components. In this chapter we present our approach to this matter which has some particular features, as it focuses on the case of multiple time series, allows for exogenous variables, and produces "revision-free" components, that is, estimates which are invariant to changes in the sample.

As is well known, any linear time series model can be written in SS form, but a not so well known fact is that this is a two-way trip, as any linear SS model can also be written in VARMAX form, see Casals, Garcia–Hiernaux, and Jerez [39]. Chapter 9 describes two algorithms used to obtain the VARMAX matrices corresponding to

---

[2]In books using a slightly different SS formulation the filter provides estimates of the contemporary states and covariances, i.e. the first and second-order moment at time t conditional on the information available up to time t. We will further discuss this issue in Chapter 4.

any linear SS model and showcases some applications in three areas: analyzing the identifiability of the parameters in a SS model, specifying a model from noisy data, and estimating the parameters of a structural SS specification so that it realizes a given VARMAX reduced form.

Chapter 10 deals with several issues relating to the aggregation and disaggregation of time series and, in particular, discusses how to specify an observable high-frequency model for a vector of time series sampled at high and low frequencies. To this end, we first study how aggregation over time affects both the dynamic components of a time series and their observability. The main conclusion is that aggregation does not affect the state variables driving the system but some of them, mainly those related to the seasonal structure, become unobservable. Building on these results, one can devise a structured specification process built on the idea that the models relating the variables in high and low sampling frequencies should be mutually consistent. After specifying such a model, the methods discussed in previous chapters provide an adequate framework for estimation, diagnostic checking, data interpolation and forecasting. An example with national accounting data illustrates the practical application of this methodology.

Finally, Chapter 11 discusses a cross-sectional extension to the classical SS formulation, in order to accommodate longitudinal or panel data. This is important if one wants to globally analyze several time series, each one originating from a different "unit." Common examples are biological variables for different subjects, or econometric variables for different firms. The assumption of independence among the units allows one to easily combine the separate likelihoods. Appropriate smoothing, filtering, and prediction recursions in this context will be discussed, as well as maximum likelihood estimates and the quantification of their uncertainty through asymptotic results. A particularly attractive and powerful feature of this methodology is the inherent ability to deal with missing or irregularly-spaced time series data, together with the ease with which smoothing and forecasting is accomplished.

The appendices provide some supplementary material. In particular: Appendix A discusses some specialized tools in numerical algebra and linear systems. Appendix B presents the asymptotic theory underlying the ML estimates for simplified versions of the SS model. Appendix C describes E^4, a free MATLAB Toolbox for time series analysis using the SS methods discussed in this book. Finally, Appendix D provides detailed instructions about how to obtain the code and data required to replicate the practical examples in this book with E^4.

Chapter 2

Linear State-Space models

2.1 The multiple error model 6
2.1.1 Model formulation 6
2.1.2 Similar transformations 8
2.1.3 Properties of the State-Space model 8
2.2 Single error models 12
2.2.1 Model formulation 12
2.2.2 State estimation in the SEM 12
2.2.3 Obtaining the SEM equivalent to a MEM 13
2.2.4 Obtaining the SEM equivalent to a general linear process 15

This chapter describes the two basic SS formulations that will be used in the remaining chapters. We will refer to the first one as a "multiple error model," or "MEM," because it allows for different errors in the state and observation equations of the model. On the other hand, the "single error model," or "SEM" has a unique vector of random disturbances, affecting both equations.

The relationship between both formulations is very rich, and can be summarized in two ideas. First, under nonrestrictive assumptions a MEM model can be written in an observationally equivalent SEM form. Second, the SEM has specific advantages for applications requiring precise estimates of the state variables. The combination of both ideas allows one to devise powerful computational procedures, as we will see e.g. in Chapters 5, 6 or 8.

The main properties of the SS model are presented using the MEM representation, bearing in mind that their extension to the SEM case is trivial. Last, we show how to write a standard linear stochastic process in SEM form.

2.1 The multiple error model

2.1.1 *Model formulation*

The basic fixed-coefficients SS formulation is given by a *state equation*[1], characterizing the system dynamics:

$$\boldsymbol{x}_{t+1} = \boldsymbol{\Phi x_t} + \boldsymbol{\Gamma u_t} + \boldsymbol{E w_t}, \tag{2.1}$$

and an *observation equation*, describing how the system output is realized as a linear combination of different dynamic components:

$$\boldsymbol{z_t} = \boldsymbol{H x_t} + \boldsymbol{D u_t} + \boldsymbol{C v_t}, \tag{2.2}$$

where $\boldsymbol{z_t} \in \mathbb{R}^m$ is a random vector of *endogenous variables* or *outputs* at time t, $\boldsymbol{x_t} \in \mathbb{R}^n$ is vector of *state variables* or simply *states*, $\boldsymbol{u_t} \in \mathbb{R}^k$ is a vector of *exogenous variables* or *inputs*, and $\boldsymbol{w_t}$ and $\boldsymbol{v_t}$ are conformable vectors of zero-mean white noise errors, such that:

$$\mathrm{E}\left(\begin{bmatrix} \boldsymbol{w_t} \\ \boldsymbol{v_t} \end{bmatrix} \begin{bmatrix} \boldsymbol{w_t} & \boldsymbol{v_t} \end{bmatrix}\right) = \begin{bmatrix} \boldsymbol{Q} & \boldsymbol{S} \\ \boldsymbol{S}^T & \boldsymbol{R} \end{bmatrix}. \tag{2.3}$$

Last, $\boldsymbol{\Phi}, \boldsymbol{\Gamma}, \boldsymbol{E}, \boldsymbol{H}, \boldsymbol{D}, \boldsymbol{C}, \boldsymbol{Q}, \boldsymbol{S}$ and $\boldsymbol{R}$ are time-invariant coefficient matrices and the covariance matrix defined in (2.3) is positive semi-definite.

Traditionally, the parameter and covariance matrices in the SS model depend on time. In (2.1)–(2.3) we dropped the time subindex. We did this to simplify the notation, because most of the theory and applications considered in this book are covered by the time-invariant framework. Occasionally we will consider some types of parameter variation, such as e.g. Example 2.1.2, Subsection 3.3.2 or Chapter 6. Whenever suitable, the text will make clear that we are using a time-varying SS model.

We will also assume that all the errors in model (2.1)–(2.2) are independent of the exogenous variables $\boldsymbol{u_t}$ and the initial state vector $\boldsymbol{x_1}$[2]. Last, we will consider that the initial state, $\boldsymbol{x_1}$, is a random vector such that its mean and covariance, conditional to the input values, are $\boldsymbol{\mu_1}$ and $\boldsymbol{P_1}$, respectively.

Example 2.1.1 Structural time series models

The SS representation can be used to formulate time series models directly. For example, Harvey [118] proposes a wide family of *structural time series models*, building on the following decomposition for $\boldsymbol{z_t}$:

$$z_t = \mu_t + \psi_t + \gamma_t + \varepsilon_t, \tag{2.4}$$

[1]There are other ways to define the state equation. For example, two common alternatives are: $\boldsymbol{x_t} = \boldsymbol{\Phi x}_{t-1} + \boldsymbol{\Gamma u_t} + \boldsymbol{E w_t}$ and $\boldsymbol{x}_{t+1} = \boldsymbol{\Phi x_t} + \boldsymbol{\Gamma u}_{t+1} + \boldsymbol{E w}_{t+1}$. These specific representations are as general as the one we use here because they can be derived from Equation (2.1) by means of a suitable change of variables.

[2]This assumption is not restrictive because, if there were any linear relationship between the model inputs and errors, we could endogeneize the former as shown in Section 3.2.

Table 2.1 *Some common specifications for the structural time series components, see [118]. The cycle parameters are such that $0 \leq \rho < 1$ and $0 \leq \lambda_c < \pi$. The white noise errors η_t, ζ_t, κ_t, κ_t^*, and ω_t are often constrained to be independent. In the models for the seasonal component, the letter S denotes the seasonal period.*

Structural component	**Model**	**Specification**
Trend (μ_t)	Random walk	$\mu_{t+1} = \mu_t + \eta_t$
	Stochastic trend	$\begin{bmatrix}\mu_{t+1}\\ \beta_{t+1}\end{bmatrix} = \begin{bmatrix}1 & 1\\ 0 & 1\end{bmatrix}\begin{bmatrix}\mu_t\\ \beta_t\end{bmatrix} + \begin{bmatrix}\eta_t\\ \zeta_t\end{bmatrix}$
	Random walk with drift	$\begin{bmatrix}\mu_{t+1}\\ \beta_{t+1}\end{bmatrix} = \begin{bmatrix}1 & 1\\ 0 & 1\end{bmatrix}\begin{bmatrix}\mu_t\\ \beta_t\end{bmatrix} + \begin{bmatrix}\eta_t\\ 0\end{bmatrix}$
	Integrated random walk	$\begin{bmatrix}\mu_{t+1}\\ \beta_{t+1}\end{bmatrix} = \begin{bmatrix}1 & 1\\ 0 & 1\end{bmatrix}\begin{bmatrix}\mu_t\\ \beta_t\end{bmatrix} + \begin{bmatrix}0\\ \zeta_t\end{bmatrix}$
Cycle (ψ_t)	Stochastic cycle	$\begin{bmatrix}\psi_{t+1}\\ \psi^*_{t+1}\end{bmatrix} = \rho\begin{bmatrix}\cos\lambda_c & \sin\lambda_c\\ -\sin\lambda_c & \cos\lambda_c\end{bmatrix}\begin{bmatrix}\psi_t\\ \psi^*_t\end{bmatrix} + \begin{bmatrix}\kappa_t\\ \kappa^*_t\end{bmatrix}$
	Nonstationary cycle	$\begin{bmatrix}\psi_{t+1}\\ \psi^*_{t+1}\end{bmatrix} = \begin{bmatrix}\cos\lambda_c & \sin\lambda_c\\ -\sin\lambda_c & \cos\lambda_c\end{bmatrix}\begin{bmatrix}\psi_t\\ \psi^*_t\end{bmatrix} + \begin{bmatrix}\kappa_t\\ \kappa^*_t\end{bmatrix}$
Seasonality (γ_t)	Dummy variable seasonality	$\gamma_{t+1} = \sum_{j=0}^{S-2} \gamma_{t-j} + \omega_t$
	Trigonometric seasonality	$\gamma_{t+1} = \sum_{j=1}^{[S/2]} (\alpha_j \cos\lambda_j t + \beta_j \sin\lambda_j t) + \omega_t;$ $\lambda_j = 2\pi j/S$

where μ_t, ψ_t, γ_t, and ε_t are, respectively, the trend, cycle, seasonal, and irregular components of the time series. Each of these *structural components* is specified by means of a state equation. Some common specifications used for this purpose are summarized in Table 2.1. Harvey [118] and related subsequent literature extend this basic model to accommodate exogenous variables, multiple time series, and cointegration constraints.

Example 2.1.2 Linear regression models

The SS representation is so flexible that it can be used either to formulate models, as in the previous example, or to accommodate as particular cases most mainstream specifications, allowing in this case for interesting variations of the basic model. Consider for example the multiple regression model:

$$z_t = \boldsymbol{u}_t^T \boldsymbol{\beta} + v_t, \tag{2.5}$$

which could be written as a particular case of (2.1)–(2.2) just by setting $\boldsymbol{H} = \boldsymbol{0}$. Alternatively, it could be formulated as:

$$\begin{aligned} \boldsymbol{\beta}_{t+1} &= \boldsymbol{\beta}_t \\ z_t &= \boldsymbol{u}_t^T \boldsymbol{\beta}_t + v_t, \end{aligned} \tag{2.6}$$

where the state equation characterizes a fixed parameter, whose value at time $t+1$ is equal to its value at time t. An immediate extension of model (2.6) would allow for a time-varying parameter. For example, we could define a regression model with random walk parameters:

$$\begin{aligned}\beta_{t+1} &= \boldsymbol{\beta_t} + w_t \\ z_t &= \boldsymbol{u}_t^T \boldsymbol{\beta_t} + v_t,\end{aligned} \tag{2.7}$$

with $\text{cov}(w_t, v_t) = 0$. Building on equation (2.7), one can immediately define a wide family of time-varying regression formulations, see Chapter 6 and the references therein.

2.1.2 *Similar transformations*

There are infinite numbers of SS models realizing the same output. To see this, consider the linear transformation of the state variables $\bar{\boldsymbol{x}}_t = \boldsymbol{T}\boldsymbol{x}_t$, where $\boldsymbol{T}$ is any arbitrary matrix such that $|\boldsymbol{T}| \neq 0$, and $|.|$ denotes the matrix determinant. It is therefore possible to compute the inverse transformation:

$$\boldsymbol{x}_t = \boldsymbol{T}^{-1}\bar{\boldsymbol{x}}_t. \tag{2.8}$$

Substituting (2.8) in (2.1)–(2.2) we obtain:

$$\boldsymbol{T}^{-1}\bar{\boldsymbol{x}}_{t+1} = \boldsymbol{\Phi}\boldsymbol{T}^{-1}\bar{\boldsymbol{x}}_t + \boldsymbol{\Gamma}\boldsymbol{u}_t + \boldsymbol{E}\boldsymbol{w}_t \tag{2.9}$$

$$\boldsymbol{z}_t = \boldsymbol{H}\boldsymbol{T}^{-1}\bar{\boldsymbol{x}}_t + \boldsymbol{D}\boldsymbol{u}_t + \boldsymbol{C}\boldsymbol{v}_t, \tag{2.10}$$

which immediately yields an SS model equivalent to (2.1)–(2.2):

$$\bar{\boldsymbol{x}}_{t+1} = \boldsymbol{T}\boldsymbol{\Phi}\boldsymbol{T}^{-1}\bar{\boldsymbol{x}}_t + \boldsymbol{T}\boldsymbol{\Gamma}\boldsymbol{u}_t + \boldsymbol{T}\boldsymbol{E}\boldsymbol{w}_t \tag{2.11}$$

$$\boldsymbol{z}_t = \boldsymbol{H}\boldsymbol{T}^{-1}\bar{\boldsymbol{x}}_t + \boldsymbol{D}\boldsymbol{u}_t + \boldsymbol{C}\boldsymbol{v}_t. \tag{2.12}$$

This change of variables in the SS model is known as a *similar transformation* because the transition matrices in (2.1)–(2.2) and (2.11)–(2.12) are similar to each other. By selecting an adequate similar transformation, one can obtain specific forms of the SS model with convenient properties, see e.g. Subsection 3.1.3.

2.1.3 *Properties of the State-Space model*

We will now define some properties of the SS model (2.1)–(2.2) that will be used throughout the rest of the book. Some of them rely on the basic concept of the *system mode*.

A *system mode* (henceforth *mode*) is an eigenvector of the transition matrix $\boldsymbol{\Phi}$. By definition, it could be interpreted as a value of the state vector which would not be altered (except for a multiplicative constant) upon propagation from t to $t+1$. The multiplicative constant is the corresponding eigenvalue, which therefore acts as an attenuation or amplification factor of the mode.

For example, consider a situation where the state transition matrix is given by

$$\boldsymbol{\Phi} = \begin{bmatrix} 1.0 & 1.0 & 0 & 0 \\ 0 & 1.0 & 0 & 0 \\ 0 & 0 & 0.7 & 0 \\ 0 & 0 & 0 & 0.5 \end{bmatrix},$$

with the eigenvalue-eigenvector pairs $\{\lambda_i, \nu_i\}$, where $\lambda_1 = 1.0 = \lambda_2$ (the eigenvalue 1.0 has multiplicity 2), $\lambda_3 = 0.7$, and $\lambda_4 = 0.5$. The corresponding eigenvectors are

$$\boldsymbol{\nu}_1 = \begin{bmatrix} 1.0 \\ 0 \\ 0 \\ 0 \end{bmatrix}, \quad \boldsymbol{\nu}_2 = \begin{bmatrix} -1.0 \\ 0 \\ 0 \\ 0 \end{bmatrix}, \quad \boldsymbol{\nu}_3 = \begin{bmatrix} 0 \\ 0 \\ 1.0 \\ 0 \end{bmatrix}, \quad \boldsymbol{\nu}_4 = \begin{bmatrix} 0 \\ 0 \\ 0 \\ 1.0 \end{bmatrix}.$$

This system would have three modes (which coincides with the number of independent eigenvectors, or geometric multiplicity of the transition matrix), two of them stationary, driven by λ_3 and λ_4, and one of them nonstationary, driven by λ_1 along the direction specified by $\boldsymbol{\nu}_1$. Thus: the first mode is $\boldsymbol{\nu}_1$, the second mode is $\boldsymbol{\nu}_3$, and the third mode is $\boldsymbol{\nu}_4$.

With this in place, we are now ready to define the following fundamental properties.

Definition 2.1 (Stability). The SS model is *stable* if all the roots of the characteristic equation $|\boldsymbol{I} - \boldsymbol{\Phi}\,\lambda\,| = 0$ lie outside the unit circle.

Note that: (a) the condition above holds if and only if all the eigenvalues of the transition matrix $\boldsymbol{\Phi}$ lie inside the unit circle, and (b) stability implies that, if the system inputs and errors are set to constant values, the state vector will converge to a steady-state equilibrium.

Definition 2.2 (Reachability / Controllability). The SS model is *reachable* if choosing adequately the value of the system inputs, $\boldsymbol{u}_t$, one can reach any arbitrary value for the state vector, starting from a null initial condition $\boldsymbol{x}_1 = \boldsymbol{0}$.

This property is also known as *controllability from the origin*. Hereafter, we will call it *controllability*, without further specification. Controllability is also linked to a dual property known as *observability*.

Definition 2.3 (Observability). The SS model is *observable* if, given the information about the inputs and outputs, one can uniquely determine the initial state.

The concepts of controllability and observability can be formalized as follows. First (and recalling that n is the dimension of the state vector) it is easy to see that, when $t = n$, the state vector can be written as:

$$\boldsymbol{x_n} = \boldsymbol{\Phi}^n \boldsymbol{x_1} + \begin{bmatrix} \boldsymbol{\Phi}^{n-1}(\boldsymbol{\Gamma E}) & \boldsymbol{\Phi}^{n-2}(\boldsymbol{\Gamma E}) & \ldots & \boldsymbol{\Phi}(\boldsymbol{\Gamma E}) & \boldsymbol{\Gamma E} \end{bmatrix} \begin{bmatrix} \boldsymbol{u_1} \\ \boldsymbol{w_1} \\ \boldsymbol{u_2} \\ \boldsymbol{w_2} \\ \vdots \\ \boldsymbol{u_{n-2}} \\ \boldsymbol{w_{n-2}} \\ \boldsymbol{u_{n-1}} \\ \boldsymbol{w_{n-1}} \end{bmatrix} \tag{2.13}$$

and, from (2.13), a necessary and sufficient condition for controllability is $\text{rank}(\boldsymbol{C_n}) = n$, where $\boldsymbol{C_n}$ is the controllability matrix:

$$\boldsymbol{C_n} = \begin{bmatrix} \boldsymbol{\Phi}^{n-1}(\boldsymbol{\Gamma E Q}^{1/2}) & \boldsymbol{\Phi}^{n-2}(\boldsymbol{\Gamma E Q}^{1/2}) & \ldots & \boldsymbol{\Phi}(\boldsymbol{\Gamma E Q}^{1/2}) & \boldsymbol{\Gamma E Q}^{1/2} \end{bmatrix} \tag{2.14}$$

so, if the rank of $\boldsymbol{C_n}$ coincides with the number of states, the image set of $\boldsymbol{C_n}$ is $\mathbb{R}^n$ and one can determine the value of $\boldsymbol{x_n}$ by choosing the sequences $\{\boldsymbol{u_t}, \boldsymbol{w_t}\}_{t=1}^n$. Similarly, it is easy to see that the observations up to $t = n$ can be written as:

$$\begin{bmatrix} \boldsymbol{z_1} \\ \boldsymbol{z_2} \\ \vdots \\ \boldsymbol{z_n} \end{bmatrix} = \begin{bmatrix} \boldsymbol{H} \\ \boldsymbol{H\Phi} \\ \vdots \\ \boldsymbol{H\Phi}^{n-1} \end{bmatrix} \boldsymbol{x_1} + \begin{bmatrix} \boldsymbol{D} & \boldsymbol{0} & \ldots & \boldsymbol{0} \\ \boldsymbol{H\Gamma} & \boldsymbol{D} & \ldots & \boldsymbol{0} \\ \vdots & \vdots & \ddots & \vdots \\ \boldsymbol{H\Phi}^{n-2}\boldsymbol{\Gamma} & \boldsymbol{H\Phi}^{n-3}\boldsymbol{\Gamma} & \ldots & \boldsymbol{D} \end{bmatrix} \begin{bmatrix} \boldsymbol{u_1} \\ \boldsymbol{u_2} \\ \vdots \\ \boldsymbol{u_n} \end{bmatrix}$$
$$+ \begin{bmatrix} \boldsymbol{0} & \boldsymbol{0} & \ldots & \boldsymbol{0} \\ \boldsymbol{HE} & \boldsymbol{0} & \ldots & \boldsymbol{0} \\ \vdots & \vdots & \ddots & \vdots \\ \boldsymbol{H\Phi}^{n-2}\boldsymbol{E} & \boldsymbol{H\Phi}^{n-3}\boldsymbol{E} & \ldots & \boldsymbol{0} \end{bmatrix} \begin{bmatrix} \boldsymbol{w_1} \\ \boldsymbol{w_2} \\ \vdots \\ \boldsymbol{w_n} \end{bmatrix} + \begin{bmatrix} \boldsymbol{C} & \boldsymbol{0} & \ldots & \boldsymbol{0} \\ \boldsymbol{0} & \boldsymbol{C} & \ldots & \boldsymbol{0} \\ \vdots & \vdots & \ddots & \vdots \\ \boldsymbol{0} & \boldsymbol{0} & \ldots & \boldsymbol{C} \end{bmatrix} \begin{bmatrix} \boldsymbol{v_1} \\ \boldsymbol{v_2} \\ \vdots \\ \boldsymbol{v_n} \end{bmatrix} \tag{2.15}$$

Therefore, a necessary and sufficient condition for observability is that $\text{rank}(\boldsymbol{O_n}) = n$, where $\boldsymbol{O_n}$ is the observability matrix:

$$\boldsymbol{O_n} = \begin{bmatrix} \boldsymbol{H}^T & (\boldsymbol{H\Phi})^T \ldots & (\boldsymbol{H\Phi}^{n-1})^T \end{bmatrix}^T. \tag{2.16}$$

Definition 2.4 (Minimality). The SS model is *minimal* if there is no alternative representation that realizes the same output, $\boldsymbol{z_t}$, with fewer states.

It can be proved that a model is minimal if and only if it is observable and controllable. Throughout this book we will assume that the SS models employed are minimal. This assumption is made without loss of generality because any given system can be transformed into an equivalent minimal representation by deleting its unobservable/non-controllable modes using the so-called *staircase algorithm*, see Rosenbrock [190].

Definition 2.5 (Canonical representation). An SS model is said to be in *canonical representation* if its matrices have a particular structure such that there is no similar transformation matrix, other than identity or trivial permutation matrices, that preserves this structure.

Canonical models assure identifiability and reduce the dimension of the parametric space, as they have the minimal number of parameters required to realize the output, z_t. A well known canonical SS representation is the *Luenberger Canonical Form*, which can be obtained from any SS model by means of a similar transformation, see Petkov, Christov and Konstantinov [181]. Chapter 9 provides further details about this form and its usefulness in deriving the VARMAX representation of an SS model.

Definition 2.6 (Detectability). The SS model is *detectable* if any unobservable mode is stable.

If a given SS model has some unobservable modes, there always exists a similar transformation that allows one to separate the observable and unobservable modes. That is:

$$\boldsymbol{\Phi}^{**} = \begin{bmatrix} \boldsymbol{\Phi}_{11}^{**} & \mathbf{0} \\ \boldsymbol{\Phi}_{21}^{**} & \boldsymbol{\Phi}_{22}^{**} \end{bmatrix} = \boldsymbol{T}\boldsymbol{\Phi}\boldsymbol{T}^{-1};\ \boldsymbol{H}^{**} = \begin{bmatrix} \boldsymbol{H}_{1}^{**} & \mathbf{0} \end{bmatrix} = \boldsymbol{H}\boldsymbol{T}^{-1} \tag{2.17}$$

where the pair $[\boldsymbol{H}_1^{**} \quad \boldsymbol{\Phi}_{11}^{**}]$ is observable and $\boldsymbol{T}$ is the matrix characterizing the similar transformation. If the system is detectable, all the eigenvalues of $\boldsymbol{\Phi}_{22}^{**}$ would then lie within the unit circle. Therefore, any unobservable states are assured to be stable.

This transformation allows us to understand the concept of observability because only the states associated to $\boldsymbol{\Phi}_{11}^{**}$ affect the system output, so we cannot estimate the unobservable states conditioned to the observed variables.

Similarly, if an SS model has some non-controllable modes, there exists a similar transformation such that:

$$\boldsymbol{\Phi}^{**} = \begin{bmatrix} \boldsymbol{\Phi}_{11}^{**} & \mathbf{0} \\ \boldsymbol{\Phi}_{21}^{**} & \boldsymbol{\Phi}_{22}^{**} \end{bmatrix};\ \boldsymbol{\Gamma}^{**} = \begin{bmatrix} \mathbf{0} \\ \boldsymbol{\Gamma}_{2}^{**} \end{bmatrix};\ \boldsymbol{E}\boldsymbol{Q}^{1/2} = \begin{bmatrix} \mathbf{0} \\ \boldsymbol{E}_{2}^{**} \end{bmatrix} \tag{2.18}$$

where the non-controllable modes are associated to $\boldsymbol{\Phi}_{11}^{**}$ and are not affected by any source of excitation. As a consequence, the non-controllable subsystem affects the system as a deterministic function of the initial state. If the eigenvalues of $\boldsymbol{\Phi}_{11}^{**}$ are less than one, we can cancel this effect by assuming that the initial state of the non-controllable subsystem is zero, see the immemorial time assumption in Definition 2.8 below.

Definition 2.7 (Stabilizability). The SS model is *stabilizable* if any non-controllable mode is stable.

Definition 2.8 (Immemorial time assumption): The SS model fulfills the immemorial time assumption if: (a) the state transition equation holds for $t = 0, -1, -2...$ and (b) $\boldsymbol{x}_r = \mathbf{0}$ with $r \to -\infty$.

Definition 2.9 (Weak stationarity): The SS model is weakly stationary (hereafter, stationary) if the first and second-order moments of its output, z_t, are time-invariant.

The following results describe the relationship between the concepts of stationarity and stability.

Result 2.1: Given a stable system without inputs, initialized with $\mathrm{E}(\boldsymbol{x}_1)$ and $\mathrm{cov}(\boldsymbol{x}_1)$, the process $\{\boldsymbol{x}_t\}$ is asymptotically weakly stationary with $\lim_{t\to\infty} \mathrm{E}(\boldsymbol{x}_t) = \mathbf{0}$ and $\lim_{t\to\infty} \mathrm{cov}(\boldsymbol{x}_t) = \boldsymbol{P}$, where the covariance $\boldsymbol{P}$ satisfies the Lyapunov algebraic equation $\boldsymbol{P} = \boldsymbol{\Phi}\boldsymbol{P}\boldsymbol{\Phi}^T + \boldsymbol{E}\boldsymbol{Q}\boldsymbol{E}^T$.

Result 2.2: If a stable system is initialized with $\mathrm{E}(\boldsymbol{x}_1) = \mathbf{0}$ and $\boldsymbol{P}_1 = \boldsymbol{\Phi}\boldsymbol{P}_1\boldsymbol{\Phi}^T + \boldsymbol{E}\boldsymbol{Q}\boldsymbol{E}^T$, then the process $\{\boldsymbol{x}_t\}$ is stationary.

Result 2.3: If a stable system without inputs was started in immemorial time, then the processes $\{\boldsymbol{x}_t\}$ and $\{\boldsymbol{z}_t\}$ are stationary within the time span of the sample.

Therefore, the concepts of stationarity and stability are different *stricto sensu*, but closely related if the system has no inputs[3]. For a formal proof of these results see Anderson and Moore [6].

2.2 Single error models

2.2.1 *Model formulation*

The SS representation defined in Section 2.1 is a *multiple-error model*, or MEM, because it has different errors affecting the state and observation equations. An alternative is a representation with a single error affecting both equations. This *single-error model*, or SEM, is given by:

$$\boldsymbol{x}_{t+1} = \boldsymbol{\Phi}\boldsymbol{x}_t + \boldsymbol{\Gamma}\boldsymbol{u}_t + \boldsymbol{K}\boldsymbol{a}_t \tag{2.19}$$

$$\boldsymbol{z}_t = \boldsymbol{H}\boldsymbol{x}_t + \boldsymbol{D}\boldsymbol{u}_t + \boldsymbol{a}_t \tag{2.20}$$

where the error term is such that $\boldsymbol{a}_t \sim \mathrm{IID}(\mathbf{0}, \boldsymbol{B})$, that is, $\boldsymbol{a}_t$ is a sequence of *independent and identically distributed* (IID) zero-mean random vectors with common covariance matrix $\boldsymbol{B}$.

2.2.2 *State estimation in the SEM*

The importance of the SEM representation lies in the fact that, under nonrestrictive assumptions, its states can be estimated with null uncertainty. To see this, note that equation (2.20) implies that $\boldsymbol{a}_t = \boldsymbol{z}_t - \boldsymbol{H}\boldsymbol{x}_t - \boldsymbol{D}\boldsymbol{u}_t$. Substituting this expression in (2.19) and rearranging the resulting terms yields:

$$\boldsymbol{x}_{t+1} = (\boldsymbol{\Phi} - \boldsymbol{K}\boldsymbol{H})\boldsymbol{x}_t + (\boldsymbol{\Gamma} - \boldsymbol{K}\boldsymbol{D})\boldsymbol{u}_t + \boldsymbol{K}\boldsymbol{z}_t. \tag{2.21}$$

Therefore, if the initial state $\boldsymbol{x}_1$ were known, then the entire sequence of states would be determined by (2.21) because all the terms on the right-hand-side of this expression would be known exactly.

[3]The assumption that the system has no inputs avoids the interference of an exogenous component that could be nonstationary. Equivalent results can be formulated under the assumption that all the inputs are stationary.

However, usually $\boldsymbol{x}_1$ is not known, but is treated as a random vector. Therefore, the propagation of (2.21) will converge exponentially fast to the exact state values as the number of samples processed increases, under the following *invertibility* condition.

Definition 2.10 (Invertibility): Model (2.19)–(2.20) is said to be invertible if all the roots of $|\boldsymbol{I}-(\boldsymbol{\Phi}-\boldsymbol{K}\boldsymbol{H})\,\lambda|=0$ lie outside the unit circle, where $\boldsymbol{K}=(\boldsymbol{\Phi}\boldsymbol{P}\boldsymbol{H}^T+\boldsymbol{E}\boldsymbol{S}\boldsymbol{C}^T)\,\boldsymbol{B}^{-1}$.

This requirement is not restrictive since the SEM representation was obtained assuming that the system is detectable. It can be proved that detectability ensures that the algebraic Riccati equation (2.25) has a strong solution, which in turn guarantees invertibility of the SEM; see Theorem 3.2. in De Souza, Gevers, and Goodwin [71].

Note that Definition 2.10 above is limited to models in SEM form because it uses the error distribution matrix $\boldsymbol{K}$ in (2.19). On the other hand, it can be extended to a MEM model just by obtaining the equivalent SEM formulation by means of the conversion procedure described in the next subsection.

2.2.3 *Obtaining the SEM equivalent to a MEM*

At first sight, the SEM (2.19)–(2.20) may seem to be a particular case of (2.1)–(2.2), constrained so that $\boldsymbol{w}_t=\boldsymbol{v}_t=\boldsymbol{a}_t$ and $\boldsymbol{C}=\boldsymbol{I}$. However, under weak assumptions model (2.1)–(2.2) can also be written as an observationally equivalent SEM, and thus both representations are equally general in practice.

Consider the following MEM for $\boldsymbol{z}_t$:

$$\bar{\boldsymbol{x}}_{t+1}=\boldsymbol{\Phi}\bar{\boldsymbol{x}}_t+\boldsymbol{\Gamma}\boldsymbol{u}_t+\boldsymbol{E}\boldsymbol{w}_t \tag{2.22}$$

$$\boldsymbol{z}_t=\boldsymbol{H}\bar{\boldsymbol{x}}_t+\boldsymbol{D}\boldsymbol{u}_t+\boldsymbol{C}\boldsymbol{v}_t, \tag{2.23}$$

where the mean and covariance of the initial state are denoted, respectively, by $\bar{\boldsymbol{\mu}}_1$ and $\bar{\boldsymbol{P}}_1$. Under these conditions, the time series $\boldsymbol{z}_t$ in (2.23) is also the output of the SEM (2.19)–(2.20) with:

$$\boldsymbol{x}_1\sim(\bar{\boldsymbol{\mu}}_1,\,\bar{\boldsymbol{P}}_1-\boldsymbol{P}) \tag{2.24}$$

$$\boldsymbol{P}=\boldsymbol{\Phi}\boldsymbol{P}\boldsymbol{\Phi}^T+\boldsymbol{E}\boldsymbol{Q}\boldsymbol{E}^T-\boldsymbol{K}\boldsymbol{B}\boldsymbol{K}^T \tag{2.25}$$

$$\boldsymbol{K}=\left(\boldsymbol{\Phi}\boldsymbol{P}\boldsymbol{H}^T+\boldsymbol{E}\boldsymbol{S}\boldsymbol{C}^T\right)\boldsymbol{B}^{-1} \tag{2.26}$$

$$\boldsymbol{B}=\boldsymbol{H}\boldsymbol{P}\boldsymbol{H}^T+\boldsymbol{C}\boldsymbol{R}\boldsymbol{C}^T. \tag{2.27}$$

This result needs only two nonrestrictive assumptions: (a) that the system (2.22)–(2.23) be detectable (see Definition 2.6), and (b) that the difference between the covariances of the initial states $\bar{\boldsymbol{P}}_1-\boldsymbol{P}$ be positive semi-definite[4]. The proof of this result can be found in Casals, Sotoca and Jerez [45]. Note that:

1. We use the same notation for the coefficient matrices $\boldsymbol{\Phi}$, $\boldsymbol{\Gamma}$, $\boldsymbol{H}$ and $\boldsymbol{D}$ in the MEM and SEM forms because they coincide when the observational equivalence between both formulations is enforced according to equations (2.24)–(2.27).

[4]This condition is sometimes justified using the immemorial time hypothesis; see Definition 2.8 and De Jong and Chu-Chun-Lin [69].

2. Expressions (2.25)–(2.27) can be interpreted as the steady-state of the Kalman filter equations, see Chapter 4.
3. The MEM (2.22)–(2.23) includes two error terms whereas the SEM (2.19)–(2.20) has only one. This means that, despite the fact that both are observationally equivalent, some information which is explicit in the covariance matrices of the MEM ($\boldsymbol{Q}$, $\boldsymbol{R}$ and $\boldsymbol{S}$) is lost when they are combined in the single error covariance matrix $\boldsymbol{B}$ of the SEM.
4. If the MEM parameter matrices are known, equations (2.24)–(2.27) can be solved for the SEM parameters and vice versa. The key to this process lies in solving equations (2.25)–(2.27) using the efficient and stable procedures described by Ionescu, Oara and Weiss [128].
5. Choosing the strong solution of the Riccati equation (2.25) has an important advantage, as it suffices to ensure that the eigenvalues of $(\boldsymbol{\Phi} - \boldsymbol{E}\boldsymbol{H})$ will lie in or within the unit circle, thus resulting in an invertible VARMAX model.

Example 2.2.3.1 Hodrick-Prescott filter

Consider a trend plus error decomposition:

$$z_t = \mu_t + \varepsilon_t \tag{2.28}$$

where the unobserved trend, μ_t, is governed by an integrated random walk process:

$$\begin{bmatrix} \mu_{t+1} \\ \beta_{t+1} \end{bmatrix} = \begin{bmatrix} 1 & 1 \\ 0 & 1 \end{bmatrix} \begin{bmatrix} \mu_t \\ \beta_t \end{bmatrix} + \begin{bmatrix} 0 \\ \zeta_t \end{bmatrix} \tag{2.29}$$

with the error covariance matrix:

$$cov \begin{bmatrix} \zeta_t \\ \varepsilon_t \end{bmatrix} = \begin{bmatrix} 1/1600 & 0 \\ 0 & 1 \end{bmatrix}. \tag{2.30}$$

Note that the noise-variance ratio of 1/1600 in (2.30) is chosen to make this model compatible with the Hodrick and Prescott [126] filter for quarterly data, see Harvey and Trimbur [120]. Solving equations (2.24)–(2.27) for this model yields the equivalent SEM:

$$z_t = \begin{bmatrix} 1 & 0 \end{bmatrix} \begin{bmatrix} x_t^1 \\ x_t^2 \end{bmatrix} + a_t \; ; \; \sigma_a^2 = 1.2509 \tag{2.31}$$

$$\begin{bmatrix} x_{t+1}^1 \\ x_{t+1}^2 \end{bmatrix} = \begin{bmatrix} 1 & 1 \\ 0 & 1 \end{bmatrix} \begin{bmatrix} x_t^1 \\ x_t^2 \end{bmatrix} + \begin{bmatrix} 0.2229 \\ 0.0224 \end{bmatrix} a_t. \tag{2.32}$$

The observational equivalence of models (2.28)–(2.30) and (2.31)–(2.32) can be tested by checking that they yield the same innovations, likelihood values, and out-of-sample point forecasts.

2.2.4 *Obtaining the SEM equivalent to a general linear process*

The procedure used to obtain the SEM representation described in Subsection 2.2.3 is very general, as it supports any model written as a linear fixed coefficient SS model. However, it requires solving the nontrivial equations (2.25)–(2.27). This can be avoided when the model for the time series is written as a general linear process or in VARMAX form because, in this case, the matrices in (2.19)–(2.20) can be obtained directly from the model parameters.

Assume that $\boldsymbol{z_t}$ follows the model:

$$\bar{\boldsymbol{F}}(\mathrm{B})\,\boldsymbol{z_t} = \bar{\boldsymbol{G}}(\mathrm{B})\,\boldsymbol{u_t} + \bar{\boldsymbol{L}}(\mathrm{B})\,\boldsymbol{a_t}, \tag{2.33}$$

where $\boldsymbol{u_t}$ and $\boldsymbol{a_t}$ are, as in previous sections, vectors of exogenous variables and white noise errors, respectively, B denotes the backshift operator, such that for any ω_t: $\mathrm{B}^i\omega_t = \omega_{t-i}$, $i = 0, \pm1, \pm2, ..., I$ and the polynomial matrices $\bar{\boldsymbol{F}}(\mathrm{B})$, $\bar{\boldsymbol{G}}(\mathrm{B})$ and $\bar{\boldsymbol{L}}(\mathrm{B})$ are defined by:

$$\bar{\boldsymbol{F}}(\mathrm{B}) = \sum_{i=0}^{p} \bar{F}_{\boldsymbol{i}}\,\mathrm{B}^i \tag{2.34}$$

$$\bar{\boldsymbol{G}}(\mathrm{B}) = \sum_{i=0}^{s} \bar{G}_{\boldsymbol{i}}\,\mathrm{B}^i \tag{2.35}$$

$$\bar{\boldsymbol{L}}(\mathrm{B}) = \sum_{i=0}^{q} \bar{L}_i\,\mathrm{B}^i \tag{2.36}$$

Note that model (2.33)–(2.36) relaxes the standard VARMAX form by not requiring the matrices $\bar{\boldsymbol{F}}_{\boldsymbol{0}}$ and $\bar{\boldsymbol{L}}_{\boldsymbol{0}}$ to be the identity. Now, following Terceiro [206], (2.33) can be written as an the equivalent SEM by pre-multiplying (2.33) by $\bar{\boldsymbol{F}}_{\boldsymbol{0}}^{-1}$, which yields:

$$\boldsymbol{F}(\mathrm{B})\,\boldsymbol{z_t} = \boldsymbol{G}(\mathrm{B})\,\boldsymbol{u_t} + \boldsymbol{L}(\mathrm{B})\,\boldsymbol{a_t} \tag{2.37}$$

with $\boldsymbol{F}(\mathrm{B}) = \bar{\boldsymbol{F}}_{\boldsymbol{0}}^{-1}\bar{\boldsymbol{F}}(B)$, $\boldsymbol{G}(\mathrm{B}) = \bar{\boldsymbol{F}}_{\boldsymbol{0}}^{-1}\bar{\boldsymbol{G}}(\mathrm{B})$, and $\boldsymbol{L}(\mathrm{B}) = \bar{\boldsymbol{F}}_{\boldsymbol{0}}^{-1}\bar{\boldsymbol{L}}(\mathrm{B})$. The matrices of the SEM equivalent to (2.37) are then:

$$\boldsymbol{\Phi} = \begin{bmatrix} -\boldsymbol{F_1} & \boldsymbol{I} & \boldsymbol{0} & \dots & \boldsymbol{0} \\ -\boldsymbol{F_2} & \boldsymbol{0} & \boldsymbol{I} & \dots & \boldsymbol{0} \\ \vdots & \vdots & \vdots & \ddots & \vdots \\ -\boldsymbol{F}_{k-1} & \boldsymbol{0} & \boldsymbol{0} & \dots & \boldsymbol{I} \\ -\boldsymbol{F_k} & \boldsymbol{0} & \boldsymbol{0} & \dots & \boldsymbol{0} \end{bmatrix} \tag{2.38}$$

$$\boldsymbol{\Gamma} = \begin{bmatrix} \boldsymbol{G_1} - \boldsymbol{F_1}\,\boldsymbol{G_0} \\ \boldsymbol{G_2} - \boldsymbol{F_2}\,\boldsymbol{G_0} \\ \vdots \\ \boldsymbol{G}_{k-1} - \boldsymbol{F}_{k-1}\,\boldsymbol{G_0} \\ \boldsymbol{G_k} - \boldsymbol{F_k}\,\boldsymbol{G_0} \end{bmatrix} \tag{2.39}$$

$$E = \begin{bmatrix} L_1 - F_1 \\ L_2 - F_2 \\ \vdots \\ L_{k-1} - F_{k-1} \\ L_k - F_k \end{bmatrix} \tag{2.40}$$

$$H = \begin{bmatrix} I & 0 & \ldots & 0 \end{bmatrix} \tag{2.41}$$

$$D = G_0, \tag{2.42}$$

where the error term coincides with the innovations in model (2.33), so that $B = \mathrm{cov}(a_t)$. The dimensions of the matrices (2.38)–(2.42) are governed by the dynamic order of the system $k = \max\{p, s, q\}$ and, according to this order, the dimension of the state vector is $n = mk$, being m the dimension of z_t.

Example 2.2.1 ARIMA models

Consider the following MA(1) and ARMA(1,1) models:

$$z_{t+1} = a_{t+1} - \theta a_t \tag{2.43}$$

$$z_{t+1} = \phi z_t + a_{t+1} - \theta a_t. \tag{2.44}$$

Applying the general transformation defined by equations (2.38)–(2.42) to these models we obtain:

$$x_{t+1} = 0 \cdot x_t - \theta a_t \tag{2.45}$$

$$z_t = x_t + a_t \tag{2.46}$$

$$x_{t+1} = \phi\, x_t + (\phi - \theta)\, a_t \tag{2.47}$$

$$z_t = x_t + a_t, \tag{2.48}$$

where (2.45)–(2.46) is equivalent to (2.43), and (2.47)–(2.48) is equivalent to (2.44).

Example 2.2.2 Transfer functions

Consider also the single-input single-output (SISO) transfer function model:

$$z_t = \frac{\omega_0 - \omega_1 \mathrm{B}}{1 - \delta_1 \mathrm{B}} u_t + (1 - \theta \mathrm{B})\, a_t. \tag{2.49}$$

The equivalent SS formulation in this case would be:

$$\begin{aligned} \begin{bmatrix} x_{t+1}^1 \\ x_{t+1}^2 \end{bmatrix} &= \begin{bmatrix} \delta_1 & 1 \\ 0 & 0 \end{bmatrix} \begin{bmatrix} x_t^1 \\ x_t^2 \end{bmatrix} + \begin{bmatrix} -\omega_1 + \delta_1 \omega_0 \\ 0 \end{bmatrix} u_t + \begin{bmatrix} \theta \\ -\theta\, \delta_1 \end{bmatrix} a_t \\ z_t &= \begin{bmatrix} 1 & 0 \end{bmatrix} \begin{bmatrix} x_t^1 \\ x_t^2 \end{bmatrix} + \omega_0 u_t + a_t. \end{aligned} \tag{2.50}$$

Example 2.2.3 VARMAX model

As a final example, consider the VAR(2) model:

$$\begin{bmatrix} \phi_{11} & \phi_{12} \\ 0 & \phi_{22} \end{bmatrix} \begin{bmatrix} z_t^1 \\ z_t^2 \end{bmatrix} = \begin{bmatrix} a_t^1 \\ a_t^2 \end{bmatrix}, \tag{2.51}$$

whose equivalent SS representation is:

$$\begin{aligned} \begin{bmatrix} x_{t+1}^1 \\ x_{t+1}^2 \end{bmatrix} &= \begin{bmatrix} -\phi_{11} & -\phi_{12} \\ 0 & -\phi_{22} \end{bmatrix} \begin{bmatrix} x_t^1 \\ x_t^2 \end{bmatrix} + \begin{bmatrix} -\phi_{11} & -\phi_{12} \\ 0 & -\phi_{22} \end{bmatrix} \begin{bmatrix} a_t^1 \\ a_t^2 \end{bmatrix} \\ \begin{bmatrix} z_t^1 \\ z_t^2 \end{bmatrix} &= \begin{bmatrix} x_t^1 \\ x_t^2 \end{bmatrix} + \begin{bmatrix} a_t^1 \\ a_t^2 \end{bmatrix}. \end{aligned} \tag{2.52}$$

Chapter 3

Model transformations

3.1 Model decomposition 19
3.1.1 Deterministic and stochastic subsystems 19
3.1.2 Implied univariate models 21
3.1.3 Block-diagonal forms 22
3.2 Model combination 24
3.2.1 Endogeneization of stochastic inputs 24
3.2.2 Colored errors 26
3.3 Change of variables in the output 28
3.3.1 Observation errors 28
3.3.2 Missing values and aggregated data 29
3.3.3 Temporal aggregation 30
3.4 Uses of these transformations 31

In this chapter we discuss different model transformations. These can be classified into three groups: model decomposition, model combination, and changes in the output variable. The last section discusses their practical applications.

3.1 Model decomposition

The expression "model decomposition" refers to the extraction of meaningful submodels from a given initial formulation.

3.1.1 Deterministic and stochastic subsystems

Assume that $\boldsymbol{z}_t$ is the observable output of the MEM (2.1)–(2.2). Consider also the additive decompositions:

$$\boldsymbol{z}_t = \boldsymbol{z}_t^d + \boldsymbol{z}_t^s \tag{3.1}$$

$$\boldsymbol{x}_t = \boldsymbol{x}_t^d + \boldsymbol{x}_t^s \tag{3.2}$$

where z_t^d denotes an additive "deterministic" component of the time series, which depends exclusively on the value of the inputs, while z_t^s refers to the "stochastic" component, which only depends on the errors. Accordingly, x_t^d and x_t^s are the deterministic and stochastic components of the state vector, respectively. Substituting (3.2) into (2.1)–(2.2) and (3.1) into (2.2) yields:

$$\boldsymbol{x}_{t+1}^d + \boldsymbol{x}_{t+1}^s = \boldsymbol{\Phi}\,(\boldsymbol{x}_t^d + \boldsymbol{x}_t^s) + \boldsymbol{\Gamma}\,\boldsymbol{u}_t + \boldsymbol{E}\,\boldsymbol{w}_t \tag{3.3}$$

$$z_t^d + z_t^s = \boldsymbol{H}(\boldsymbol{x}_t^d + \boldsymbol{x}_t^s) + \boldsymbol{D}\,\boldsymbol{u}_t + \boldsymbol{C}\,\boldsymbol{v}_t, \tag{3.4}$$

and (3.3)–(3.4) can be immediately decomposed into two different SS models. The deterministic subsystem, which realizes z_t^d, is:

$$\boldsymbol{x}_{t+1}^d = \boldsymbol{\Phi}\,\boldsymbol{x}_t^d + \boldsymbol{\Gamma}\,\boldsymbol{u}_t \tag{3.5}$$

$$z_t^d = \boldsymbol{H}\,\boldsymbol{x}_t^d + \boldsymbol{D}\,\boldsymbol{u}_t, \tag{3.6}$$

and the stochastic subsystem, whose output is z_t^s, is given by:

$$\boldsymbol{x}_{t+1}^s = \boldsymbol{\Phi}\,\boldsymbol{x}_t^s + \boldsymbol{E}\,\boldsymbol{w}_t \tag{3.7}$$

$$z_t^s = \boldsymbol{H}\,\boldsymbol{x}_t^s + \boldsymbol{C}\,\boldsymbol{v}_t. \tag{3.8}$$

This deterministic/stochastic decomposition is a standard result of linear systems theory, see e.g., Van Overschee and De Moor [215] or Casals, Jerez, and Sotoca [42]. The terms "deterministic" and "stochastic" may be confusing for some readers, who would probably find the terms "input-driven" and "error-driven" to be more accurate. Taking this remark into account, we will maintain the original terminology. In Chapter 8 we will show how this decomposition can be applied to extract the deterministic components from a time series.

The following example illustrates this decomposition on the basis of a simple transfer function model.

Example 3.1.1 Obtaining the deterministic and stochastic subsystems implied by a transfer function

Consider the following SISO transfer function:

$$z_t = \frac{0.5}{1-0.7\mathrm{B}}u_t + \frac{1-0.8\mathrm{B}}{1-\mathrm{B}}a_t. \tag{3.9}$$

With this notation, writing the models for the deterministic and stochastic components is trivial, since taking into account expression (3.1), it is obvious that:

$$z_t^d = \frac{0.5}{1-0.7\mathrm{B}}u_t \tag{3.10}$$

$$z_t^s = \frac{1-0.8\mathrm{B}}{1-\mathrm{B}}a_t. \tag{3.11}$$

On the other hand, following the procedure described in Subsection 2.2.4, it is easy to obtain the SS representation of model (3.9):

$$\begin{aligned}\begin{bmatrix} x_{t+1}^1 \\ x_{t+1}^2 \end{bmatrix} &= \begin{bmatrix} 1.7 & 1 \\ -0.7 & 0 \end{bmatrix} \begin{bmatrix} x_t^1 \\ x_t^2 \end{bmatrix} + \begin{bmatrix} 0.35 \\ -0.35 \end{bmatrix} u_t + \begin{bmatrix} 0.2 \\ -0.14 \end{bmatrix} a_t \\ z_t &= \begin{bmatrix} 1 & 0 \end{bmatrix} \begin{bmatrix} x_t^1 \\ x_t^2 \end{bmatrix} + 0.5 u_t + a_t, \end{aligned} \tag{3.12}$$

which can be decomposed into the following additive deterministic and stochastic subsystems:

$$\begin{aligned}\begin{bmatrix} x_{t+1}^{1,d} \\ x_{t+1}^{2,d} \end{bmatrix} &= \begin{bmatrix} 1.7 & 1 \\ -0.7 & 0 \end{bmatrix} \begin{bmatrix} x_t^{1,d} \\ x_t^{2,d} \end{bmatrix} + \begin{bmatrix} 0.35 \\ -0.35 \end{bmatrix} u_t \\ z_t^d &= \begin{bmatrix} 1 & 0 \end{bmatrix} \begin{bmatrix} x_t^{1,d} \\ x_t^{2,d} \end{bmatrix} + 0.5 u_t \end{aligned} \tag{3.13}$$

$$\begin{aligned}\begin{bmatrix} x_{t+1}^{1,s} \\ x_{t+1}^{2,s} \end{bmatrix} &= \begin{bmatrix} 1.7 & 1 \\ -0.7 & 0 \end{bmatrix} \begin{bmatrix} x_t^{1,s} \\ x_t^{2,s} \end{bmatrix} + \begin{bmatrix} 0.2 \\ -0.14 \end{bmatrix} a_t \\ z_t^s &= \begin{bmatrix} 1 & 0 \end{bmatrix} \begin{bmatrix} x_t^{1,s} \\ x_t^{2,s} \end{bmatrix} + a_t. \end{aligned} \tag{3.14}$$

3.1.2 *Implied univariate models*

Consider a minimal system written in the SEM form (2.19)–(2.20). If we combine the state equation (2.19) with the *i*-th row of the observation equation (2.20), we obtain the implied univariate model for the *i*-th variable in $\boldsymbol{z}_t$:

$$\boldsymbol{x}_{t+1} = \boldsymbol{\Phi x}_t + \boldsymbol{\Gamma u}_t + \boldsymbol{K a}_t \tag{3.15}$$

$$z_t^i = \boldsymbol{H}^i \boldsymbol{x}_t + \boldsymbol{D}^i \boldsymbol{u}_t + a_t^i, \tag{3.16}$$

where z_t^i, $\boldsymbol{H}^i$, $\boldsymbol{D}^i$, and a_t^i denote the *i*-th row of the corresponding vectors and matrices. Note that the *i*-th variable z_t^i may be unaffected by some states in $\boldsymbol{x}_t$, so that minimality of (2.1)–(2.2) does not ensure that model (3.15)–(3.16) is minimal, see Definition 2.4. To obtain the minimal representation we could apply the aforementioned staircase algorithm to the implied univariate model, see Rosenbrock [190], to delete any unobservable/non-controllable modes and, accordingly, to obtain a minimal representation for z_t^i. Note also that Chapter 9 presents two procedures to obtain the VARMAX representation equivalent to an SS model. Applying them to model (3.15)–(3.16) would result in the equivalent ARIMAX form.

The following example illustrates these ideas by decomposing a simple bivariate model.

Example 3.1.2 Obtaining the ARIMA models implied by a VARMAX process

Consider the VARMA model:

$$\begin{bmatrix} 1-0.8\mathrm{B} & 0.2\mathrm{B} \\ -0.3\mathrm{B} & 1 \end{bmatrix} \begin{bmatrix} z_t^1 \\ z_t^2 \end{bmatrix} = \begin{bmatrix} 1 & 0 \\ 0 & 1-0.5\mathrm{B} \end{bmatrix} \begin{bmatrix} a_t^1 \\ a_t^2 \end{bmatrix}, \tag{3.17}$$

whose equivalent SEM representation is given by:

$$\begin{aligned} \begin{bmatrix} x_{t+1}^1 \\ x_{t+1}^2 \end{bmatrix} &= \begin{bmatrix} 0.8 & -0.2 \\ 0.3 & 0 \end{bmatrix} \begin{bmatrix} x_t^1 \\ x_t^2 \end{bmatrix} + \begin{bmatrix} 0.8 & -0.2 \\ 0.3 & -0.5 \end{bmatrix} \begin{bmatrix} a_t^1 \\ a_t^2 \end{bmatrix} \\ \begin{bmatrix} z_t^1 \\ z_t^2 \end{bmatrix} &= \begin{bmatrix} x_t^1 \\ x_t^2 \end{bmatrix} + \begin{bmatrix} a_t^1 \\ a_t^2 \end{bmatrix}. \end{aligned} \tag{3.18}$$

In this case, the implied SS model for the first output would be given by the state equation in (3.18) and the first row of its observation equation, that is:

$$z_t^1 = x_t^1 + a_t^1. \tag{3.19}$$

According to (3.18), the implied model for the second output would be given by the same state equation and the second row of its observation equation:

$$z_t^2 = x_t^2 + a_t^2. \tag{3.20}$$

Finally, applying the procedures discussed in Chapter 9, we would obtain the ARIMA processes implied by models (3.17) or (3.18):

$$\left(1-0.800\mathrm{B}+0.060\mathrm{B}^2\right) z_t^1 = \left(1-0.111\mathrm{B}+0.048\mathrm{B}^2\right) \varepsilon_t^1, \quad \sigma_{\varepsilon^1}^2 = 1.035, \tag{3.21}$$

$$\left(1-0.800\mathrm{B}+0.060\mathrm{B}^2\right) z_t^2 = \left(1-1.004\mathrm{B}+0.332\mathrm{B}^2\right) \varepsilon_t^2, \quad \sigma_{\varepsilon^2}^2 = 1.204, \tag{3.22}$$

where the errors ε_t^1 and ε_t^2 are functions of a_t^1 and a_t^2, see Chapter 9.

The code required to replicate this example can be found in the file `Example0312.m`, see Appendix D.

3.1.3 Block-diagonal forms

As we saw in Section 2.1, if $\boldsymbol{z_t}$ is realized by (2.1)–(2.2), then it is also realized by the model:

$$\bar{\boldsymbol{x}}_{t+1} = \bar{\boldsymbol{\Phi}}\,\bar{\boldsymbol{x}}_t + \bar{\boldsymbol{\Gamma}}\boldsymbol{u_t} + \bar{\boldsymbol{E}}\boldsymbol{w_t} \tag{3.23}$$

$$\boldsymbol{z_t} = \bar{\boldsymbol{H}}\,\bar{\boldsymbol{x}}_t + \boldsymbol{D}\boldsymbol{u_t} + \boldsymbol{C}\boldsymbol{v_t}, \tag{3.24}$$

where the states in (3.23)–(3.24) are related to those in (2.1)–(2.2) by the similar transformation $\boldsymbol{T}\boldsymbol{x_t} = \bar{\boldsymbol{x}}_t$, see Subsection 2.1.2.

The transformation matrix $\boldsymbol{T}$ is typically chosen so that the resulting model has some specific advantage. In particular, we could determine $\boldsymbol{T}$ so that the transformed transition matrix $\bar{\Phi}$ is *block-diagonal*[1]. The advantage of this specific decomposition lies in the fact that the states of the transformed model are uniquely associated with the eigenvalues of the transition matrix which characterize the dynamic structure of the model, as well as its properties in the frequency domain.

Specifically, if $\lambda_{j,k} = a_j \pm b_j i$ is a pair of complex conjugate eigenvalues of Φ, the corresponding states will generate a peak in the pseudo-spectrum of the series at the frequency $f_{j,k} = (2\pi)^{-1} \arctan(b_j/a_j)$ cycles per time unit, and the corresponding values of the period (p) and damping factor (df) would then be $p = 1/f_{j,k}$ and $df = \|\lambda_{j,k}\|$, where $\|.\|$ denotes modulus. In this way, a block-diagonal model decomposes the response to shocks in $\boldsymbol{u_t}$ and $\boldsymbol{w_t}$ into the basic movements corresponding to the different eigenvalues of the transition matrix, which represent independent reactions of the state variables to an excitation in any input of the state equation. The following example clarifies these ideas.

Example 3.1.3 An ARIMA model in block-diagonal form

Consider the ARIMA(2,1,1) model:

$$(1+0.75\mathrm{B}+0.25\mathrm{B}^2)(1-\mathrm{B})z_t = (1-0.5\mathrm{B})a_t, \tag{3.25}$$

which has a unit root and two stationary AR roots: $-1.5 \pm 1.323i$. Applying the method described in Subsection 2.2.4, we can obtain the equivalent SEM:

$$\begin{bmatrix} x^1_{t+1} \\ x^2_{t+1} \\ x^3_{t+1} \end{bmatrix} = \begin{bmatrix} 0.25 & 1 & 0 \\ 0.50 & 0 & 1 \\ 0.25 & 0 & 0 \end{bmatrix} \begin{bmatrix} x^1_t \\ x^2_t \\ x^3_t \end{bmatrix} + \begin{bmatrix} -0.25 \\ 0.50 \\ 0.25 \end{bmatrix} a_t \tag{3.26}$$

$$z_t = \begin{bmatrix} 1 & 0 & 0 \end{bmatrix} \begin{bmatrix} x^1_t \\ x^2_t \\ x^3_t \end{bmatrix} + a_t, \tag{3.27}$$

in which the states do not have a clear interpretation. However, the previously described transformation provides the equivalent block-diagonal form:

$$\begin{bmatrix} \bar{x}^1_{t+1} \\ \bar{x}^2_{t+1} \\ \bar{x}^3_{t+1} \end{bmatrix} = \begin{bmatrix} 1 & 0 & 0 \\ 0 & -0.375 & 0.909 \\ 0 & -0.120 & -0.375 \end{bmatrix} \begin{bmatrix} \bar{x}^1_t \\ \bar{x}^2_t \\ \bar{x}^3_t \end{bmatrix} + \begin{bmatrix} -0.319 \\ 0.594 \\ 0.022 \end{bmatrix} a_t \tag{3.28}$$

$$z_t = \begin{bmatrix} -0.785 & -0.833 & -0.238 \end{bmatrix} \begin{bmatrix} \bar{x}^1_t \\ \bar{x}^2_t \\ \bar{x}^3_t \end{bmatrix} + a_t, \tag{3.29}$$

[1]The most famous block-diagonalizing algorithm is the Jordan decomposition which, on the other hand, is very unstable, see Golub and Van Loan [109]. We find more convenient a simpler and more stable method which consists of applying the real Schur decomposition to Φ and diagonalizing the resulting real Schur matrix by solving a system of Sylvester equations, see Petkov, Christov and Konstantinov [181]; see Appendix A.

which clearly shows that the first state corresponds to the unit root, while the second and third states describe the harmonic cycle generated by the complex AR roots. Therefore, block-diagonalizing an SS model yields an equivalent form in which the states can be interpreted as unobserved structural components of the time series, see Casals, Jerez, and Sotoca [41].

The code required to replicate this example can be found in the file `Example0313.m`, see Appendix D.

3.2 Model combination

The expression "model combination" refers to the composition of different models into a single SS formulation. In practice, model combination is a useful technique for implementing a complex model by defining simpler pieces and then joining them together.

3.2.1 Endogeneization of stochastic inputs

Following Terceiro [206], Chapter 3, consider the MEM formulation:

$$\boldsymbol{x}_{t+1}^{z} = \boldsymbol{\Phi}^{z}\,\boldsymbol{x}_{t}^{z} + \boldsymbol{\Gamma}\,\boldsymbol{u}_{t} + \boldsymbol{E}^{z}\,\boldsymbol{w}_{t}^{z} \tag{3.30}$$

$$\boldsymbol{z}_{t} = \boldsymbol{H}^{z}\,\boldsymbol{x}_{t}^{z} + \boldsymbol{D}\,\boldsymbol{u}_{t} + \boldsymbol{C}^{z}\,\boldsymbol{v}_{t}^{z}, \tag{3.31}$$

and an SS model for the inputs in (3.30)–(3.31):

$$\boldsymbol{x}_{t+1}^{u} = \boldsymbol{\Phi}^{u}\,\boldsymbol{x}_{t}^{u} + \boldsymbol{E}^{u}\,\boldsymbol{w}_{t}^{u} \tag{3.32}$$

$$\boldsymbol{u}_{t} = \boldsymbol{H}^{u}\,\boldsymbol{x}_{t}^{u} + \boldsymbol{C}^{u}\,\boldsymbol{v}_{t}^{u}, \tag{3.33}$$

where the errors in (3.30)–(3.31) and (3.32)–(3.33) are assumed to be mutually independent.

Building on this notation, our goal consists of formulating a single model for the output vector $\left[\boldsymbol{z}_t^T\ \boldsymbol{u}_t^T\right]^T$. We will refer to this representation as an "endogeneized" or "nested in inputs" model. To this end, we substitute (3.33) in (3.30)–(3.31), obtaining:

$$\boldsymbol{x}_{t+1}^{z} = \boldsymbol{\Phi}^{z}\,\boldsymbol{x}_{t}^{z} + \boldsymbol{\Gamma}\,(\boldsymbol{H}^{u}\,\boldsymbol{x}_{t}^{u} + \boldsymbol{C}^{u}\,\boldsymbol{v}_{t}^{u}) + \boldsymbol{E}^{z}\,\boldsymbol{w}_{t}^{z} \tag{3.34}$$

$$\boldsymbol{z}_{t} = \boldsymbol{H}^{z}\,\boldsymbol{x}_{t}^{z} + \boldsymbol{D}\,(\boldsymbol{H}^{u}\,\boldsymbol{x}_{t}^{u} + \boldsymbol{C}^{u}\,\boldsymbol{v}_{t}^{u}) + \boldsymbol{C}^{z}\,\boldsymbol{v}_{t}^{z}. \tag{3.35}$$

Finally, the state equation of the endogeneized model results from combining expressions (3.34) and (3.32) into a single state equation:

$$\begin{bmatrix} \boldsymbol{x}_{t+1}^{z} \\ \boldsymbol{x}_{t+1}^{u} \end{bmatrix} = \begin{bmatrix} \boldsymbol{\Phi}^{z} & \boldsymbol{\Gamma}\boldsymbol{H}^{u} \\ 0 & \boldsymbol{\Phi}^{u} \end{bmatrix} \begin{bmatrix} \boldsymbol{x}_{t}^{z} \\ \boldsymbol{x}_{t}^{u} \end{bmatrix} + \begin{bmatrix} \boldsymbol{E}^{z} & \boldsymbol{\Gamma}\boldsymbol{C}^{u} & 0 \\ 0 & 0 & \boldsymbol{E}^{u} \end{bmatrix} \begin{bmatrix} \boldsymbol{w}_{t}^{z} \\ \boldsymbol{v}_{t}^{u} \\ \boldsymbol{w}_{t}^{u} \end{bmatrix}, \tag{3.36}$$

whose observation equation is obtained by combining equations (3.35) and (3.33):

$$\begin{bmatrix} \boldsymbol{z}_{t} \\ \boldsymbol{u}_{t} \end{bmatrix} = \begin{bmatrix} \boldsymbol{H}^{z} & \boldsymbol{D}\boldsymbol{H}^{u} \\ 0 & \boldsymbol{H}^{u} \end{bmatrix} \begin{bmatrix} \boldsymbol{x}_{t}^{z} \\ \boldsymbol{x}_{t}^{u} \end{bmatrix} + \begin{bmatrix} \boldsymbol{C}^{z} & \boldsymbol{D}\boldsymbol{C}^{u} \\ 0 & \boldsymbol{C}^{u} \end{bmatrix} \begin{bmatrix} \boldsymbol{v}_{t}^{z} \\ \boldsymbol{v}_{t}^{u} \end{bmatrix}. \tag{3.37}$$

This model combination procedure, together with the deterministic-stochastic decomposition presented in Subsection 3.1.1, has a clear implication: there is no loss of generality in defining an SS model without exogenous inputs, $\boldsymbol{u_t}$. If the inputs are deterministic or purely exogenous, they can be "deleted" by decomposing the model into its input and error driven terms; see Subsection 3.1.1. On the other hand, if we have models for both $\boldsymbol{z_t}$ and $\boldsymbol{u_t}$, they can be combined into a single formulation without exogenous variables as just described.

Therefore, in the remainder of this book we will avoid unnecessary complexity by considering a purely stochastic SS formulation most often:

$$\boldsymbol{x}_{t+1} = \boldsymbol{\Phi} \boldsymbol{x_t} + \boldsymbol{E} \boldsymbol{w_t} \tag{3.38}$$

$$\boldsymbol{z_t} = \boldsymbol{H} \boldsymbol{x_t} + \boldsymbol{C} \boldsymbol{v_t}. \tag{3.39}$$

Example 3.2.1 Combining a SISO model with the model for the input

Consider a SISO transfer function:

$$z_t = \frac{0.3+0.6\mathrm{B}}{1-0.7\mathrm{B}} u_t + \frac{1-0.8\,\mathrm{B}}{1-\mathrm{B}} a_t^z \tag{3.40}$$

whose input follows the ARIMA(1,1) model:

$$u_t = \frac{1-0.9\mathrm{B}}{1-\mathrm{B}} a_t^u \tag{3.41}$$

and such that:

$$\operatorname{cov}\left(\begin{bmatrix} a_t^z \\ a_t^u \end{bmatrix} \right) = \begin{bmatrix} 1.0 & 0 \\ 0 & 0.5 \end{bmatrix}. \tag{3.42}$$

Following the procedure described above, the first step needed to combine these two models would be to formulate them in SS form. Accordingly, the SEM representation for model (3.40) is:

$$\begin{aligned} \begin{bmatrix} x_{t+1}^{1,z} \\ x_{t+1}^{2,z} \end{bmatrix} &= \begin{bmatrix} 1.70 & 1 \\ -0.70 & 0 \end{bmatrix} \begin{bmatrix} x_t^{1,z} \\ x_t^{2,z} \end{bmatrix} + \begin{bmatrix} 0.81 \\ -0.81 \end{bmatrix} u_t + \begin{bmatrix} 0.20 \\ -0.14 \end{bmatrix} a_t^z \\ z_t &= \begin{bmatrix} 1 & 0 \end{bmatrix} \begin{bmatrix} x_t^{1,z} \\ x_t^{2,z} \end{bmatrix} + 0.30 u_t + a_t^z, \end{aligned} \tag{3.43}$$

while the SS form for the input model (3.41) is:

$$\begin{aligned} x_{t+1}^{1,u} &= x_t^{1,u} + 0.1 a_t^u \\ u_t &= x_t^{1,u} + a_t. \end{aligned} \tag{3.44}$$

Finally, combining (3.43) and (3.44) according to the procedure explained above results in a joint SS model for both the input and output variables:

$$\begin{bmatrix} x_{t+1}^1 \\ x_{t+1}^2 \\ x_{t+1}^3 \end{bmatrix} = \begin{bmatrix} 1.70 & 1 & 0.81 \\ -0.70 & 0 & -0.81 \\ 0 & 0 & 0.1 \end{bmatrix} \begin{bmatrix} x_t^1 \\ x_t^2 \\ x_t^3 \end{bmatrix} + \begin{bmatrix} 0.20 & 0.81 \\ -0.14 & -0.81 \\ 0 & 1 \end{bmatrix} \begin{bmatrix} a_t^z \\ a_t^u \end{bmatrix}$$

$$\begin{bmatrix} z_t \\ u_t \end{bmatrix} = \begin{bmatrix} 1 & 0 & 0.30 \\ 0 & 0 & 1 \end{bmatrix} \begin{bmatrix} x_t^1 \\ x_t^2 \\ x_t^3 \end{bmatrix} + \begin{bmatrix} 1 & 0.30 \\ 0 & 1 \end{bmatrix} \begin{bmatrix} a_t^z \\ a_t^u \end{bmatrix}. \tag{3.45}$$

This can also be written in an equivalent VARMA form by applying the procedures described in Chapter 9:

$$\begin{bmatrix} 1-1.70\mathrm{B}+0.70\mathrm{B}^2 & 0 \\ 0 & 1-\mathrm{B} \end{bmatrix} \begin{bmatrix} z_t \\ u_t \end{bmatrix} = \begin{bmatrix} 1-1.50\mathrm{B}+0.56\mathrm{B}^2 & 0.78\mathrm{B}-0.708\mathrm{B}^2 \\ 0 & 1-0.9\mathrm{B} \end{bmatrix} \begin{bmatrix} a_t^1 \\ a_t^2 \end{bmatrix} \tag{3.46}$$

with:

$$\operatorname{cov}\left(\begin{bmatrix} a_t^1 \\ a_t^2 \end{bmatrix} \right) = \begin{bmatrix} 1.05 & 0.15 \\ 0.15 & 0.50 \end{bmatrix}. \tag{3.47}$$

Note that the error covariance (3.42) is diagonal, while that of the endogeneized model, (3.47), is not. This happens because the latter corresponds to the reduced-form VARMAX model (3.46) which captures the instantaneous relationship between z_t and u_t as a nonzero error covariance. On the other hand, models (3.40) and (3.45) are structural forms, which represent this relationship by means of a parameter in the model structure. Accordingly, the errors in the structural form of this model are uncorrelated.

The code required to replicate this example can be found in the file `Example0321.m`, see Appendix D.

3.2.2 *Colored errors*

Consider an SS model in SEM form:

$$\boldsymbol{x}_{t+1}^{z} = \boldsymbol{\Phi}^{z}\, \boldsymbol{x}_t^{z} + \boldsymbol{E}^{z}\, \boldsymbol{a}_t^{a} \tag{3.48}$$

$$\boldsymbol{z}_t = \boldsymbol{H}^{z}\, \boldsymbol{x}_t^{z} + \boldsymbol{a}_t^{a}, \tag{3.49}$$

where the errors $\boldsymbol{a}_t^{a}$ are colored noise realized by the model:

$$\boldsymbol{x}_{t+1}^{a} = \boldsymbol{\Phi}^{a}\, \boldsymbol{x}_t^{a} + \boldsymbol{E}^{a}\, \boldsymbol{a}_t \tag{3.50}$$

$$\boldsymbol{a}_t^{a} = \boldsymbol{H}^{a}\, \boldsymbol{x}_t^{a} + \boldsymbol{a}_t, \tag{3.51}$$

and $\boldsymbol{a}_t$ is a vector of zero-mean white noise errors. Once again, we will formulate a single model combining the dynamic structure of $\boldsymbol{z}_t$ with that of the colored errors $\boldsymbol{a}_t^{a}$. We refer to this formulation as a "nested in errors" model. Substituting (3.51) in (3.48) and (3.49) yields:

$$\boldsymbol{x}_{t+1}^{z} = \boldsymbol{\Phi}^{z}\, \boldsymbol{x}_t^{z} + \boldsymbol{E}^{z}\, (\boldsymbol{H}^{a}\, \boldsymbol{x}_t^{a} + \boldsymbol{a}_t) \tag{3.52}$$

$$z_t = H^z x_t^z + H^a x_t^a + a_t. \tag{3.53}$$

Last, the state equation of the nested model results from combining equations (3.52) and (3.50) into a single state equation:

$$\begin{bmatrix} x_{t+1}^z \\ x_{t+1}^a \end{bmatrix} = \begin{bmatrix} \Phi^z & E^z H^a \\ 0 & \Phi^a \end{bmatrix} \begin{bmatrix} x_t^z \\ x_t^a \end{bmatrix} + \begin{bmatrix} E^z \\ E^a \end{bmatrix} a_t, \tag{3.54}$$

and the nested observation equation results from rearranging terms in (3.53):

$$z_t = \begin{bmatrix} H^z & H^a \end{bmatrix} \begin{bmatrix} x_t^z \\ x_t^a \end{bmatrix} + a_t. \tag{3.55}$$

This procedure is useful in defining complex models by combining different multiplicative "building blocks." The following example clarifies this idea.

Example 3.2.2 Defining a model as the product of several dynamic factors

Consider the AR(4) model:

$$\left(1 - 2.10\mathrm{B} + 1.58\mathrm{B}^2 - 0.62\mathrm{B}^3 + 0.14\mathrm{B}^4\right) z_t = a_t, \tag{3.56}$$

where the roots of the polynomial on the left-hand side are $1 \pm 2i$, $1/0.7$ and 1. Assume now that we want to define model (3.56) in the equivalent factor form[2]:

$$\left(1 - 0.4\mathrm{B} + 0.2\mathrm{B}^2\right)(1 - 0.7\mathrm{B})(1 - \mathrm{B})\, z_t = a_t. \tag{3.57}$$

The starting point for this is to obtain the SS representations for the factors. Accordingly, the model for the factors $(1-\mathrm{B})(1-0.7\mathrm{B})\, z_t = a_t^a$ is:

$$\begin{aligned} \begin{bmatrix} x_{t+1}^{1,z} \\ x_{t+1}^{2,z} \end{bmatrix} &= \begin{bmatrix} 1.70 & 1 \\ -0.70 & 0 \end{bmatrix} \begin{bmatrix} x_t^{1,z} \\ x_t^{2,z} \end{bmatrix} + \begin{bmatrix} 1.7 \\ -0.7 \end{bmatrix} a_t^a \\ z_t &= \begin{bmatrix} 1 & 0 \end{bmatrix} \begin{bmatrix} x_t^{1,z} \\ x_t^{2,z} \end{bmatrix} + a_t^a, \end{aligned} \tag{3.58}$$

and the model for the factor $\left(1 - 0.4\mathrm{B} + 0.2\mathrm{B}^2\right) a_t^a = a_t$ is:

$$\begin{aligned} \begin{bmatrix} x_{t+1}^{1,a} \\ x_{t+1}^{2,a} \end{bmatrix} &= \begin{bmatrix} 0.4 & 1 \\ -0.2 & 0 \end{bmatrix} \begin{bmatrix} x_t^{1,a} \\ x_t^{2,a} \end{bmatrix} + \begin{bmatrix} 0.4 \\ -0.2 \end{bmatrix} a_t \\ a_t^a &= \begin{bmatrix} 1 & 0 \end{bmatrix} \begin{bmatrix} x_t^{1,a} \\ x_t^{2,a} \end{bmatrix} + a_t. \end{aligned} \tag{3.59}$$

---

[2]There are many reasons why one may want to work with a model in factor form. For example, one may want to estimate its real roots and standard errors separately, in order to infer if this could be a unit root by means of a Dickey and Fuller [77] type statistic. It would also be useful to define a model with multiple seasonal factors such as, e.g., the ARIMA process describing an hourly time series with daily and weekly seasonal cycles, whose periods would then be 24 and 24×7 observations per cycle respectively.

Under these conditions, the SS representation of (3.57) is obtained by combining (3.58) and (3.59) into the composite formulation:

$$\begin{bmatrix} x_{t+1}^1 \\ x_{t+1}^2 \\ x_{t+1}^3 \\ x_{t+1}^4 \end{bmatrix} = \begin{bmatrix} 1.7 & 1 & 1.7 & 0 \\ -0.7 & 0 & -0.7 & 0 \\ 0 & 0 & 0.4 & 1 \\ 0 & 0 & -0.2 & 0 \end{bmatrix} \begin{bmatrix} x_t^1 \\ x_t^2 \\ x_t^3 \\ x_t^4 \end{bmatrix} + \begin{bmatrix} 1.7 \\ -0.7 \\ 0.4 \\ -0.2 \end{bmatrix} a_t \tag{3.60}$$

$$z_t = \begin{bmatrix} 1 & 0 & 1 & 0 \end{bmatrix} \begin{bmatrix} x_t^1 \\ x_t^2 \\ x_t^3 \\ x_t^4 \end{bmatrix} + a_t.$$

The example in Subsection 5.4.2 illustrates how these formulations allow one to capture a complex dynamic structure by means of a nonstandard formulation.

The code required to replicate this example can be found in the file `Example0322.m`, see Appendix D.

3.3 Change of variables in the output

The expression "change of variables in the output" refers to a situation where the model output cannot be observed directly, usually because of certain imperfections in the data, such as observation errors, missing values, or aggregation constraints.

3.3.1 Observation errors

Assume that the output of a MEM model in the form (3.38)–(3.39), $\boldsymbol{z}_t$, is affected by a white noise error, so that the observed variable is $\bar{\boldsymbol{z}}_t$ such that:

$$\bar{\boldsymbol{z}}_t = \boldsymbol{z}_t + \tilde{\boldsymbol{z}}_t, \tag{3.61}$$

where $\tilde{\boldsymbol{z}}_t$ is an uncorrelated observation error. In this case, the derivation of a model for $\bar{\boldsymbol{z}}_t$ follows immediately by adding $\tilde{\boldsymbol{z}}_t$ to both sides of the observation equation (3.39):

$$\boldsymbol{z}_t + \tilde{\boldsymbol{z}}_t = \boldsymbol{H}\boldsymbol{x}_t + \boldsymbol{C}\boldsymbol{v}_t + \tilde{\boldsymbol{z}}_t, \tag{3.62}$$

which simplifies to:

$$\bar{\boldsymbol{z}}_t = \boldsymbol{H}\boldsymbol{x}_t + \begin{bmatrix} \boldsymbol{C} & \boldsymbol{I} \end{bmatrix} \begin{bmatrix} \boldsymbol{v}_t \\ \tilde{\boldsymbol{z}}_t \end{bmatrix}. \tag{3.63}$$

Note that the case of colored observational noise can be treated by combining the previous change of variable with the procedure described in Subsection 3.2.2.

Example 3.3.1 Defining an AR(2) model with white noise observation errors

Consider the AR(2) model:

$$\left(1 - 0.4\mathrm{B} + 0.2\mathrm{B}^2\right) z_t = a_t, \tag{3.64}$$

whose output is affected by an observation error, such that $\bar{z}_t = z_t + \tilde{z}_t$. The errors a_t and $\tilde{z}_t$ are zero-mean white noise sequences such that:

$$\operatorname{cov}\left(\begin{bmatrix} a_t \\ \tilde{z}_t \end{bmatrix}\right) = \begin{bmatrix} 1.0 & 0 \\ 0 & 0.5 \end{bmatrix}. \tag{3.65}$$

Following the procedure described in Subsection 2.2.4, the SS model for the unobserved variable z_t would be:

$$\begin{bmatrix} x_{t+1}^{1,z} \\ x_{t+1}^{2,z} \end{bmatrix} = \begin{bmatrix} 0.4 & 1 \\ -0.2 & 0 \end{bmatrix} \begin{bmatrix} x_t^{1,z} \\ x_t^{2,z} \end{bmatrix} + \begin{bmatrix} 0.4 \\ -0.2 \end{bmatrix} a_t \tag{3.66}$$

$$z_t = \begin{bmatrix} 1 & 0 \end{bmatrix} \begin{bmatrix} x_t^{1,z} \\ x_t^{2,z} \end{bmatrix} + a_t,$$

and, following the procedure described above, the MEM model for the error-contaminated variable has the same state equation[3], but the following observation equation:

$$\bar{z}_t = \begin{bmatrix} 1 & 0 \end{bmatrix} \begin{bmatrix} x_t^{1,z} \\ x_t^{2,z} \end{bmatrix} + \begin{bmatrix} 1 & 1 \end{bmatrix} \begin{bmatrix} a_t \\ \tilde{z}_t \end{bmatrix}, \tag{3.67}$$

which can also be written in an equivalent ARMA form by applying the procedures described in Chapter 9 as:

$$(1 - 0.400\text{B} + 0.200\text{B}^2)\,\bar{z}_t = (1 - 0.145\text{B} + 0.064\text{B}^2)\,\bar{a}_t \quad \sigma_{\bar{a}}^2 = 1.561. \tag{3.68}$$

The code required to replicate this example can be found in the file `Example0331.m`, see Appendix D.

3.3.2 *Missing values and aggregated data*

Consider a MEM in the form (3.38)–(3.39), and assume now that we do not observe the m components of $\boldsymbol{z_t}$ directly, but the $\bar{m}$ components of $\bar{\boldsymbol{z}}_t$ instead, such that:

$$\bar{\boldsymbol{z}}_t = \bar{\boldsymbol{H}}_t\, \boldsymbol{z}_t, \tag{3.69}$$

where $\bar{\boldsymbol{H}}_t$ is a $\bar{m} \times m$ matrix of known coefficients which defines how the vector $\boldsymbol{z_t}$ is observed at time t. Some particular cases of interest are:

1. $\bar{\boldsymbol{H}}_t = \boldsymbol{I}_m$, so that $\bar{\boldsymbol{z}}_t = \boldsymbol{z}_t$;
2. $\bar{\boldsymbol{H}}_t = \boldsymbol{0}$, in which case the t-th observation is missing;
3. $\bar{\boldsymbol{H}}_t = \begin{bmatrix} \boldsymbol{I}_{\bar{m}} & \boldsymbol{0} \end{bmatrix}$, where the last $m - \bar{m}$ components of the output are not observed;

[3]This happens because a white noise observation error does not add new dynamics to the system. A colored error would add its own dynamics to those of the unobserved variable and change the state equation accordingly.

4. $\bar{H}_t = [1, 1, \ldots, 1]$, which implies that we only observe the sum of the m components in z_t.

Building on model (3.38)–(3.39) and the time-varying observation restriction (3.69), it is easy to see that the model for $\bar{z}_t$ will have the state equation (3.38), and an observation equation that is obtained by pre-multiplying (3.39) by $\bar{H}_t$:

$$\bar{z}_t = \bar{H}_t H x_t + \bar{H}_t C v_t. \tag{3.70}$$

The fact that the change in variables does not affect the state equation means that the system dynamics are not altered by any linear constraints on the observation of its output. On the other hand, if the observation equation is changed, some dynamic components of model (3.38)–(3.39) may lose observability in model (3.38)–(3.70).

Finally, note that equation (3.70) has time-varying parameters. As is well known, see e.g. Anderson and Moore [6], most SS methods can be applied to systems with time-varying coefficients, see e.g. Chapter 6. In this book we opted to simplify the notation by dropping the time subindex in the matrices of the SS model. Despite this concession to clarity, most of the procedures described here can be used in the time-varying parameter case (with the obvious modifications) and are therefore able to deal with missing and aggregated data.

3.3.3 *Temporal aggregation*

Consider the MEM (3.38)–(3.39), defined for $t = 1, 2, \ldots, T$. Without loss of generality, we will assume that z_t is a quarterly time series observable only at times $t = 4, 8, 12, \ldots$ with an aggregation constraint determined by its nature as either a flow or a stock variable. In the first case, the sample would include the values $z_t + z_{t-1} + z_{t-2} + z_{t-3}$ $(t = 4, 8, 12, \ldots)$. On the other hand, if z_t is a stock, the observed values would be z_t $(t = 4, 8, 12, \ldots)$.

Following Terceiro [206], the model for the aggregated values of a flow variable can be formulated by augmenting the state vector by as many lags of z_t as needed to fit the aggregation rule. By doing so, we obtain:

$$\begin{bmatrix} x_{t+1} \\ z_t \\ z_{t-1} \\ z_{t-2} \\ z_{t-3} \end{bmatrix} = \begin{bmatrix} \Phi & 0 & 0 & 0 & 0 \\ H & 0 & 0 & 0 & 0 \\ 0 & I & 0 & 0 & 0 \\ 0 & 0 & I & 0 & 0 \\ 0 & 0 & 0 & I & 0 \end{bmatrix} \begin{bmatrix} x_t \\ z_{t-1} \\ z_{t-2} \\ z_{t-3} \\ z_{t-4} \end{bmatrix} + \begin{bmatrix} E & 0 \\ 0 & C \\ 0 & 0 \\ 0 & 0 \\ 0 & 0 \end{bmatrix} \begin{bmatrix} w_t \\ v_t \end{bmatrix} \tag{3.71}$$

$$\begin{bmatrix} z_t \\ z_{t-1} \\ z_{t-2} \\ z_{t-3} \end{bmatrix} = \begin{bmatrix} H & 0 & 0 & 0 & 0 \\ 0 & I & 0 & 0 & 0 \\ 0 & 0 & I & 0 & 0 \\ 0 & 0 & 0 & I & 0 \end{bmatrix} \begin{bmatrix} x_t \\ z_{t-1} \\ z_{t-2} \\ z_{t-3} \\ z_{t-4} \end{bmatrix} + \begin{bmatrix} C \\ 0 \\ 0 \\ 0 \end{bmatrix} v_t, \tag{3.72}$$

where the time index runs in multiples of four, that is, $t = 4, 8, 12, \ldots$. Building on equations (3.71)–(3.72), it is trivial to obtain the corresponding model for a flow variable by multiplying the observation equation (3.72) by $\begin{bmatrix} I & I & I & I \end{bmatrix}$. The resulting observation equation would then be:

$$z_t + z_{t-1} + z_{t-2} + z_{t-3} = \begin{bmatrix} H & I & I & I & 0 \end{bmatrix} \begin{bmatrix} x_t \\ z_{t-1} \\ z_{t-2} \\ z_{t-3} \\ z_{t-4} \end{bmatrix} + C v_t. \qquad (3.73)$$

On the other hand, the model for a stock variable is obtained by multiplying equation (3.72) by $\begin{bmatrix} I & 0 & 0 & 0 \end{bmatrix}$, resulting in

$$z_t = \begin{bmatrix} H & 0 & 0 & 0 & 0 \end{bmatrix} \begin{bmatrix} x_t \\ z_{t-1} \\ z_{t-2} \\ z_{t-3} \\ z_{t-4} \end{bmatrix} + C v_t. \qquad (3.74)$$

The treatment for temporally aggregated data just outlined is adequate for many uses, but has a clear inconvenience: it increases the size of the state vector. In Chapter 10 we will see an efficient alternative which does not increase the size of the state vector and, therefore, is more stable and suitable for computationally-intensive tasks.

3.4 Uses of these transformations

The model transformations discussed in this chapter illustrate the flexibility of the SS representation, but their practical use may be unclear at first sight. In the rest of this book, and throughout the SS literature, we find many examples of their applicability, such as the following:

1. The endogeneization procedure described in Subsection 3.2.1 provides an immediate treatment for estimating a model with: (a) stochastic inputs, see Subsection 5.3.2, or (b) missing covariates in panel data models, see Chapter 11.
2. The deterministic-stochastic decomposition and the block-diagonal representation presented in subsections 3.1.1 and 3.1.3, respectively, are the foundations of the signal extraction procedure described in Chapter 8.
3. SS methods for data aggregated over space and time, see subsections 3.3.2 and 3.3.3, are the basis of many benchmarking and nowcasting procedures routinely applied by statistical agencies, see e.g., Durbin and Quenneville [83] or Nunes [169].
4. Finally, the observation equation transformations discussed in subsections 3.3.1 and 3.3.2 are the basis of the methods for modeling samples with missing values or observation errors, see e.g., Jones [132].

These are just a few examples which illustrate the flexibility of the SS model in allowing one to address sophisticated problems that are difficult to solve using standard statistical approaches.

Chapter 4

Filtering and Smoothing

4.1 The conditional moments of a State-Space model 33
4.2 The Kalman Filter 34
4.3 Decomposition of the smoothed moments 36
4.4 Smoothing for a general State-Space model 37
4.5 Smoothing for fixed-coefficients and single-error models 39
4.6 Uncertainty of the smoothed estimates in a SEM 40
4.7 Examples 42
4.7.1 Recursive Least Squares 42
4.7.2 Cleaning the Wolf sunspot series 42
4.7.3 Extracting the Hodrick-Prescott trend 44

In this chapter we present the main state estimation algorithms which are known as "filtering" and "fixed-interval smoothing." Both focus on estimating the sequence of states, but are based on different information sets, and constitute the foundation of most SS procedures.

4.1 The conditional moments of a State-Space model

Consider an SS model, such as the general MEM (2.1)–(2.2) or the SEM (2.19)–(2.20) and denote:

$$\boldsymbol{x}_{t|j} = \mathrm{E}\left(\boldsymbol{x}_t \,|\, \boldsymbol{\Omega}_j\right) \tag{4.1}$$

$$\boldsymbol{P}_{t|j} = \mathrm{E}\left[(\boldsymbol{x}_t - \boldsymbol{x}_{t|j})(\boldsymbol{x}_t - \boldsymbol{x}_{t|j})^T \,|\, \boldsymbol{\Omega}_j\right], \tag{4.2}$$

where $\boldsymbol{\Omega}_j$ is the information set containing all the input and output measures up to time *j*. With this notation, the sequences of filtered and fixed-interval smoothed states are, respectively[1]:

$$\boldsymbol{x}_{t+1|t}\,, \quad \boldsymbol{P}_{t+1|t}\,, \qquad t = 1,\ldots,N, \tag{4.3}$$

[1] In many works the results of filtering are denoted as: $\boldsymbol{x}_{t|t}$ and $\boldsymbol{P}_{t|t}$ because they use the alternative state equation $\boldsymbol{x}_t = \boldsymbol{\Phi}\boldsymbol{x}_{t-1} + \boldsymbol{\Gamma}\boldsymbol{u}_t + \boldsymbol{E}\boldsymbol{w}_t$.

$$\boldsymbol{x}_{t|N}, \quad \boldsymbol{P}_{t|N}, \qquad t = 1,\ldots,N. \tag{4.4}$$

Note that extending these sequences out of sample would provide forecasts for the state values, as well as for the corresponding error covariances,

$$\boldsymbol{x}_{t|N}, \quad \boldsymbol{P}_{t|N}, \qquad t = N+1, N+2, \ldots. \tag{4.5}$$

Building on these results and the observation equation of the model, it is a straightforward exercise to compute forecasts and error covariances for the outputs of the SS model, $\boldsymbol{z_t}$. For example, assuming $\boldsymbol{u}_{N+h}$ is observed (and non-stochastic), the relevant equations for h-step ahead forecasting in the case of MEM (2.1)–(2.2) would be:

$$\mathrm{E}\left(\boldsymbol{z}_{N+h} \left| \boldsymbol{\Omega_j} \right.\right) = \boldsymbol{H}\boldsymbol{x}_{N+h|N} + \boldsymbol{D}\boldsymbol{u}_{N+h},$$

$$\mathrm{E}\left[(\boldsymbol{z}_{N+h} - \boldsymbol{z}_{N+h|N})(\boldsymbol{z}_{N+h} - \boldsymbol{x}_{N+h|N})^T \left| \boldsymbol{\Omega_j} \right.\right] = \boldsymbol{H}\boldsymbol{P}_{N+h|N}\boldsymbol{H}^T + \boldsymbol{C}\boldsymbol{R}\boldsymbol{C}^T,$$

where $\boldsymbol{x}_{N+h|N}$ and $\boldsymbol{P}_{N+h|N}$ are obtained by iterative application (for $h = 1, 2, \ldots$, etc.) of Kalman filter equations (4.6) and (4.9) given below, starting from a filtered state that has been evaluated up to time $t = N$.

Filtering and smoothing procedures are applied in this book with different goals in mind. For example, we use the Kalman filter to compute the innovations required to evaluate the likelihood for an SS model, see Terceiro [206] and Chapter 5, while smoothing algorithms are applied to estimate unobserved components in structural time-series models, as in Harvey [118], and to deal with samples containing missing values or aggregated data, see e.g., Ansley and Kohn [7], Kohn and Ansley [137] or Naranjo, Trindade and Casella [165].

4.2 The Kalman Filter

Building on an SS model in the form (2.1)–(2.2), the Kalman Filter (KF), see Kalman [134], computes one-step-ahead estimates of the states as well as the corresponding error covariance matrix by propagating the following equations:

$$\boldsymbol{x}_{t+1|t} = \boldsymbol{\Phi}\,\boldsymbol{x}_{t|t-1} + \boldsymbol{\Gamma}\,\boldsymbol{u}_t + \boldsymbol{K}_t\,\tilde{\boldsymbol{z}}_{t|t-1} \tag{4.6}$$

$$\tilde{\boldsymbol{z}}_{t|t-1} = \boldsymbol{z}_t - \boldsymbol{H}\,\boldsymbol{x}_{t|t-1} - \boldsymbol{D}\,\boldsymbol{u}_t \tag{4.7}$$

$$\boldsymbol{B}_t = \boldsymbol{H}\,\boldsymbol{P}_{t|t-1}\,\boldsymbol{H}^T + \boldsymbol{C}\boldsymbol{R}\boldsymbol{C}^T \tag{4.8}$$

$$\boldsymbol{P}_{t+1|t} = \boldsymbol{\Phi}\,\boldsymbol{P}_{t|t-1}\,\boldsymbol{\Phi}^T + \boldsymbol{E}\,\boldsymbol{Q}\,\boldsymbol{E}^T - \boldsymbol{K}_t\,\boldsymbol{B}_t\,\boldsymbol{K}_t^T \tag{4.9}$$

$$\boldsymbol{K}_t = (\boldsymbol{\Phi}\,\boldsymbol{P}_{t|t-1}\,\boldsymbol{H}^T + \boldsymbol{E}\boldsymbol{S}\boldsymbol{C}^T)\,\boldsymbol{B}_t^{-1}, \tag{4.10}$$

see Appendix A, starting from suitable initial conditions, $\boldsymbol{x}_1$ and $\boldsymbol{P}_1$, see e.g., De Jong and Chu-Chun-Lin [69], or Casals and Sotoca [43]. Note that:

1. The updating equation (4.6) transforms the *ex-ante* estimate for the state at time t, $\boldsymbol{x}_{t|t-1}$, in the *ex-post* value, $\boldsymbol{x}_{t+1|t}$, according to a linear function of the innovation, $\tilde{\boldsymbol{z}}_{t|t-1}$, so the larger the value of this forecast error, the larger the update will be.

2. Equation (4.7) computes the one-step-ahead forecast error for z_t, also known as "innovation," on the basis of the *ex-ante* state estimate, $x_{t|t-1}$. The corresponding error covariance is given by the matrix B_t in equation (4.8).
3. Expression (4.9) is known as the Riccati equation of the KF. It provides the covariance matrix for the sequence of KF states and, as we shall see later, it has a stationary solution under nonrestrictive assumptions. This fact plays an important role in many SS computational procedures.
4. The Riccati equation (4.9) does not depend on the sample values, z_t. This fact has two interesting implications. First, no observation helps more than any other in decreasing uncertainty about the states. Second, equations (4.8)–(4.9) and (4.6)–(4.7) can be propagated independently.
5. Last, the matrix K_t defined in equation (4.10) is known as the KF gain. Its appearance in the KF can be justified in many ways. First, it can be shown that the sequence (4.10) minimizes the trace of the *ex-post* covariance $P_{t+1|t}$, see Jazwinsky [130] and Appendix A, as well as that of any linear combination of the variances in the main diagonal of $P_{t+1|t}$, see Demetry [72]. Under normality, it provides the maximum likelihood estimates of the states.

Note that equation (4.7) can be trivially rewritten as

$$z_t = H x_{t|t-1} + D u_t + \tilde{z}_{t|t-1}.$$

Therefore, expressions (4.6) and this restructured (4.7) can be interpreted, respectively, as the state and observation equations of the time-varying parameter MEM:

$$\begin{aligned} x_{t+1|t} &= \Phi x_{t|t-1} + \Gamma u_t + K_t \tilde{z}_{t|t-1} \\ z_t &= H x_{t|t-1} + D u_t + \tilde{z}_{t|t-1}. \end{aligned} \tag{4.11}$$

The errors in (4.11) are the innovations of the KF and, because of this, it is known in the literature as the "innovations representation," see e.g. Harvey [118]. The difference between this model and the previously defined SEM, (2.19)–(2.20), is that the error distribution matrix in (4.11), K_t, is time-varying, while the corresponding matrix in the SEM, K, is constant. It can be shown that the matrix K in the SEM is the steady-state solution for the sequence K_t. Because of this, the SEM is also known in the literature as the "steady-state innovations model," see e.g. Anderson and Moore [6].

In a stationary system with no inputs, adequate initial conditions are given by the steady-state solution of equations (4.6)–(4.10):

$$x_1 = 0 \tag{4.12}$$

$$P_1 = \Phi P_1 \Phi^T + EQE^T, \tag{4.13}$$

where the Lyapunov equation (4.13) has a unique positive semi-definite solution, see Akaike [3] and Anderson and Moore [6]. When the model has exogenous inputs, the initialization of the KF is more complex and needs to take into account the deterministic or stochastic structure of $\boldsymbol{u_t}$. We will discuss this issue further in Section 5.2.

Finally, when the system is nonstationary the uncertainty in the prior distribution of the initial state is infinite. The literature refers to this case as "diffuse initial conditions," and it has different consequences on filtering and fixed-interval smoothing algorithms. First, it is impossible to initialize the standard KF with exact diffuse initial conditions[2]. Therefore, in the nonstationary case one can adopt two basic strategies: (a) using the information filter, see Anderson and Moore [6], which propagates the system states and the inverse of their covariances so the exact diffuse initial conditions are $\boldsymbol{x_1} = \mathbf{0}$ and $\boldsymbol{P_1^{-1}} = \mathbf{0}$; and (b) developing specialized algorithms to cope with nonstationary states for each specific application. An example of the latter approach is the likelihood evaluation procedure presented in Subsection 5.1.3.

Fixed-interval smoothing, on the other hand, allows for diffuse initial conditions. To this end, in Section 4.3 we will characterize the effect of initial conditions on the smoothed moments. This result will then be used for the exact smoother described in Section 4.4.

4.3 Decomposition of the smoothed moments

Consider the SS model:

$$\bar{\boldsymbol{x}}_{t+1} = \boldsymbol{\Phi}\,\bar{\boldsymbol{x}}_t + \boldsymbol{\Gamma}\,\boldsymbol{u}_t + \boldsymbol{E}\,\boldsymbol{w}_t \tag{4.14}$$

$$\bar{\boldsymbol{z}}_t = \boldsymbol{H}\,\bar{\boldsymbol{x}}_t + \boldsymbol{D}\,\boldsymbol{u}_t + \boldsymbol{C}\,\boldsymbol{v}_t, \tag{4.15}$$

where the states and observations correspond to model (2.1)–(2.2) with the initial conditions $\bar{\boldsymbol{x}}_1 = \mathbf{0}$ and $\bar{\boldsymbol{P}}_1 = \mathbf{0}$. Note that $\bar{\boldsymbol{z}}_t$ is not observed. Propagating state equations (2.1) and (4.14), it follows that:

$$\boldsymbol{x}_t = \boldsymbol{\Phi}^{t-1}\,\boldsymbol{x}_1 + \bar{\boldsymbol{x}}_t, \tag{4.16}$$

where $\bar{\boldsymbol{x}}_t$ is independent of $\boldsymbol{x}_1$. Thus the expectation of (4.16) conditional on the information set $\boldsymbol{\Omega}_N$ is:

$$\boldsymbol{x}_{t|N} = \boldsymbol{\Phi}^{t-1}\,\boldsymbol{x}_{1|N} + \bar{\boldsymbol{x}}_{t|N}\,. \tag{4.17}$$

Also, from (2.1)–(2.2) and (4.14)–(4.15) it is easy to prove, see Rosenberg [189], that:

$$\tilde{\boldsymbol{z}}_t = \boldsymbol{H}\,\bar{\bar{\boldsymbol{\Phi}}}_{t-1}\boldsymbol{x}_1 + \tilde{\bar{\boldsymbol{z}}}_t, \tag{4.18}$$

where $\tilde{\boldsymbol{z}}_t = \boldsymbol{z}_t - \boldsymbol{z}_{t|t-1}$ is the sequence of KF innovations corresponding to (2.1)–(2.2). The values $\tilde{\bar{\boldsymbol{z}}}_t$, defined accordingly by $\tilde{\bar{\boldsymbol{z}}}_t = \bar{\boldsymbol{z}}_t - \bar{\boldsymbol{z}}_{t|t-1}$, are uncorrelated innovations, given that $\bar{\boldsymbol{x}}_1 = \mathbf{0}$ and $\bar{\boldsymbol{P}}_1 = \mathbf{0}$ are the proper initial conditions for

[2]The literature often suggests initializing the KF with $\boldsymbol{x_1} = \mathbf{0}$ and $\boldsymbol{P_1} = k\boldsymbol{I}$, where k is an arbitrary scalar whose value is "large enough" to approximate an infinite covariance matrix. As Ansley and Kohn [8] point out, such an initialization yields substantial estimation errors and numerical instability.

(4.14)–(4.15). Finally, the matrix sequence $\bar{\bar{\boldsymbol{\Phi}}}_t$ is defined recursively via $\bar{\bar{\boldsymbol{\Phi}}}_t = (\boldsymbol{\Phi} - \boldsymbol{K}_t \boldsymbol{H})\,\bar{\bar{\boldsymbol{\Phi}}}_{t-1}$ with $\bar{\bar{\boldsymbol{\Phi}}}_1 = \boldsymbol{I}$.

Equation (4.18) can be written for the entire sample as:

$$\tilde{\boldsymbol{z}} = \boldsymbol{X}\,\boldsymbol{x}_1 + \bar{\tilde{\boldsymbol{z}}}, \tag{4.19}$$

where $\boldsymbol{X}$ is the block-diagonal matrix whose t-th block is $\boldsymbol{H}\,\bar{\bar{\boldsymbol{\Phi}}}_{t-1}$, and the mN-dimensional vectors $\tilde{\boldsymbol{z}}$ and $\bar{\tilde{\boldsymbol{z}}}$ contain the respective innovations resulting from a KF initialized with null values for the mean and covariance of the initial state, hereafter KF($\boldsymbol{0}$,$\boldsymbol{0}$). Note that this initialization assures that $\bar{\tilde{\boldsymbol{z}}}$ is independent of $\boldsymbol{x}_1$. The problem reduces then to obtaining the conditional expectation $\boldsymbol{x}_{t|N}$ on the right-hand side of (4.17), as well as its covariance, $\boldsymbol{P}_{t|N}$, taking into account (4.19). The following theorem shows how to achieve this.

Theorem 4.1 *The smoothed moments of the state vector can be written as:*

$$\boldsymbol{x}_{t|N} = \left[\boldsymbol{\Phi}^{t-1} - E(\bar{\boldsymbol{x}}_t \bar{\tilde{\boldsymbol{z}}}^T)\boldsymbol{B}^{-1}\boldsymbol{X}\right]\boldsymbol{x}_{1|N} + E(\bar{\boldsymbol{x}}_t \bar{\tilde{\boldsymbol{z}}}^T)\boldsymbol{B}^{-1}\tilde{\boldsymbol{z}} \tag{4.20}$$

$$\boldsymbol{P}_{t|N} = \left[\boldsymbol{\Phi}^{t-1} - E(\bar{\boldsymbol{x}}_t \bar{\tilde{\boldsymbol{z}}}^T)\boldsymbol{B}^{-1}\boldsymbol{X}\right]\boldsymbol{P}_{1|N}\left[\boldsymbol{\Phi}^{t-1} - E(\bar{\boldsymbol{x}}_t \bar{\tilde{\boldsymbol{z}}}^T)\boldsymbol{B}^{-1}\boldsymbol{X}\right]^T + \bar{\boldsymbol{P}}_{t|N}, \tag{4.21}$$

where $\boldsymbol{B}$ *is a block-diagonal matrix that contains the covariance matrices of* $\bar{\tilde{\boldsymbol{z}}}_t$*, and* $\bar{\boldsymbol{P}}_{t|N}$ *is the smoothed second-order moment of the states in (4.14)–(4.15).*

A proof of this result can be found in the Appendix of Casals, Jerez, and Sotoca [40].

Note that equations (4.20)–(4.21) apply to both, stationary and nonstationary systems, as the only terms affected by the covariance of the initial state $\bar{\boldsymbol{P}}_1$ are $\boldsymbol{x}_{1|N}$ and $\boldsymbol{P}_{1|N}$, and this dependence occurs through $\bar{\boldsymbol{P}}_1^{-1}$, which is finite, see [66] and equations (A.1)–(A.2) in the Appendix of Casals, Jerez, and Sotoca [40]. The computation of $\mathrm{E}(\bar{\boldsymbol{x}}_t \bar{\tilde{\boldsymbol{z}}}^T)\boldsymbol{B}^{-1}\boldsymbol{X}$ and $\mathrm{E}(\bar{\boldsymbol{x}}_t \bar{\tilde{\boldsymbol{z}}}^T)\boldsymbol{B}^{-1}\tilde{\boldsymbol{z}}$ in (4.20)–(4.21) depends on the specific smoothing algorithm to be used. This issue is discussed further in the next Section.

4.4 Smoothing for a general State-Space model

Results (4.20)–(4.21) can be combined with any standard smoother to derive an algorithm which can be applied to nonstationary systems. In this section we will do this using an efficient and stable method due to De Jong [67].

The smoothing algorithm proposed by De Jong [67] first runs a KF($\boldsymbol{0}$,$\boldsymbol{0}$) through the entire sample and then performs a backward recursion given by:

$$\boldsymbol{x}_{t|N} = \boldsymbol{x}_{t|t-1} + \boldsymbol{P}_{t|t-1}\,\boldsymbol{r}_{t-1} \tag{4.22}$$

$$\boldsymbol{P}_{t|N} = \boldsymbol{P}_{t|t-1} - \boldsymbol{P}_{t|t-1}\,\boldsymbol{R}_{t-1}\boldsymbol{P}_{t|t-1} \tag{4.23}$$

$$\boldsymbol{r}_{t-1} = \boldsymbol{H}^T\,\boldsymbol{B}_t^{-1}\tilde{\boldsymbol{z}}_t + \bar{\boldsymbol{\Phi}}_t^T\boldsymbol{r}_t, \qquad \text{with} \quad \boldsymbol{r}_N = \boldsymbol{0} \tag{4.24}$$

$$\boldsymbol{R}_{t-1} = \boldsymbol{H}^T\,\boldsymbol{B}_t^{-1}\boldsymbol{H} + \bar{\boldsymbol{\Phi}}_t^T\,\boldsymbol{R}_t\,\bar{\boldsymbol{\Phi}}_t, \qquad \text{with} \quad \boldsymbol{R}_N = \boldsymbol{0} \tag{4.25}$$

$$\bar{\boldsymbol{\Phi}}_t = \boldsymbol{\Phi} - \boldsymbol{K}_t\,\boldsymbol{H}, \tag{4.26}$$

where $\boldsymbol{x}_{t|t-1}$ and $\boldsymbol{P}_{t|t-1}$ were computed in the forward step, $\boldsymbol{B_t}$ is the t-th diagonal block of $\boldsymbol{B}$, and $\boldsymbol{K_t}$ is the KF gain. In this framework, the terms corresponding to (4.20)–(4.21) are given by:

$$\mathrm{E}(\bar{\boldsymbol{x}}_t \tilde{\bar{\boldsymbol{z}}}^T)\boldsymbol{B}^{-1}\tilde{\boldsymbol{z}} = \boldsymbol{x}_{t|t-1} + \boldsymbol{P}_{t|t-1}\boldsymbol{r}_{t-1} \tag{4.27}$$

$$\bar{\boldsymbol{P}}_{t|N} = \boldsymbol{P}_{t|t-1} - \boldsymbol{P}_{t|t-1}\boldsymbol{R}_{t-1}\boldsymbol{P}_{t|t-1} \tag{4.28}$$

$$\boldsymbol{\Phi}^{t-1} - \mathrm{E}(\bar{\boldsymbol{x}}_t \tilde{\bar{\boldsymbol{z}}}^T)\boldsymbol{B}^{-1}\boldsymbol{X} = \left(\boldsymbol{I} - \boldsymbol{P}_{t|t-1}\boldsymbol{R}_{t-1}\right)\bar{\bar{\boldsymbol{\Phi}}}_t, \tag{4.29}$$

and combining (4.22)–(4.26) with (4.27)–(4.29) yields the following algorithm:

Forward step : Propagate a KF(**0**,**0**) and the following equations:

$$\bar{\bar{\boldsymbol{\Phi}}}_{t+1} = (\boldsymbol{\Phi} - \boldsymbol{K_t}\boldsymbol{H})\,\bar{\bar{\boldsymbol{\Phi}}}_t, \qquad \text{with} \quad \bar{\bar{\boldsymbol{\Phi}}}_{\boldsymbol{1}} = \boldsymbol{I} \tag{4.30}$$

$$\boldsymbol{X_t} = \boldsymbol{H}\,\bar{\bar{\boldsymbol{\Phi}}}_t \tag{4.31}$$

$$\boldsymbol{W_t} = \boldsymbol{W}_{t-1} + \boldsymbol{X}_t^T\boldsymbol{B}_t^{-1}\boldsymbol{X_t}, \qquad \text{with} \quad \boldsymbol{W_0} = \boldsymbol{0} \tag{4.32}$$

$$\boldsymbol{w_t} = \boldsymbol{w}_{t-1} + \boldsymbol{X}_t^T\boldsymbol{B}_t^{-1}\tilde{\boldsymbol{z}}_t, \qquad \text{with} \quad \boldsymbol{w_0} = \boldsymbol{0}. \tag{4.33}$$

Backward step : Propagate (4.24)–(4.26) and

$$\boldsymbol{V}_{t|N} = \left(\boldsymbol{I} - \boldsymbol{P}_{t|t-1}\boldsymbol{R}_{t-1}\right)\bar{\bar{\boldsymbol{\Phi}}}_t \tag{4.34}$$

$$\boldsymbol{x}_{t|N} = \boldsymbol{x}_{t|t-1} + \boldsymbol{P}_{t|t-1}\boldsymbol{r}_{t-1} + \boldsymbol{V}_{t|N}\boldsymbol{x}_{1|N} \tag{4.35}$$

$$\boldsymbol{P}_{t|N} = \boldsymbol{P}_{t|t-1} - \boldsymbol{P}_{t|t-1}\boldsymbol{R}_{t-1}\boldsymbol{P}_{t|t-1} + \boldsymbol{V}_{t|N}\boldsymbol{P}_{1|N}\boldsymbol{V}_{t|N}^T \tag{4.36}$$

where:

$$\boldsymbol{x}_{1|N} = \boldsymbol{P}_{1|N}\left(\bar{\boldsymbol{P}}_{\boldsymbol{1}}^{-1}\bar{\boldsymbol{x}}_{\boldsymbol{1}} + \boldsymbol{w_N}\right) \tag{4.37}$$

$$\boldsymbol{P}_{1|N} = \left(\bar{\boldsymbol{P}}_{\boldsymbol{1}}^{-1} + \boldsymbol{W_N}\right)^{-1}. \tag{4.38}$$

Comparing (4.35)–(4.36) with (4.22)–(4.23), we see immediately that the distortion due to the arbitrary (**0**,**0**) initialization is corrected by the terms $\boldsymbol{V}_{t|N}\boldsymbol{x}_{1|N}$ and $\boldsymbol{V}_{t|N}\boldsymbol{P}_{1|N}\boldsymbol{V}_{t|N}^T$. This algorithm could of course have been implemented with any initial conditions. Null values were chosen only because they save computing time by simplifying the forward KF recursion.

These results can be easily applied to solve any other signal extraction problem such as, e.g., fixed-point or fixed-interval smoothing, by modifying the conditioning information set in (4.20)–(4.21), and then combining the resulting equations with a standard algorithm.

4.5 Smoothing for fixed-coefficients and single-error models

To simplify notation, we have thus far assumed that the SS model is time-invariant. However, algorithm (4.30)–(4.38) does not require fixed parameters, and can therefore be generalized to time-varying systems just by adding a time subindex to the non-constant matrices. On the other hand, if the system is indeed time-invariant, further computational gains can be obtained by using a KF($\mathbf{0}$,$\bar{\boldsymbol{P}}$) instead of a KF($\mathbf{0}$,$\mathbf{0}$), where $\bar{\boldsymbol{P}}$ denotes the stationary solution of the Riccati equation in the KF. In this case, algorithm (4.30)–(4.38) simplifies to the following procedure:

Initial step : Compute $\bar{\boldsymbol{P}}$ such that:

$$\bar{\boldsymbol{P}} = \boldsymbol{\Phi}\bar{\boldsymbol{P}}\boldsymbol{\Phi}^T + \boldsymbol{E}\boldsymbol{Q}\boldsymbol{E}^T - \bar{\boldsymbol{K}}\bar{\boldsymbol{B}}\bar{\boldsymbol{K}}^T, \tag{4.39}$$

where

$$\bar{\boldsymbol{B}} = \boldsymbol{H}\bar{\boldsymbol{P}}\boldsymbol{H}^T + \boldsymbol{C}\boldsymbol{R}\boldsymbol{C}^T \tag{4.40}$$

$$\bar{\boldsymbol{K}} = \left(\boldsymbol{\Phi}\bar{\boldsymbol{P}}\boldsymbol{H}^T + \boldsymbol{E}\boldsymbol{S}\boldsymbol{C}^T\right)\bar{\boldsymbol{B}}^{-1}. \tag{4.41}$$

If the system is detectable, (4.39) has a unique and positive semi-definite strong solution, see Theorem 3.1 and Lemma 3.1 in De Souza, Gevers, and Goodwin [71] and Appendix A.10.

Forward step: The KF($\mathbf{0}$,$\bar{\boldsymbol{P}}$) in the forward step simplifies to computing:

$$\boldsymbol{x}_{t+1|t} = \boldsymbol{\Phi}\,\boldsymbol{x}_{t|t-1} + \boldsymbol{\Gamma}\,\boldsymbol{u}_t + \bar{\boldsymbol{K}}\,\tilde{\boldsymbol{z}}_t \tag{4.42}$$

$$\tilde{\boldsymbol{z}}_t = \boldsymbol{z}_t - \boldsymbol{H}\,\boldsymbol{x}_{t|t-1} - \boldsymbol{D}\,\boldsymbol{u}_t, \tag{4.43}$$

as well as equations (4.30)–(4.33) with the substitutions $\boldsymbol{K}_t = \bar{\boldsymbol{K}}$ and $\boldsymbol{B}_t^{-1} = \bar{\boldsymbol{B}}^{-1}$.

Backward step: Propagate (4.24)–(4.26) with $\boldsymbol{K}_t = \bar{\boldsymbol{K}}$ and $\boldsymbol{B}_t^{-1} = \bar{\boldsymbol{B}}^{-1}$, as well as the expressions:

$$\boldsymbol{x}_{t|N} = \boldsymbol{x}_{t|t-1} + \boldsymbol{P}_{t|t-1}\boldsymbol{r}_{t-1} + \boldsymbol{V}_{t|N}\left[\left(\bar{\boldsymbol{P}}_1 - \bar{\boldsymbol{P}}\right)^{-1} + \boldsymbol{W}_N\right]^{-1} \cdot \left[\left(\bar{\boldsymbol{P}}_1 - \bar{\boldsymbol{P}}\right)^{-1}\bar{\boldsymbol{x}}_1 + \boldsymbol{w}_N\right] \tag{4.44}$$

$$\boldsymbol{P}_{t|N} = \boldsymbol{P}_{t|t-1} - \boldsymbol{P}_{t|t-1}\boldsymbol{R}_{t-1}\boldsymbol{P}_{t|t-1} + \boldsymbol{V}_{t|N}\left[\left(\bar{\boldsymbol{P}}_1 - \bar{\boldsymbol{P}}\right)^{-1} + \boldsymbol{W}_N\right]^{-1}\boldsymbol{V}_{t|N}^T, \tag{4.45}$$

instead of (4.35)–(4.36). Note that $(\bar{\boldsymbol{P}}_1 - \bar{\boldsymbol{P}})^{-1}$ is the inverse covariance for the initial state of the equivalent SEM derived in the initial step.

The computational efficiency of this recursion is due to a positive trade-off between the setup cost required to compute $\bar{\boldsymbol{P}}$ from (4.39)–(4.41), and the reduced computation and memory requirements in the forward and backward steps. This follows because of three facts: (a) propagating (4.42)–(4.43) requires less computation than the analogous KF($\mathbf{0}$,$\mathbf{0}$); (b) in the time-invariant case $\mathrm{E}(\tilde{\boldsymbol{z}}_t\,\tilde{\boldsymbol{z}}_t^T) = \bar{\boldsymbol{B}}$ for all t, so this matrix needs to be inverted only once, and (c) since $\boldsymbol{K_t} = \bar{\boldsymbol{K}}$, $\boldsymbol{B}_t^{-1} = \bar{\boldsymbol{B}}^{-1}$, and $\boldsymbol{P}_{t|t-1} = \bar{\boldsymbol{P}}$, for all t, the computational load and memory requirements of the procedure are reduced.

Further gains can be obtained by writing the model in the equivalent SEM form, as shown in Chapter 2. In this case, the previous algorithm is further simplified because the solution of the Riccati equation is $\bar{\boldsymbol{P}} = \boldsymbol{0}$. In the next subsection we will discuss other advantages of this formulation.

4.6 Uncertainty of the smoothed estimates in a SEM

The following proposition defines an important property of the smoothed states of fixed-coefficients SEM:

Proposition 4.1 *In the SEM (2.19)–(2.20), if the eigenvalues of* $\boldsymbol{\Phi} - \boldsymbol{K}\boldsymbol{H}$ *are either on or inside the unit circle, the smoothing covariance matrix* $\boldsymbol{P}_{t|N}$ *converges to zero as t increases. When all these eigenvalues are strictly inside the unit circle, this convergence is exponentially fast.*

Therefore, under the conditions of Proposition 4.1, the smoothed state estimates converge to exact values. (Note that these conditions are closely related to the invertibility of the SS model; see Definition 2.10.) The practical significance of this result is twofold. First, the smoothed states can be treated as actual data because they have null variances. Second, they will not be revised if the sample increases. This happens because in filtering and smoothing, revisions are proportional to the uncertainty affecting the estimates; therefore an estimate with null uncertainty will never be updated.

The important property embodied by Proposition 4.1 can be derived as a Corollary of Theorem 4.2 in De Souza, Gevers, and Goodwin [71], where detectability is shown to be a necessary and sufficient condition for convergence of the difference Riccati equation to the unique strong solution of the algebraic Riccati equation. For our purposes, the following proof is simpler and more direct.

Proof of Proposition 4.1. Ignoring exogenous inputs without loss of generality, equation (2.20) implies that $\boldsymbol{a}_t = \boldsymbol{z}_t - \boldsymbol{H}\boldsymbol{x}_t$. Substituting this expression in (2.19) yields

$$\boldsymbol{x}_{t+1} = (\boldsymbol{\Phi} - \boldsymbol{K}\boldsymbol{H})\boldsymbol{x}_t + \boldsymbol{K}\boldsymbol{z}_t. \tag{4.46}$$

Therefore, and if the initial state $\boldsymbol{x}_1$ were known, the entire sequence of states would be exactly determined by (4.46), because the inputs on the right-hand side of this expression would also be known. However, usually $\boldsymbol{x}_1$ is unknown, and is treated as a random variable. In this case, the covariance matrix of the smoothed estimates (by definition conditional on the entire sample) would be

$$\boldsymbol{P}_{t|N} = (\boldsymbol{\Phi} - \boldsymbol{K}\boldsymbol{H})^{t-1}\boldsymbol{P}_{1|N}\left[(\boldsymbol{\Phi} - \boldsymbol{K}\boldsymbol{H})^{t-1}\right]^T, \tag{4.47}$$

where the eigenvalues of $\boldsymbol{\Phi} - \boldsymbol{K}\boldsymbol{H}$ coincide with the reciprocals of the roots of the MA terms implied by the SS model (see Definition 2.10). If the model is strictly invertible, then the eigenvalues of $\Phi - \boldsymbol{K}\boldsymbol{H}$ lie strictly inside the unit circle, and, therefore, $\boldsymbol{P}_{t|N}$ converges to zero at an exponential rate as t increases. On the other hand, if some eigenvalues of $\Phi - \boldsymbol{K}\boldsymbol{H}$ lie exactly on the unit circle, this convergence to exact values will still occur, but is not exponentially fast. ■

The convergence of $\boldsymbol{P}_{t|N}$ depends on the structure of the model. In particular:

1. When the model has a pure p-th order autoregressive structure, the trace of $\boldsymbol{P}_{t|N}$ collapses to zero after processing p observations. This property holds even when the process has unit AR roots, and means that the states in the AR-SEM can be uniquely determined from a given amount of information about the system outputs.
2. If the model has MA roots and all of them are invertible, then $\mathrm{tr}(\boldsymbol{P}_{t|N})$ converges exponentially fast to zero as t increases. The decay rate of the coefficients in the corresponding AR representation governs this convergence, so an MA root close to the unit circle slows it. Note that this type of convergence also holds for models with strictly non-invertible MA roots, as they can be written in an equivalent strictly invertible form.
3. When there are MA roots on the unit circle, $\boldsymbol{P}_{t|N}$ converges to zero asymptotically. In this case $\boldsymbol{P}_{t|N}$ converges to a value which depends on the sample size, and tends to zero as the sample increases. Therefore, $\mathrm{tr}(\boldsymbol{P}_{t|N})$ converges to zero as $t \to \infty$.

The following corollaries provide further insight into the implications of Proposition 4.1.

Corollary 4.1 *Under the conditions of Proposition 4.1, the KF covariances, $\mathbf{P}_{t|t-1}$, converge to zero.*

Proof. The proof is identical to that of Proposition 4.1, replacing the subindex N in (4.47) by $t-1$.

Corollary 4.2 *Under the conditions of Proposition 4.1, the smoothed and KF state estimates converge to the same values, i.e., $\|\boldsymbol{x}_{t|N} - \boldsymbol{x}_{t|t-1}\| \to 0$ as $t \to \infty$, where $\|.\|$ denotes vector norm.*

Proof. Immediate from Proposition 4.1 and Corollary 4.1.

Therefore, one-sided KF estimates also converge to exact values (Corollary 4.1) and, consequently, to the smoothed estimates of the states (Corollary 4.2). For online applications, the recursive structure of the KF would be a clear advantage, in particular if the recursion is simplified in the light of Corollary 4.1. On the other hand, using filtering instead of smoothing implies some inefficiency, as $\boldsymbol{P}_{t|N}$ converges to zero faster than $\boldsymbol{P}_{t|t-1}$.

4.7 Examples

4.7.1 Recursive Least Squares

As we saw in Example 2.1.2, the constant-coefficient linear regression model could be formulated as:

$$\beta_{t+1} = \beta_t \tag{4.48}$$

$$z_t = \boldsymbol{u}_t^T \beta_t + v_t \tag{4.49}$$

Identifying terms between (4.48)–(4.49) and the MEM (2.1)–(2.2) we see that $\boldsymbol{\Phi} = \boldsymbol{I}$, $\boldsymbol{\Gamma} = \mathbf{0}$, $\boldsymbol{E} = \mathbf{0}$, $\boldsymbol{H} = \boldsymbol{u}_t^T$, $\boldsymbol{D} = \mathbf{0}$, and $\boldsymbol{C} = \boldsymbol{I}$, with $\boldsymbol{Q} = \mathbf{0}$, $\boldsymbol{S} = \mathbf{0}$, $\boldsymbol{R} = \sigma_v^2$ and $\boldsymbol{x_t} = \boldsymbol{\beta_t}$. Now, applying the KF equations (4.7)–(4.10) to model (4.48)–(4.49) yields:

$$\tilde{z}_{t|t-1} = \boldsymbol{z_t} - \boldsymbol{u}_t^T \hat{\beta}_t \tag{4.50}$$

$$\hat{\beta}_{t+1} = \hat{\beta}_t + \boldsymbol{k_t}\, \tilde{z}_{t|t-1} \tag{4.51}$$

$$\boldsymbol{P}_{t+1|t} = \boldsymbol{P}_{t|t-1} - \boldsymbol{k_t} \left(\boldsymbol{u}_t^T \boldsymbol{P}_{t|t-1} \boldsymbol{u}_t + \sigma_v^2\right) \boldsymbol{k_t}^T \tag{4.52}$$

$$\boldsymbol{k_t} = \left(\boldsymbol{P}_{t|t-1} \boldsymbol{u_t}^T\right) \left(\boldsymbol{u}_t^T \boldsymbol{P}_{t|t-1} \boldsymbol{u}_t + \sigma_v^2\right)^{-1}, \tag{4.53}$$

which simplify trivially to the equations of the “Recursive Least Squares” estimator:

$$\hat{\beta}_{t+1} = \hat{\beta}_t + \left(\boldsymbol{P}_{t|t-1} \boldsymbol{u_t}^T\right) \left(\boldsymbol{u}_t^T \boldsymbol{P}_{t|t-1} \boldsymbol{u}_t + \sigma_v^2\right)^{-1} \left(\boldsymbol{z_t} - \boldsymbol{u}_t^T \hat{\beta}_t\right) \tag{4.54}$$

$$\boldsymbol{P}_{t+1|t} = \boldsymbol{P}_{t|t-1} - \boldsymbol{P}_{t|t-1} \boldsymbol{u_t}^T \left(\boldsymbol{u}_t^T \boldsymbol{P}_{t|t-1} \boldsymbol{u}_t + \sigma_v^2\right)^{-1} \boldsymbol{u}_t \boldsymbol{P}_{t|t-1}. \tag{4.55}$$

Equation (4.54) computes the *ex-post* estimate $\hat{\beta}_{t+1}$ as a linear combination of the *ex-ante* estimate $\hat{\beta}_t$ and the innovation $\boldsymbol{z_t} - \boldsymbol{u}_t^T \hat{\beta}_t$. Starting from suitable initial conditions, this equation can be run for each observation in the sample, thus allowing one to recursively estimate the vector of regression parameters. Meanwhile, note that (4.55) furnishes the covariances for $\hat{\beta}_{t+1}$. It is interesting to see that this equation is not coupled with (4.54).

Note that equations (4.54)–(4.55) are not computable, as they depend on the error variance σ_v^2, which is most often unknown. This issue can be solved easily by the following changes of variables:

$$\boldsymbol{P}_{t+1|t} = \frac{1}{\sigma_v^2} \boldsymbol{V}_{t+1} \quad , \quad \boldsymbol{P}_{t|t-1} = \frac{1}{\sigma_v^2} \boldsymbol{V}_t, \tag{4.56}$$

which provide the computable form of the Recursive Least Squares estimator of β.

4.7.2 Cleaning the Wolf sunspot series

This example illustrates the use of the smoothing algorithm described in previous sections to filter out the observational noise from a variable in an error-in-variables model. To this end, consider Wolf’s annual series of sunspot numbers, from 1700 to 1988, taken from Tong [210]. This dataset draws on records compiled by human

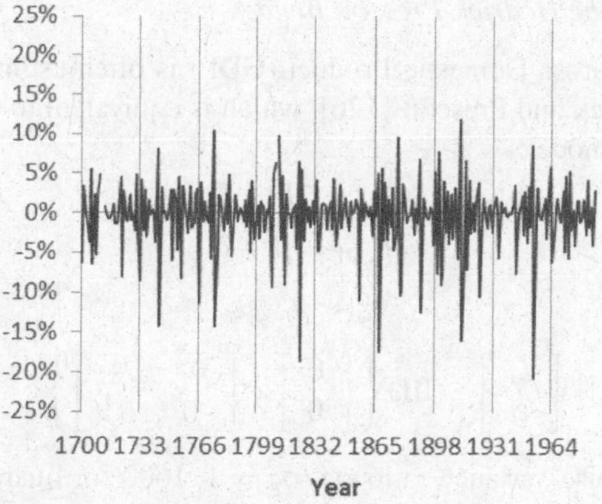

Figure 4.1 *Percent relative corrections on the original Wolf numbers:* $(z_t^*/z_t - 1) \times 100$.

observers using optical devices of varying quality, so it seems natural to assume that the data is affected by observational errors.

On the other hand, previous analyses have consistently found that this series displays a harmonic cycle with a period of about 11 years. Building on these ideas, we fitted the following AR(2) plus white noise errors model to the original data :

$$(1 - \phi_1 \mathrm{B} - \phi_2 \mathrm{B}^2) z_t^* = \mu + w_t, \qquad \{w_t\} \sim \mathrm{IID}(0, \sigma_w^2) \tag{4.57}$$

$$z_t = z_t^* + v_t, \qquad \{v_t\} \sim \mathrm{IID}(0, \sigma_v^2), \tag{4.58}$$

where the sequences z_t and z_t^* are, respectively, the square root of the Wolf number at year t and the underlying "decontaminated" value. Maximizing the Gaussian likelihood results in the estimates and standard errors displayed in Table 4.1. Note

Table 4.1 *Estimates and standard errors for the AR(2) plus noise model fitted to the Wolf sunspot series.*

Parameter	ϕ_1	ϕ_2	μ	σ_w^2	σ_v^2
estimate	1.444	-0.743	1.476	2.205	0.147
std. error	0.048	0.047	0.145	-	-

that the primary AR(2) structure has complex roots, implying a damped cycle with a period of 11.07 years. Using model (4.57)–(4.58) it is easy to obtain smoothed estimates for the underlying decontaminated series z_t^*. The percent relative corrections, $(z_t^*/z_t - 1) \times 100$, implied by this model are shown in Figure 4.1.

The code required to replicate this example can be found in the file `Example0472.m`, see Appendix D.

4.7.3 Extracting the Hodrick-Prescott trend

The potential Gross Domestic Product (GDP) is often estimated with the filter proposed by Hodrick and Prescott [126], which is equivalent to the smoothed trend resulting from the model:

$$\mu_{t+1} = \mu_t + \beta_t \tag{4.59}$$

$$\beta_{t+1} = \beta_t + \zeta_t \tag{4.60}$$

$$z_t = \mu_t + \varepsilon_t \tag{4.61}$$

$$\begin{bmatrix} \zeta_t \\ \varepsilon_t \end{bmatrix} \sim \text{IID}\left\{ \begin{bmatrix} 0 \\ 0 \end{bmatrix}, \begin{bmatrix} \sigma_\zeta^2 & 0 \\ 0 & \sigma_\varepsilon^2 \end{bmatrix} \right\}, \tag{4.62}$$

with a signal-to-noise variance ratio $\sigma_\zeta^2/\sigma_\varepsilon^2 = 1/1600$ for quarterly data, see Harvey and Trimbur [120]. A typical analysis in this framework consists of estimating the unknown variances, taking into account the variance ratio constraint, and then obtaining smoothed estimates of the trend component μ_t.

This procedure was applied to the quarterly and seasonally adjusted series of US GDP (GDP_t) in constant 2000 US Dollar, from 1947/Q1 to 2008/Q3. We fitted model (4.59)–(4.62) to the transformed variable $z_t = \log(GDP_t) \times 100$, and obtained the variance estimates of $\hat{\sigma}_\varepsilon^2 = 1.875$ and $\hat{\sigma}_\zeta^2 = 1.875/1600$. Figure 4.2 shows the estimates of the change in HP trend (the unobserved component β_t). This series is often used to pinpoint turning points in the economic cycle.

The code required to replicate this example can be found in the file `Example0473.m`, see Appendix D.

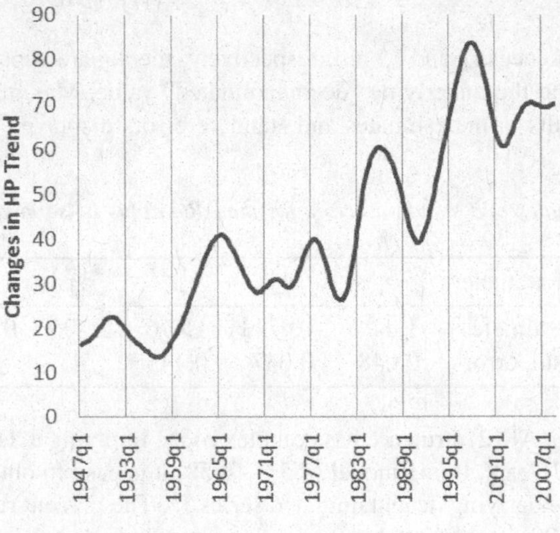

Figure 4.2 *Changes in the HP trend.*

Chapter 5

Likelihood computation for fixed-coefficients models

5.1	Maximum likelihood estimation	45
	5.1.1 Problem statement	45
	5.1.2 Prediction error decomposition	46
	5.1.3 Initialization of the Kalman filter in the stationary case	47
5.2	The likelihood for a nonstationary model	48
	5.2.1 Diffuse likelihood	49
	5.2.2 Minimally conditioned likelihood	50
	5.2.3 Likelihood computation for a fixed-parameter SEM	52
	5.2.4 Initialization of the Kalman filter in the nonstationary case	52
5.3	The likelihood for a model with inputs	54
	5.3.1 Models with deterministic inputs	54
	5.3.2 Models with stochastic inputs	55
5.4	Examples	57
	5.4.1 Models for the Airline Passenger series	57
	5.4.2 Modeling the series of Housing Starts and Sales	60

In this chapter we discuss how to compute the Gaussian likelihood for a linear SS model. As we will see, the presence of unit roots and exogenous inputs critically affects its evaluation.

5.1 Maximum likelihood estimation

5.1.1 Problem statement

Consider a MEM without inputs:

$$\boldsymbol{x}_{t+1} = \boldsymbol{\Phi}\boldsymbol{x}_t + \boldsymbol{E}\boldsymbol{w}_t \tag{5.1}$$

$$\boldsymbol{z}_t = \boldsymbol{H}\boldsymbol{x}_t + \boldsymbol{C}\boldsymbol{v}_t, \tag{5.2}$$

with corresponding error covariances:

$$\mathrm{E}\left(\begin{bmatrix}\boldsymbol{w}_t \\ \boldsymbol{v}_t\end{bmatrix}\begin{bmatrix}\boldsymbol{w}_t & \boldsymbol{v}_t\end{bmatrix}\right) = \begin{bmatrix}\boldsymbol{Q} & \boldsymbol{S} \\ \boldsymbol{S}^T & \boldsymbol{R}\end{bmatrix}. \tag{5.3}$$

Our goal is to estimate a parameter vector $\boldsymbol{\theta}$ which includes all the unknown elements within the matrices $\boldsymbol{\Phi}$, $\boldsymbol{E}$, $\boldsymbol{H}$, $\boldsymbol{C}$, $\boldsymbol{Q}$, $\boldsymbol{R}$ and $\boldsymbol{S}$. Assuming that the sample information is $\boldsymbol{Z} = (\boldsymbol{z}_1, \boldsymbol{z}_2, ..., \boldsymbol{z}_N)$, with N being the sample size, the maximum likelihood (ML) estimator of $\boldsymbol{\theta}$ is

$$\hat{\boldsymbol{\theta}} = \arg\max_{\boldsymbol{\theta}} \; [P(\boldsymbol{Z};\boldsymbol{\theta})], \tag{5.4}$$

where $P(\boldsymbol{Z};\boldsymbol{\theta})$ is the joint probability density function of the observations. If $\boldsymbol{Z}$ is treated as given, this is the so-called *likelihood function*. The estimates obtained by maximizing this function have many convenient asymptotic properties, which are discussed in Appendix B.

In this framework, an ML estimation method will include two basic elements:

1. A procedure to evaluate the likelihood function for any tentative value of the parameter vector $\boldsymbol{\theta}$. In this chapter we will discuss how to do this in the case of a fixed-coefficients SS model. Chapter 6 covers the likelihood evaluation for some specific time-varying parameter models.
2. An algorithm to compute the ML estimate $\boldsymbol{\theta}$. This can be done with many methods, most of them being variations on the basic Newton-Raphson algorithm, see e.g., Dennis and Schnabel [74], requiring the first and second-order derivatives of the log-likelihood function. These can be computed either analytically, see appendices B to D in Terceiro [206], or by numerical approximation, see Dennis and Schnabel [74]. A popular method in the statistics literature, because it obviates the evaluation of derivatives, is the so-called *EM Algorithm*, which will be discussed in detail in Chapter 11.

Bearing in mind that the numerical optimization problem is substantial, but has been satisfactorily resolved in the literature, we will focus our attention on discussing how to evaluate the Gaussian likelihood for a SS model. To this end, the basic idea is to write the likelihood function in a specific form known as "prediction error decomposition."

5.1.2 Prediction error decomposition

When the observations are not independent, one can write $P(\boldsymbol{Z};\boldsymbol{\theta})$ by applying successively the definition of conditional density:

$$P(\boldsymbol{Z};\boldsymbol{\theta}) = \prod_{t=1}^{N} P(\boldsymbol{z}_t \,|\, \boldsymbol{Z}^{t-1};\boldsymbol{\theta}), \tag{5.5}$$

where $\boldsymbol{Z}^{t-1} = (\boldsymbol{z}_1, \boldsymbol{z}_2, ..., \boldsymbol{z}_{t-1})$. It is convenient to simplify the mathematical structure of (5.5) by taking logs so, instead of maximizing (5.5) with respect to $\boldsymbol{\theta}$, we will consider the equivalent problem of minimizing the negative log-likelihood:

$$\ell(\boldsymbol{Z};\boldsymbol{\theta}) = -\log P(\boldsymbol{Z};\boldsymbol{\theta}) = -\sum_{t=1}^{N} \log P(\boldsymbol{z}_t \,|\, \boldsymbol{Z}^{t-1};\boldsymbol{\theta}). \tag{5.6}$$

and, assuming normality for $\boldsymbol{w}_t$, $\boldsymbol{v}_t$ and $\boldsymbol{x}_1$, the general term $P(\boldsymbol{z}_t \,|\, \boldsymbol{Z}^{t-1};\boldsymbol{\theta})$ can be computed by applying a KF to (5.1)–(5.2) for any given value of $\boldsymbol{\theta}$. In particular, the negative Gaussian log-likelihood (5.6) can be written as:

$$\ell(\boldsymbol{Z};\boldsymbol{\theta}) = \frac{1}{2}\sum_{t=1}^{N} (m\log(2\pi) + \log|\boldsymbol{B}_t| + \tilde{\boldsymbol{z}}_t^T \boldsymbol{B}_t^{-1} \tilde{\boldsymbol{z}}_t), \tag{5.7}$$

where the sequence of innovations is defined as: $\tilde{\boldsymbol{z}}_t = \boldsymbol{z}_t - \mathrm{E}(\boldsymbol{z}_t \,|\, \boldsymbol{Z}^{t-1};\boldsymbol{\theta})$ and $\boldsymbol{B}_t = \mathrm{cov}(\tilde{\boldsymbol{z}}_t \,|\, \boldsymbol{Z}^{t-1};\boldsymbol{\theta})$. The function (5.7) is known as the "prediction error decomposition form" of the log-likelihood, see Schweppe [192]. The reason for this terminology stems from the fact that $\tilde{\boldsymbol{z}}_t$ can be interpreted as a vector of one-step-ahead prediction errors which, under normality, follows the distribution, $\tilde{\boldsymbol{z}}_t \sim NID(\boldsymbol{0}, \boldsymbol{B}_t)$. The term *NID* will henceforth denote *Normal and Independently Distributed.*

Evaluation of the log-likelihood (5.7) requires computation of the sequences $\tilde{\boldsymbol{z}}_t$ and $\boldsymbol{B}_t$, obtained by propagating the KF :

$$\hat{\boldsymbol{x}}_{t+1|t} = \boldsymbol{\Phi}\hat{\boldsymbol{x}}_{t|t-1} + \boldsymbol{K}_t\tilde{\boldsymbol{z}}_t \tag{5.8}$$

$$\tilde{\boldsymbol{z}}_t = \boldsymbol{z}_t - \boldsymbol{H}\hat{\boldsymbol{x}}_{t|t-1} \tag{5.9}$$

$$\boldsymbol{B}_t = \boldsymbol{H}\boldsymbol{P}_{t|t-1}\boldsymbol{H}^T + \boldsymbol{C}\boldsymbol{R}\boldsymbol{C}^T \tag{5.10}$$

$$\boldsymbol{K}_t = (\boldsymbol{\Phi}\boldsymbol{P}_{t|t-1}\boldsymbol{H}^T + \boldsymbol{E}\boldsymbol{S}\boldsymbol{C}^T)\boldsymbol{B}_t^{-1} \tag{5.11}$$

$$\boldsymbol{P}_{t+1|t} = \boldsymbol{\Phi}\boldsymbol{P}_{t|t-1}\boldsymbol{\Phi}^T + \boldsymbol{E}\boldsymbol{Q}\boldsymbol{E}^T - \boldsymbol{K}_t\boldsymbol{B}_t\boldsymbol{K}_t^T. \tag{5.12}$$

Note that expressions (5.8)–(5.12) can be immediately generalized to the time-varying parameter case just by adding time subindex t to the system matrices $\boldsymbol{\Phi}$, $\boldsymbol{E}$, $\boldsymbol{H}$, $\boldsymbol{C}$, $\boldsymbol{Q}$, $\boldsymbol{R}$ and $\boldsymbol{S}$. This is useful, for example, in estimating random coefficients models or periodic models, see Chapter 6. Given any value for the parameter vector $\boldsymbol{\theta}$, it is relatively easy to: (a) generate the corresponding SS matrices, (b) propagate the KF (5.8)–(5.12), and (c) use the resulting outputs to evaluate (5.7).

5.1.3 *Initialization of the Kalman filter in the stationary case*

The KF (5.8)–(5.12) is a recursive algorithm which requires initial conditions for the state vector and its covariance matrix. Determining these conditions is a complex problem which has been tackled in many works such as: Akaike [3], Anderson and Moore [6], De Jong[66], Ansley and Kohn [8], or De Jong and Chu-Chun-Lin [69]. We will now address this problem assuming that the state equation (5.1) is stable, see Definition 2.1. Under this assumption, adequate initial conditions for the KF are $\boldsymbol{\mu}$ and $\boldsymbol{P}_1$ such that:

$$\boldsymbol{\mu} = \mathrm{E}(\boldsymbol{x}_1) = \boldsymbol{0} \tag{5.13}$$

$$P_1 = \Phi P_1 \Phi^T + EQE^T, \tag{5.14}$$

where the algebraic Lyapunov equation (5.14) has a unique positive semi-definite solution, see Akaike [3], Anderson and Moore [6] and Golub and Van Loan [109]. To solve it, we need to rewrite (5.14) as a system of linear equations:

$$\text{vec}(P_1) = (\Phi \otimes \Phi)\text{vec}(P_1) + \text{vec}(EQE^T), \tag{5.15}$$

where $\otimes$ denotes the Kronecker product and vec(.) is the *vectorization* operator, which stacks the columns of its argument into a single vector. The solution of (5.15) is then[1]:

$$\text{vec}(P_1) = (I - \Phi \otimes \Phi)^{-1}\text{vec}(EQE^T). \tag{5.16}$$

Note that the eigenvalues of $I - \Phi \otimes \Phi$ are $1 - \lambda_i \lambda_j$, for all $i, j = 1, 2, ..., n$, where λ_i is the i-th eigenvalue of Φ. Therefore, the stability assumption ensures that equation (5.15) has a unique solution. Moreover, since the matrix EQE^T is positive semidefinite, then P_1 is positive semi-definite in general. If the system is controllable[2], P_1 would be strictly positive definite.

5.2 The likelihood for a nonstationary model

In the nonstationary case at least one eigenvalue of Φ in (5.1) has a unit modulus. Accordingly the system is not stable and, under the immemorial time hypothesis (see Definition 2.9 in Chapter 2), the uncertainty associated with the initial state is not finite. As a consequence, the initialization of the KF becomes more complex.

Many papers deal with this problem which is generically described as "diffuse initialization" see De Jong [68], Ansley and Kohn [9], Koopman and Durbin [142] and Koopman [139]. As we saw in Section 4.2, this situation could be tackled in general by initializing the KF with an arbitrarily large covariance, or by using the information filter, see Anderson and Moore [6]. Ansley and Kohn [8] point out that the former method induces numerical errors, and the latter cannot be applied to all cases. Alternatively, one could develop specialized algorithms for each specific application. In this line, the literature provides three main approaches to evaluate the likelihood for a model with nonstationary states:

1. Ansley and Kohn [8] propose a data transformation which cancels the nonstationary states. This approach has been further developed in other works, such as Kohn and Ansley [137] or Ansley and Kohn [9] and, in the univariate case, in Bell and Hillmer [19] or Gomez and Maravall [110].
2. Koopman [139] derives an exact solution to initialize a KF with diffuse initial conditions by: (a) setting $\mu = \mathbf{0}$ and $P_1 = P_* + kP_\infty$ $(k \to \infty)$, where the matrices P_* and P_∞ are known, and (b) transforming the KF variables so that they do not depend on k.

[1] There are computationally efficient procedures that do not require solving a system of dimension n^2 such as the one in (5.15). The basic idea consists of using a real Schur decomposition of Φ that computes P_1 by solving a system of dimension 1×1 or 2×2, see , Christov and Konstantinov [181] and Appendix A.

[2] See Chapter 2, Definition 2.2

3. De Jong, [66, 68], suggest initializing the KF with $\boldsymbol{\mu} = \mathbf{0}$ and $\boldsymbol{P_1} = \mathbf{0}$, and then adjusting the log-likelihood to compensate for these arbitrary initial conditions.

These methods have specific problems and limitations. In particular, Ansley and Kohn [8] recognize that method 1 needs a complex and nonstandard filtering, and requires the data transformation to be independent of the parameter values. This requirement is not fulfilled, for example, when one wants to estimate structural time series models such as those described by Harvey [118]. On the other hand, methods 2 and 3 have two problems: (a) these methods do not allow for stochastic inputs, and (b) its likelihood depends in some cases on the scale of the state vector. In Subsection 5.2.1 we provide further details about this issue.

Subsection 5.2.1 describes the diffuse likelihood method 3 in detail. Subsection 5.2.2 presents the *minimally conditioned likelihood*, a procedure which draws many ideas from the diffuse likelihood but avoids its scaling problem. Last, Subsection 5.3.1 discusses the effect of exogenous inputs on the calculation of the likelihood.

5.2.1 *Diffuse likelihood*

The diffuse likelihood can be written as:

$$\ell_\infty(\boldsymbol{Z};\boldsymbol{\theta}) = \ell(\boldsymbol{Z};\boldsymbol{\theta}) - \log|\boldsymbol{P_1}|, \tag{5.17}$$

where $\ell_\infty(\boldsymbol{Z};\boldsymbol{\theta})$ denotes the diffuse log-likelihood of model (5.1)–(5.2) and $\ell(\boldsymbol{Z};\boldsymbol{\theta})$ is the corresponding standard log-likelihood, see De Jong [68]. On this basis, De Jong [68] proposes an evaluation algorithm based on the so-called diffuse KF, while Koopman [139] suggests using two specialized filters for the diffuse and non-diffuse components, respectively.

The first term on the right-hand side of (5.17), $\ell(\boldsymbol{Z};\boldsymbol{\theta})$, can be computed following De Jong [66], who assumes that the initial state is such that $\boldsymbol{x_1} \sim N(\boldsymbol{\mu}, \boldsymbol{P_1})$. Under these conditions:

$$\begin{aligned}\ell(\boldsymbol{Z};\boldsymbol{\theta}) = \log|\boldsymbol{P}_1| + \boldsymbol{\mu}^T \boldsymbol{P}_1^{-1}\boldsymbol{\mu} + \sum_{t=1}^{N} \log|\boldsymbol{B}_t| + \sum_{t=1}^{N} \tilde{\boldsymbol{z}}_t^T \boldsymbol{B}_t^{-1}\tilde{\boldsymbol{z}}_t + \log|\boldsymbol{P}_1^{-1} + \boldsymbol{W}_N| \\ - (\boldsymbol{P}_1^{-1}\boldsymbol{\mu} + \boldsymbol{w}_N)^T(\boldsymbol{P}_1^{-1} + \boldsymbol{W}_N)^{-1}(\boldsymbol{P}_1^{-1}\boldsymbol{\mu} + \boldsymbol{w}_N),\end{aligned} \tag{5.18}$$

where $\tilde{\boldsymbol{z}}_t$ and $\boldsymbol{B_t}$ are, respectively, the innovations and their covariance matrices resulting from a KF(**0**,**0**) applied to (5.1)–(5.2). On the other hand, the values of $\boldsymbol{w_N}$ and $\boldsymbol{W_N}$ in (5.18) can be computed as follows:

$$\boldsymbol{w_t} = \boldsymbol{w}_{t-1} + \bar{\boldsymbol{\Phi}}_{t-1}^T \boldsymbol{H}^T \boldsymbol{B_t}^{-1}\tilde{\boldsymbol{z}}_t \tag{5.19}$$

$$\boldsymbol{W_t} = \boldsymbol{W}_{t-1} + \bar{\boldsymbol{\Phi}}_{t-1}^T \boldsymbol{H}^T \boldsymbol{B_t}^{-1}\boldsymbol{H}\bar{\boldsymbol{\Phi}}_{t-1} \tag{5.20}$$

$$\bar{\boldsymbol{\Phi}}_t = (\boldsymbol{\Phi} - \boldsymbol{K_t}\boldsymbol{H})\bar{\boldsymbol{\Phi}}_{t-1}, \tag{5.21}$$

where $\boldsymbol{w_t}$ and $\boldsymbol{W_t}$ should be initialized with null values, $\bar{\boldsymbol{\Phi}}_\mathbf{0} = \boldsymbol{I}$, and $\boldsymbol{K_t}$ is the gain of a KF(**0**,**0**) applied to (5.1)–(5.2). Note that:

1. The terms $\log|\boldsymbol{P_1}|$, $\boldsymbol{\mu}^T\boldsymbol{P_1}^{-1}\boldsymbol{\mu}$, $\log|\boldsymbol{P_1}^{-1}+\boldsymbol{W_N}|$, $\boldsymbol{P_1}^{-1}\boldsymbol{\mu}+\boldsymbol{w_N}$, and $(\boldsymbol{P_1}^{-1}+\boldsymbol{W_N})^{-1}$ in (5.18) depend on the mean and covariance of the initial state, $\boldsymbol{\mu}$ and $\boldsymbol{P_1}$. When the system is stationary, all of them can be computed.
2. When there are some unit roots, the first term in (5.18) diverges to infinity, whence it is ignored for computing the diffuse likelihood. The other terms are not problematic because they depend on $\boldsymbol{P_1}^{-1}$ which is finite[3].
3. The diffuse approach has a clear shortcoming, as the likelihood value depends on the scale of the state vector in some models. The following example illustrates this situation.

Example 5.2.1 Scale-dependency of the diffuse likelihood

Consider the observationally equivalent models:

$$\begin{aligned} x_{t+1} &= x_t + w_t \\ z_t &= \alpha x_t + v_t, \end{aligned} \tag{5.22}$$

and

$$\begin{aligned} x^*_{t+1} &= x^*_t + \alpha\, w_t \\ z_t &= x^*_t + v_t, \end{aligned} \tag{5.23}$$

where α is an arbitrary constant and the system needs some identification constraints such as $\mathrm{var}(w_t) = 1$ and $\mathrm{cov}\,(w_t, v_t) = 0$. According to (5.17), the diffuse likelihood for models (5.22) and (5.23) are, respectively:

$$\ell^1_\infty(\boldsymbol{Z};\alpha) = \ell(\boldsymbol{Z};\boldsymbol{\theta}) - \frac{1}{2}\log|\mathrm{cov}\,(\boldsymbol{x_1})| \tag{5.24}$$

$$\ell^2_\infty(\boldsymbol{Z};\alpha) = \ell(\boldsymbol{Z};\boldsymbol{\theta}) - \frac{1}{2}\log|\mathrm{cov}\,(\alpha\boldsymbol{x_1})|, \tag{5.25}$$

and these values do not coincide because $\ell^1_\infty(\boldsymbol{Z};\alpha) - \ell^2_\infty(\boldsymbol{Z};\alpha) = \log(\alpha)$. Therefore, the values of the diffuse likelihood corresponding to observationally equivalent representations, such as (5.22) and (5.23), can be different. Note also that the first-order derivatives of the likelihood are also scale-dependent. The procedure described in the following subsection solves these problems.

5.2.2 *Minimally conditioned likelihood*

An alternative to the diffuse likelihood would be a Gaussian likelihood conditional on the minimum subsample required to eliminate the effect of the diffuse states. This strategy is closely related to the diffuse likelihood, but does not suffer

[3]If all the modes in MEM (5.1)–(5.2) were unstable, $\boldsymbol{P_1}^{-1}$ would be null. On the other hand, if the model has some stable modes, the matrix $\boldsymbol{P_1}^{-1}$ converges to a nonzero value, see De Jong and Chu-Chun-Lin [69].

from its shortcomings. Following Casals, Sotoca and Jerez [47], the diffuse and minimally conditioned likelihoods are related by the following expression:

$$\ell^*(\boldsymbol{Z};\boldsymbol{\theta}) = \ell_\infty(\boldsymbol{Z};\boldsymbol{\theta}) - \log\left|\boldsymbol{O}_1^T\boldsymbol{O}_1\right|, \tag{5.26}$$

where $\ell^*(\boldsymbol{Z};\boldsymbol{\theta})$ denotes the minimally conditioned log-likelihood and $\ell_\infty(\boldsymbol{Z};\boldsymbol{\theta})$ is the diffuse function. The difference between both functions is therefore the term $-\log\left|\boldsymbol{O}_1^T\boldsymbol{O}_1\right|$ which plays a very important role, as it frees the minimally conditioned likelihood from any scale dependency. To compute this correction term, one needs to:

1. Write observation equation (5.2) in matrix form:

$$\boldsymbol{Z} = \boldsymbol{O}\boldsymbol{x}_1 + \boldsymbol{Z}^*, \tag{5.27}$$

   where $\boldsymbol{Z}^*$ is the part of the sample which does not depend on $\boldsymbol{x}_1$, and $\boldsymbol{O}$ is the extended observability matrix, defined as:

$$\boldsymbol{O} = \begin{bmatrix} \boldsymbol{H} \\ \boldsymbol{H\Phi} \\ \vdots \\ \boldsymbol{H\Phi}^{n-1} \end{bmatrix}. \tag{5.28}$$

2. Separate the diffuse and non-diffuse components of the initial state by means of the similar transformation:

$$\boldsymbol{T}\boldsymbol{x}_1 = \begin{bmatrix} \boldsymbol{x}_1^N \\ \boldsymbol{x}_1^S \end{bmatrix}, \tag{5.29}$$

   where $\boldsymbol{T}$ is the matrix characterizing the transformation such that $\boldsymbol{T\Phi T}^{-1} = \text{diag}(\boldsymbol{\Phi}^N, \boldsymbol{\Phi}^S)$, $\boldsymbol{x}_1^N$ is a vector including the diffuse states, such that $\text{cov}(\boldsymbol{x}_1^N) \to \infty$, and $\boldsymbol{x}_1^S$ is a vector of stationary components. Denoting $\boldsymbol{T}^{-1} = [\boldsymbol{M}\ \boldsymbol{N}]$, we can write equation (5.29) as:

$$\boldsymbol{x}_1 = \boldsymbol{M}\boldsymbol{x}_1^N + \boldsymbol{N}\boldsymbol{x}_1^S \tag{5.30}$$

3. Apply the decomposition (5.30) to (5.27), which yields:

$$\boldsymbol{Z} = \boldsymbol{O}_N\boldsymbol{x}_1^N + \boldsymbol{Z}^S, \tag{5.31}$$

   where $\boldsymbol{O}_N = \boldsymbol{OM}$ is the extended observability matrix corresponding to the diffuse initial states and $\boldsymbol{Z}^S$ is the part of the sample which does not depend on the diffuse states.

4. Compute the correction term $-\log\left|\boldsymbol{O}_1^T\boldsymbol{O}_1\right|$, where $\boldsymbol{O}_1$ includes the first columns of $\boldsymbol{O}_N$ so that $\text{rank}(\boldsymbol{O}_N) = k$, where k is the number of diffuse states in the system. This specific choice implies that the likelihood computation is conditioned on the first k observations that can reduce the covariance matrix of $\boldsymbol{x}_1^N$ (conditional on the first k observations) to a finite value.

Casals, Sotoca and Jerez [47] proved that the minimally conditioned likelihood: (a) can be efficiently computed by two equivalent methods which, in comparison with other approaches, (b) only requires standard filtering, so it is able to accommodate missing data and cointegration constraints. Last, (c) the choice made for the correction term $-\log\left|O_1^T O_1\right|$ assures that the values provided by the minimally conditioned likelihood coincide with those that would be obtained by differencing the data and then computing the likelihood for the stationary model; a desirable consistency property.

5.2.3 *Likelihood computation for a fixed-parameter SEM*

Likelihood evaluation can be substantially simplified when the model is a constant-parameter SEM. First, in an SEM $\boldsymbol{C} = \boldsymbol{I}$ and there is only one error covariance, $\boldsymbol{Q}$, so in this case the log-likelihood function (5.18) simplifies to:

$$\ell(\boldsymbol{Z};\boldsymbol{\theta}) = \log|\boldsymbol{P}_1| + \boldsymbol{\mu}^T \boldsymbol{P}_1^{-1}\boldsymbol{\mu} + N\log|\boldsymbol{Q}| + \sum_{t=1}^{N} \tilde{\boldsymbol{z}}_t^T \boldsymbol{Q}^{-1}\tilde{\boldsymbol{z}}_t + \log|\boldsymbol{P}_1^{-1} + \boldsymbol{W}_N|$$
$$- (\boldsymbol{P}_1^{-1}\boldsymbol{\mu} + \boldsymbol{w}_N)^T (\boldsymbol{P}_1^{-1} + \boldsymbol{W}_N)^{-1} (\boldsymbol{P}_1^{-1}\boldsymbol{\mu} + \boldsymbol{w}_N). \quad (5.32)$$

The KF$(\mathbf{0},\mathbf{0})$ thus simplifies drastically, because $\boldsymbol{P} = \mathbf{0}$ is a solution of the Riccati equation (5.12), which implies that $\boldsymbol{P}_{t+1|tt} = \mathbf{0}$, $\boldsymbol{B}_t = \boldsymbol{Q}$, and $\boldsymbol{K}_t = \boldsymbol{E}$. As these KF variables have exact and constant values which do not need to be updated, the resulting likelihood computation algorithm is faster and more stable than those previously described, see Casals, Sotoca and Jerez [45].

5.2.4 *Initialization of the Kalman filter in the nonstationary case*

To evaluate (5.18) we need to determine $\boldsymbol{P}_1^{-1}$. To this end, we will separate the stable and non-stable modes by applying a similar transformation to model (5.1)–(5.2):

$$\begin{bmatrix} \boldsymbol{x}_{t+1}^N \\ \boldsymbol{x}_{t+1}^S \end{bmatrix} = \boldsymbol{T}\boldsymbol{x}_t, \quad (5.33)$$

such that the state equation is decomposed in two blocks:

$$\begin{bmatrix} \boldsymbol{x}_{t+1}^N \\ \boldsymbol{x}_{t+1}^S \end{bmatrix} = \begin{bmatrix} \boldsymbol{\Phi}^N & \mathbf{0} \\ \mathbf{0} & \boldsymbol{\Phi}^S \end{bmatrix} \begin{bmatrix} \boldsymbol{x}_t^N \\ \boldsymbol{x}_t^S \end{bmatrix} + \begin{bmatrix} \boldsymbol{E}^N \\ \boldsymbol{E}^S \end{bmatrix} \boldsymbol{w}_t, \quad (5.34)$$

where the vectors $\boldsymbol{x}_{t+1}^N$ and $\boldsymbol{x}_{t+1}^S$ correspond, respectively, to the nonstationary and stationary modes of the system. Note that this transformation is a particular case of the block-diagonalization procedure presented in Subsection 3.1.3 and Appendix A. De Jong and Chu-Chun-Lin [69] originally suggested using the Jordan decomposition to obtain (5.34). In our opinion, the procedures discussed in Subsection 3.1.3 are more efficient and stable.

Assume now that the model has been initialized with arbitrary values. The covariance of the state vector can then be propagated according to the following equations:

$$\boldsymbol{P}_{t+1}^N = \boldsymbol{\Phi}^N \boldsymbol{P}_t^N (\boldsymbol{\Phi}^N)^T + \boldsymbol{E}^N \boldsymbol{Q} (\boldsymbol{E}^N)^T \quad (5.35)$$

$$P_{t+1}^{S} = \Phi^{S} P_t^{S} (\Phi^{S})^T + E^{S} Q (E^{S})^T \tag{5.36}$$

$$P_{t+1}^{NS} = \Phi^{N} P_t^{NS} (\Phi^{S})^T + E^{N} Q (E^{S})^T, \tag{5.37}$$

where P_t^N, P_t^S, and P_t^{NS} are, respectively, the covariances of the nonstationary and stationary states, as well as the cross-covariance between both types of states. As the eigenvalues of the nonstationary subsystem are equal to one, the variances of all its modes diverge to infinity, so that $\lim_{t\to\infty} P_t^N = \infty$ and, accordingly, $\lim_{t\to\infty} (P_t^N)^{-1} = \mathbf{0}$. On the other hand, P_t^S and P_t^{NS} converge to the following equilibrium values:

$$P^{S} = \Phi^{S} P^{S} (\Phi^{S})^T + E^{S} Q (E^{S})^T \tag{5.38}$$

$$P^{NS} = \Phi^{N} P^{NS} (\Phi^{S})^T + E^{N} Q (E^{S})^T. \tag{5.39}$$

In their otherwise excellent paper, De Jong and Chu-Chun-Lin [69] assumed that the covariance between the stationary and nonstationary states converges to zero. To see that this is not true, apply the vectorization operator to (5.39):

$$\mathrm{vec}(P^{NS}) = (I - \Phi^{S} \otimes \Phi^{N})^{-1} vec[E^{N} Q (E^{S})^T]. \tag{5.40}$$

Taking into account that the eigenvalues of $I - \Phi^{S} \otimes \Phi^{N}$ are $1 - \lambda_i^N \lambda_j^S$, where λ_i^N and λ_j^S are the i-th and j-th eigenvalues of Φ^N and Φ^S, respectively, and bearing in mind that there are no reciprocal eigenvalues in these matrices, the value P^{NS} is the unique solution of (5.40), see Section A.7. Therefore, the following convergence result holds:

$$\lim_{t\to\infty} \mathrm{cov}\left(\begin{bmatrix} x_t^N \\ x_t^S \end{bmatrix} \right) = \begin{bmatrix} \infty & (P^{NS})^T \\ P^{NS} & P^{S} \end{bmatrix}. \tag{5.41}$$

Applying the matrix block-inversion lemma to equation (5.41), we obtain:

$$\lim_{t\to\infty} \left(\mathrm{cov}\left(\begin{bmatrix} x_t^N \\ x_t^S \end{bmatrix} \right) \right)^{-1} = \begin{bmatrix} \mathbf{0} & \mathbf{0} \\ \mathbf{0} & (P^{S})^{-1} \end{bmatrix}, \tag{5.42}$$

and using $T^{-1} = [M \; N]$ and $[M \; N]^{-1} = \begin{bmatrix} U \\ V \end{bmatrix}$ we have:

$$\lim_{t\to\infty} [\mathrm{cov}(x_t)]^{-1} = \lim_{t\to\infty} \mathrm{cov}\left(T^{-1} \begin{bmatrix} x_t^N \\ x_t^S \end{bmatrix} (T^{-1})^T \right)^{-1} = V^T (P^S)^{-1} V, \tag{5.43}$$

so that, under the immemorial time hypothesis, expression (5.43) provides an adequate initial condition for $[\mathrm{cov}(x_t)]^{-1}$.

On the other hand, the state vector should be initialized to null values. This choice is adequate because: (a) in a system without inputs the vector of stationary states (x_t^S) converges to zero, and (b) the variances of the nonstationary states (x_t^N) diverge to infinity, so that the particular choice made for their initial values is irrelevant. Thus, and since any initial condition is adequate for these states, the simplest choice would be $\mu = \mathbf{0}$.

5.3 The likelihood for a model with inputs

Consider now a MEM with exogenous inputs:

$$\boldsymbol{x}_{t+1} = \boldsymbol{\Phi}\boldsymbol{x}_t + \boldsymbol{\Gamma}\boldsymbol{u}_t + \boldsymbol{E}\boldsymbol{w}_t \tag{5.44}$$

$$\boldsymbol{z}_t = \boldsymbol{H}\boldsymbol{x}_t + \boldsymbol{D}\boldsymbol{u}_t + \boldsymbol{C}\boldsymbol{v}_t, \tag{5.45}$$

where the unknown elements in $\boldsymbol{\Gamma}$ and $\boldsymbol{D}$ are added to the vector $\boldsymbol{\theta}$ to be estimated, and the sample information is now given by $\boldsymbol{Z} = (\boldsymbol{z}_1, \boldsymbol{z}_2, ..., \boldsymbol{z}_N)$ and $\boldsymbol{U} = (\boldsymbol{u}_1, \boldsymbol{u}_2, ..., \boldsymbol{u}_N)$. Under these conditions, the log-likelihood function in (5.6) becomes:

$$\ell(\boldsymbol{\theta}) = -\log P(\boldsymbol{Z}, \boldsymbol{U}; \boldsymbol{\theta}) = -\sum_{t=1}^{N} \log P(\boldsymbol{z}_t \,|\, \boldsymbol{Z}^{t-1}, \boldsymbol{U}; \boldsymbol{\theta}). \tag{5.46}$$

Evaluating (5.46) requires propagating a KF analogous to (5.8)–(5.12), but replacing (5.8) and (5.9) by the following expressions:

$$\hat{\boldsymbol{x}}_{t+1|t} = \boldsymbol{\Phi}\hat{\boldsymbol{x}}_{t|t-1} + \boldsymbol{\Gamma}\boldsymbol{u}_t + \boldsymbol{K}_t\tilde{\boldsymbol{z}}_t \tag{5.47}$$

$$\tilde{\boldsymbol{z}}_t = \boldsymbol{z}_t - \boldsymbol{H}\hat{\boldsymbol{x}}_{t|t-1} - \boldsymbol{D}\boldsymbol{u}_t. \tag{5.48}$$

The likelihood function can be written in a form analogous to (5.18) using the following result:

$$\ell(\boldsymbol{Z} \,|\, \boldsymbol{U}) = \ell(\boldsymbol{x}_1 \,|\, \boldsymbol{U}) + \ell(\boldsymbol{Z} \,|\, \boldsymbol{U}, \boldsymbol{x}_1) - \ell(\boldsymbol{x}_1 \,|\, \boldsymbol{Z}, \boldsymbol{U}) \tag{5.49}$$

and, under normality, evaluation of (5.49) depends crucially on the values of $\boldsymbol{\mu} = \mathrm{E}(\boldsymbol{x}_1 \,|\, \boldsymbol{U})$ and $\boldsymbol{P}_1 = \mathrm{cov}(\boldsymbol{x}_1 \,|\, \boldsymbol{U})$. Following De Jong [66], minus twice the log-likelihood (discarding constants) is then:

$$\begin{aligned} \ell(\boldsymbol{Z}, \boldsymbol{U}; \boldsymbol{\theta}) = {} & \log|\boldsymbol{P}_1| + \boldsymbol{\mu}^T \boldsymbol{P}_1^{-1} \boldsymbol{\mu} + \sum_{t=1}^{N} \log|\boldsymbol{B}_t| + \sum_{t=1}^{N} \tilde{\boldsymbol{z}}_t^T \boldsymbol{B}_t^{-1} \tilde{\boldsymbol{z}}_t + \log|\boldsymbol{P}_1^{-1} + \boldsymbol{W}_N| \\ & - (\boldsymbol{P}_1^{-1}\boldsymbol{\mu} + \boldsymbol{w}_N)^T (\boldsymbol{P}_1^{-1} + \boldsymbol{W}_N)^{-1} (\boldsymbol{P}_1^{-1}\boldsymbol{\mu} + \boldsymbol{w}_N), \end{aligned} \tag{5.50}$$

where $\tilde{\boldsymbol{z}}_t$ and $\boldsymbol{B}_t$ are the innovations and their covariances resulting from a KF(**0**,**0**) given by (5.47)–(5.48) and (5.10)–(5.12). Again, the values of $\boldsymbol{w}_N$ and $\boldsymbol{W}_N$ in (5.50) can be computed via (5.19)–(5.21).

5.3.1 Models with deterministic inputs

The literature typically assumes that the inputs are: (a) exactly known within the sample period, (b) exogenous to the system, and (c) deterministic, which affects the expected value of state vector, but not its covariance matrix. Under these assumptions, the initial conditions for the KF depend on the stationarity of the system.

Stationary systems. If the inputs are constant and known, a stable and adequate initialization for the state vector in the stationary case would be given by:

$$\boldsymbol{\mu} = (\boldsymbol{I} - \boldsymbol{\Phi})^{-1}\boldsymbol{\Gamma}\boldsymbol{u}_1 \tag{5.51}$$

$$\boldsymbol{P}_1 = \boldsymbol{\Phi}\boldsymbol{P}_1\boldsymbol{\Phi}^T + \boldsymbol{E}\boldsymbol{Q}\boldsymbol{E}^T, \tag{5.52}$$

where we assume that $\boldsymbol{u}_1 = \boldsymbol{u}_0 = \boldsymbol{u}_{-1}...$.

Nonstationary systems. When the model has inputs we cannot apply directly the concepts of stationarity and stability defined in Chapter 2. This is because a stable system could have nonstationary inputs which would then realize nonstationary states. Bearing in mind this disclaimer, we will decompose the state equation (5.44) into the corresponding "stationary" and "nonstationary" subsystems as in (5.34):

$$\begin{bmatrix} \boldsymbol{x}_{t+1}^N \\ \boldsymbol{x}_{t+1}^S \end{bmatrix} = \begin{bmatrix} \boldsymbol{\Phi}^N & \boldsymbol{0} \\ \boldsymbol{0} & \boldsymbol{\Phi}^S \end{bmatrix} \begin{bmatrix} \boldsymbol{x}_t^N \\ \boldsymbol{x}_t^S \end{bmatrix} + \begin{bmatrix} \boldsymbol{\Gamma}^N \\ \boldsymbol{\Gamma}^S \end{bmatrix} \boldsymbol{u}_t + \begin{bmatrix} \boldsymbol{E}^N \\ \boldsymbol{E}^S \end{bmatrix} \boldsymbol{w}_t. \tag{5.53}$$

To avoid discontinuities, we will also assume that the pair $(\boldsymbol{\Phi}^N, \boldsymbol{E}^N)$ is controllable. This is enough to avoid a situation where some modes of the nonstationary subsystem have a persistent excitation from $\boldsymbol{u}_t$, thus causing them to diverge to infinity. Under these assumptions, and keeping in mind that the initial state vector has been transformed according to:

$$\boldsymbol{T}\boldsymbol{x}_1 = \begin{bmatrix} \boldsymbol{x}_1^N \\ \boldsymbol{x}_1^S \end{bmatrix}, \tag{5.54}$$

we can multiply the expected value of the initial state by $\boldsymbol{T}^{-1} = [\boldsymbol{M}\ \boldsymbol{N}]$, obtaining:

$$\boldsymbol{\mu} = \boldsymbol{M}\mathrm{E}(\boldsymbol{x}_1^N) + \boldsymbol{N}\mathrm{E}(\boldsymbol{x}_1^S). \tag{5.55}$$

Under (5.53) and the immemorial time hypothesis, the value of $\mathrm{E}(\boldsymbol{x}_1^N)$ is again irrelevant, and so we can take it to be null. On the other hand, the initial condition for the stationary states is $\mathrm{E}(\boldsymbol{x}_1^S) = (\boldsymbol{I} - \boldsymbol{\Phi}^S)^{-1}\boldsymbol{\Gamma}^S\boldsymbol{u}_1$. Accordingly, expression (5.55) is equivalent to:

$$\boldsymbol{\mu} = \boldsymbol{N}\mathrm{E}(\boldsymbol{x}_1^S) = \boldsymbol{N}(\boldsymbol{I} - \boldsymbol{\Phi}^S)^{-1}\boldsymbol{\Gamma}^S\boldsymbol{u}_1, \tag{5.56}$$

which determines the desired initial condition.

5.3.2 *Models with stochastic inputs*

In many cases the exogenous variables should be regarded as stochastic. The literature (mostly in engineering) does not pay enough attention to this issue, probably due to the fact that in most physical systems inputs can be safely assumed to be either deterministic or controllable. However, it is very relevant in other situations such as, e.g., when modeling an economic time series.

The key for a proper initialization of the KF with stochastic inputs consists of decomposing the state equation (5.44) into its deterministic and stochastic subsystems, as shown in Subsection 3.1.1. After doing so, the state equations (3.5) and (3.7) can be jointly written as:

$$\begin{bmatrix} \boldsymbol{x}^d_{t+1} \\ \boldsymbol{x}^s_{t+1} \end{bmatrix} = \begin{bmatrix} \boldsymbol{\Phi} & \boldsymbol{0} \\ \boldsymbol{0} & \boldsymbol{\Phi} \end{bmatrix} \begin{bmatrix} \boldsymbol{x}^d_t \\ \boldsymbol{x}^s_t \end{bmatrix} + \begin{bmatrix} \boldsymbol{\Gamma} \\ \boldsymbol{0} \end{bmatrix} \boldsymbol{u}_t + \begin{bmatrix} \boldsymbol{0} \\ \boldsymbol{E} \end{bmatrix} \boldsymbol{w}_t, \tag{5.57}$$

and the problem reduces then to initializing both subsystems. Initialization for the stochastic system was discussed in Sections 5.1 and 5.2. As for the deterministic subsystem, its initial conditions depend on whether the dynamic structure of the inputs is known.

The dynamic structure of the inputs is known. In this case, a proper initialization would require determining $\boldsymbol{\mu} = \mathrm{E}(\boldsymbol{x}_1 \,|\boldsymbol{U})$ and $\boldsymbol{P}_1 = \mathrm{cov}(\boldsymbol{x}_1 \,|\boldsymbol{U})$. These exact expressions were derived by Casals and Sotoca [43] under the immemorial time hypothesis[4].

The mean and covariance of the initial state vector conditional on the inputs have different expressions depending on the particular stochastic structure of the inputs. For example, if the inputs follow a white noise process, it is easy to see that $\boldsymbol{\mu} = \mathrm{E}(\boldsymbol{x}_1 \,|\boldsymbol{U}) = \boldsymbol{0}$ and $\boldsymbol{P}_1 = \mathrm{cov}(\boldsymbol{x}_1 \,|\boldsymbol{U}) = \boldsymbol{\Phi}\boldsymbol{P}_1\boldsymbol{\Phi}^T + \boldsymbol{E}\boldsymbol{Q}\boldsymbol{E}^T + \boldsymbol{\Gamma}\boldsymbol{\Sigma}_a\boldsymbol{\Gamma}^T$, where $\boldsymbol{\Sigma}_a$ is the covariance matrix of the white noise.

The dynamic structure of the inputs is unknown. The previous framework can be extended to the case when the model for the inputs is unknown. The resulting initial conditions are approximate but, on the other hand, their computational cost is smaller than that of the exact method. To this end, we use again the decomposition (5.57) assuming that the states of both subsystems are independent, that is, $\mathrm{cov}(\boldsymbol{x}^s_t, \boldsymbol{x}^d_t) = \boldsymbol{0}$ for all t, see Van Overschee and De Moor [215][5]. The observation equations corresponding to the states in (5.57) are:

$$\boldsymbol{z}^d_t = \boldsymbol{H}\boldsymbol{x}^d_t + \boldsymbol{D}\boldsymbol{u}_t \tag{5.58}$$

$$\boldsymbol{z}^s_t = \boldsymbol{H}\boldsymbol{x}^s_t + \boldsymbol{v}_t. \tag{5.59}$$

From (5.57), (5.58) and (5.59) it is immediate that:

$$\boldsymbol{\mu} = \boldsymbol{\mu}^d + \boldsymbol{\mu}^s \tag{5.60}$$

$$\boldsymbol{P}_1 = \boldsymbol{P}^d_1 + \boldsymbol{P}^s_1, \tag{5.61}$$

where $\boldsymbol{\mu} = \mathrm{E}(\boldsymbol{x}_1 \,|\boldsymbol{U})$, $\boldsymbol{\mu}^d = \mathrm{E}(\boldsymbol{x}^d_1 |\boldsymbol{U})$, and $\boldsymbol{\mu}^s = \mathrm{E}(\boldsymbol{x}^s_1 \,|\boldsymbol{U})$. Then, the problem reduces to estimating $\boldsymbol{\mu}^d$, $\boldsymbol{\mu}^s$, $\boldsymbol{P}^d_1$, and $\boldsymbol{P}^s_1$.

[4] In the nonstationary case this procedure may yield indeterminate results such as, e.g., $(0)(\infty)$. However, this is irrelevant for likelihood computation because (5.18) does not depend on these conditional moments, but on $\boldsymbol{P}_1^{-1}$, $\boldsymbol{P}_1^{-1}\boldsymbol{\mu}$, and $\boldsymbol{\mu}^T\boldsymbol{P}_1^{-1}\boldsymbol{\mu}$ instead, and these terms always converge to finite and easy to compute values.

[5] This hypothesis is not too restrictive, as the independence between $\boldsymbol{x}_1$ and $\boldsymbol{w}_t$ implies that $\mathrm{cov}(\boldsymbol{x}^s_1, \boldsymbol{x}^d_1) = \boldsymbol{0}$ is a sufficient condition for $\mathrm{cov}(\boldsymbol{x}^s_t, \boldsymbol{x}^d_t) = \boldsymbol{0}$ for all t.

For the stochastic system, the immemorial time argument assures that an adequate initialization is $\boldsymbol{\mu}^s = \mathbf{0}$, because $\mathrm{E}(\boldsymbol{x}_1^s|U) = \mathrm{E}(\boldsymbol{x}_1^s)$. The solution for $\boldsymbol{P}_1^s$ is the same as described for models without inputs (see Subsections 5.1.3 and 5.2.4). In the deterministic subsystem (5.58), we assume that the initial state is a fixed and unknown value with $\boldsymbol{P}_1^d = \mathbf{0}$. ML estimates of $\boldsymbol{\mu}^d$ can be computed as $\boldsymbol{\mu}^d = \boldsymbol{W}_N^{-1}\boldsymbol{w}_N$, where $\boldsymbol{W}_N$ and $\boldsymbol{w}_N$ are by-products of the propagation of (5.19)–(5.21). In this case, the likelihood (5.50) can be concentrated with respect to $\boldsymbol{\mu}^d$ and $\boldsymbol{P}_1^d = \mathbf{0}$.

Therefore, the approximation consists in: (a) computing $\boldsymbol{\mu}^s$ using the immemorial time argument, (b) computing $\boldsymbol{\mu}^d$ by ML, and (c) using (5.60) and (5.61) to calculate the approximate initial conditions for the whole system. The resulting initial conditions are again different, depending on the stationary or nonstationary nature of the model.

1. If the model is stationary, both the deterministic and stochastic subsystems are stationary. Hence $\boldsymbol{M} = \mathbf{0}$, see Expression (5.55), $\boldsymbol{\Phi}^S = \boldsymbol{\Phi}$, and the approximation to the initial conditions is given by:

$$\boldsymbol{\mu} = \boldsymbol{W}_N^{-1}\boldsymbol{w}_N \tag{5.62}$$

$$\boldsymbol{P}_1 = \boldsymbol{\Phi}\boldsymbol{P}_1\boldsymbol{\Phi}^T + \boldsymbol{E}\boldsymbol{Q}\boldsymbol{E}^T. \tag{5.63}$$

   Note that expressions (5.62) and (5.63) do not depend on the parameters of the model for the inputs.

2. When the model is nonstationary, the conditional mean of the initial state is $\boldsymbol{\mu} = \boldsymbol{W}_N^{-1}\boldsymbol{w}_N$, and the inverse of the corresponding covariance is $\boldsymbol{P}_1^{-1} = \mathbf{0}$. This result coincides with the initial conditions obtained by De Jong and Chu-Chun-Lin [69] for the same case. If $\boldsymbol{P}_1^{-1} = \mathbf{0}$, the value of $\boldsymbol{\mu}$ is irrelevant.

3. The initial conditions for a partially nonstationary model are $\boldsymbol{\mu} = \boldsymbol{W}_N^{-1}\boldsymbol{w}_N$ and $\boldsymbol{P}_1^{-1} = \boldsymbol{V}^T(\boldsymbol{P}^S)^{-1}\boldsymbol{V}$, where $\boldsymbol{P}^S$ is calculated according to the Lyapunov equation given by (5.38), and $\boldsymbol{V}$ is defined in (5.43).

The key idea in deriving suitable initial conditions then, consists in computing the decomposition (5.57)–(5.59) followed by an application of the immemorial time hypothesis. On the other hand, the correct initialization of a model with stochastic inputs consists of obtaining ML estimates of the initial state for the system.

5.4 Examples

5.4.1 Models for the Airline Passenger series

Examples 5.4.1 and 5.4.2 are based on the famous monthly time series of international airline passengers, which is shown in Figure 5.1.

Example 5.4.1 Estimating the Airline model in stationary and nonstationary form

This example emphasizes the estimation of seasonal ARIMA models and illustrates the fact that SS methods allow one to obtain identical estimates for a model

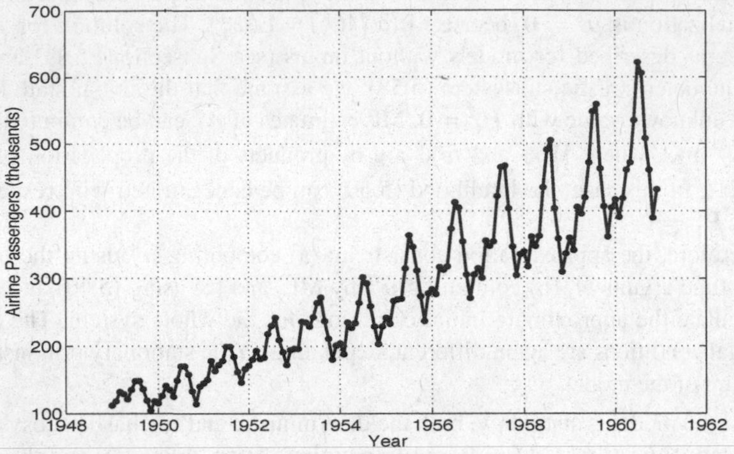

Figure 5.1 *International Airline Passengers (thousands) from 1949 to 1960. Source: see Box, Jenkins and Reinsel [30].*

in stationary and nonstationary forms. While both representations are strictly equivalent, the latter has clear advantages for some specific applications, which will be discussed later.

As is well known, the (log) airline series is adequately represented by an $\text{IMA}(1,1)\times(1,1)_{12}$ model:

$$(1-\text{B})(1-\text{B}^{12})\log y_t = (1-\theta\text{B})(1-\Theta\text{B}^{12})\,a_t \tag{5.64}$$

where y_t denotes the number of airline passengers in month t. On the other hand, if we denote $z_t = (1-\text{B})(1-\text{B}^{12})\log y_t$, the nonstationary model (5.64) could be written in the equivalent stationary $\text{MA}(1)\times\text{MA}_{12}$ form:

$$z_t = (1-\theta\text{B})(1-\Theta\text{B}^{12})\,a_t \tag{5.65}$$

where the parameters and errors in both representations should be the same.

Table 5.1 summarizes the main estimation results obtained for both representations. The table reports a single set of values because the figures obtained for both forms are identical up to the fourth decimal place. This high level of numerical consistency is only assured when the nonstationary model is estimated by optimizing the minimally conditioned likelihood described in Subsection 5.2.2. Maximizing the likelihood evaluated according to the diffuse filters of De Jong [68] or Koopman [139] (see Subsection 5.2.1) often produces estimates which differ from those of the stationary model. The code required to replicate this example can be found in the file `Example0541.m`, see Appendix D.

Example 5.4.2 A transfer function model with calendar effects and missing values

Table 5.1 *Estimation results for the airline models (5.64) and (5.65).*

Parameter	Estimate	Standard error
$\hat{\theta}$	0.4018	0.0766
$\hat{\Theta}$	0.5569	0.0738
$\hat{\sigma}_a$	0.0367	–
Log-likelihood	−244.6965	–

The previous example showed that the same basic model can be estimated in both stationary and nonstationary forms, providing results which are close to identical if the likelihood computation procedure is suitably chosen. On the other hand, it is natural to ask: why should we take on the additional cost in terms of complexity and computational burden to estimate a model in nonstationary form? The answer to this question is that this representation is more convenient for many purposes. Without loss of generality, we will discuss this idea using the equivalent models (5.64) and (5.65).

Assume for example that the goal of the modeling exercise is to forecast the nonstationary variable y_t. In this case, estimating the nonstationary form (5.64) is more convenient, as it immediately provides point and interval forecasts for $\log y_t$, and hence y_t. In comparison, the stationary form (5.65) provides forecasts for $z_t = (1-\mathrm{B})(1-\mathrm{B}^{12})\log y_t$, requiring a more complex treatment to recover the original metrics.

Another situation where the nonstationary form is more practical is when one wants to perform the trend-cycle-seasonal decomposition of y_t, see Chapter 8. In this case, the nonstationary form (5.64) can be immediately decomposed into the additive components of $\log y_t$ whose anti-logs provide the corresponding multiplicative decomposition. Once again, the same results could be obtained on the basis of the stationary formulation (5.65) by means of a more complex procedure.

A third case where the nonstationary representation has clear advantages is when the sample contains missing values. They could arise, either by a discontinuity in the sampling, or as an effective treatment for additive impulse-type outliers see Gomez, Maravall and Peña [111]. In this case one should never difference the data because differencing propagates the missing values, thus destroying valuable information. Therefore, working with the nonstationary model is a must in this case.

To illustrate these ideas we will fit the following transfer function to the airline series:

$$\begin{aligned}\log y_t &= \omega_0^L L_t + \omega_0^W W_t + \omega_0^E E_t + N_t \\ (1-\mathrm{B})(1-\mathrm{B}^{12})N_t &= (1-\theta\mathrm{B})(1-\Theta\mathrm{B}^{12})a_t,\end{aligned} \tag{5.66}$$

where L_t is the number of labor days (Monday to Friday), W_t, is the number of weekend days (Saturday and Sunday) in a month, E_t, is a dummy variable whose value is one if Easter occurs in month t and zero otherwise, and N_t denotes an autocorrelated error term which follows an $IMA(1,1)\times(1,1)_{12}$ process. Therefore, the explanatory variables in this model are devised to capture calendar effects, see Cleveland and Devlin [54].

Estimating model (5.66) by ML provides the following results:

$$\log y_t = \underset{(.014)}{.039}L_t + \underset{(.014)}{.049}W_t + \underset{(.010)}{.028}E_t + \hat{N}_t$$
$$(1-\mathrm{B})(1-\mathrm{B}^{12})\hat{N}_t = (1-\underset{(.082)}{.222}\mathrm{B})(1-\underset{(.071)}{.533}\mathrm{B}^{12})\hat{a}_t \quad , \quad \hat{\sigma}_a = .033 \tag{5.67}$$

and Figure 5.2 displays the residuals $\hat{a}_t$.

Building on these results we detected three additive outliers, in observations number 29, 54 and 62. The standard treatment for this situation consists of adding three impulse-type intervention variables to the specification; see Box and Tiao [31]. According to the previous discussion, we re-estimated the model (5.66) treating these three observations as missing values. This provided the following estimation results:

$$\log y_t^* = \underset{(.012)}{.034}L_t + \underset{(.012)}{.044}W_t + \underset{(.008)}{.023}E_t + \hat{N}_t$$
$$(1-\mathrm{B})(1-\mathrm{B}^{12})\hat{N}_t = (1-\underset{(.085)}{.082}\mathrm{B})(1-\underset{(.074)}{.484}\mathrm{B}^{12})\hat{a}_t^* \quad , \quad \hat{\sigma}_{a^*} = .029, \tag{5.68}$$

where the "asterisk" (*) symbol denotes the variable with missing values. Note that the residuals of this model, displayed in Figure 5.3, have fewer outliers than those in Figure 5.2, so this atypical intervention procedure has been effective.

The code required to replicate this example can be found in the file `Example0542.m`, see Appendix D.

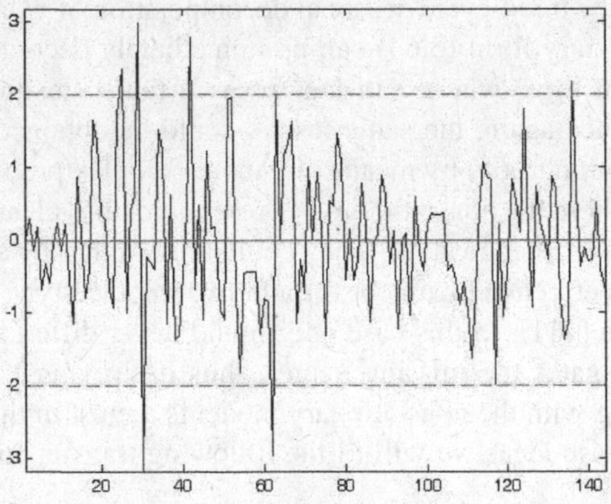

Figure 5.2 *Standardized residuals of the transfer function (5.67).*

5.4.2 *Modeling the series of Housing Starts and Sales*

Examples 5.4.3 and 5.4.4 are based on the monthly U.S. single-family housing starts (ST_t) and sales (SS_t) series, see Figure 5.4. This dataset has been analyzed in many works, such as those of Hillmer and Tiao [124], Ahn and Reinsel [1] or

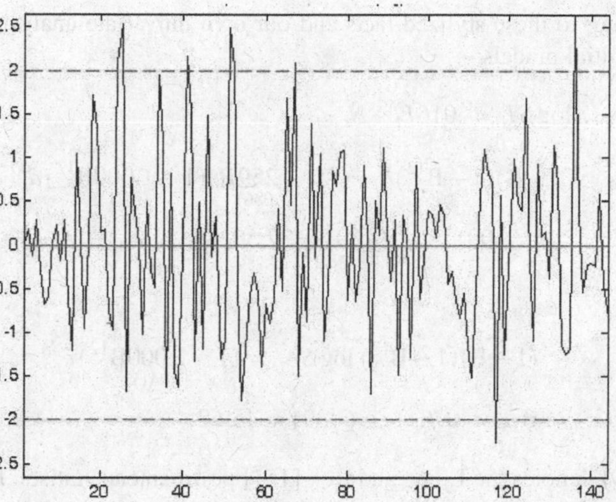

Figure 5.3 *Standardized residuals of the transfer function (5.68).*

Tiao [208]. Some stylized facts found in previous analyses are: (a) both series are adequately represented by an airline model (b) their seasonal structure is probably deterministic, and (c) they are cointegrated.

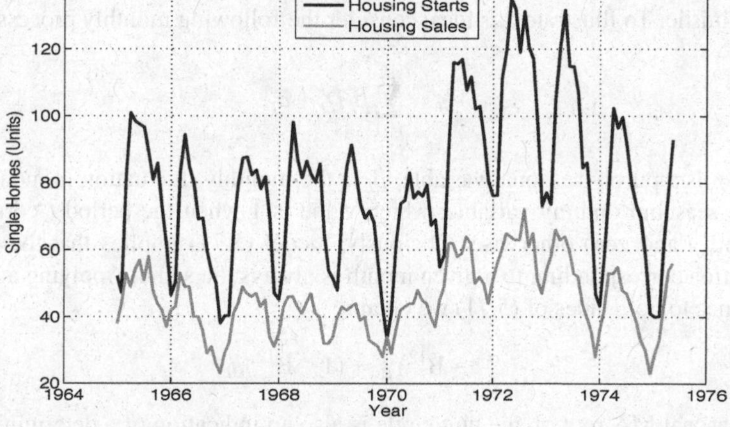

Figure 5.4 *Monthly U.S. single-family Housing Starts and Housing Sales, in thousands of units, from January 1965 through May 1975.*

Example 5.4.3 Univariate modeling

According to these stylized facts and our own univariate analysis, we fitted the following initial models:

$$\log ST_t = \underset{(.006)}{.016} L_t + \hat{N}_t$$
$$(1-\mathrm{B})(1-\mathrm{B}^{12})\hat{N}_t = (1 - \underset{(.086)}{.259}\mathrm{B})(1 - \underset{(.041)}{1.000}\mathrm{B}^{12})\,\hat{a}_t^{ST}$$
$$\hat{\sigma}_{ST} = .081 \quad , \quad Q(39) = 37.50, \tag{5.69}$$

and

$$(1-\mathrm{B})(1-\mathrm{B}^{12}) \log SA_t = (1 - \underset{(.041)}{1.000}\mathrm{B}^{12})\,\hat{a}_t^{SA}$$
$$\hat{\sigma}_{SA} = .078 \quad , \quad Q(39) = 45.62, \tag{5.70}$$

where $Q(39)$ denotes the Ljung and Box [151] portmanteau statistic for the null that the model errors are white noise, computed with the first 39 sample auto-correlations.

Note that the housing starts series depends on the number of work days (Monday to Friday) in each month, while the housing sales data does not display any significant calendar effect.

Another relevant feature of models (5.69) and (5.70) is that the seasonal MA parameter is unity in both cases, so the corresponding polynomial has twelve roots on the unit circle. This is a frequent result in two cases: first, when a seasonal difference is applied to a time series which does not need it; in this case the misspecified seasonal difference tends to be compensated by an equivalent seasonal MA polynomial; second, a unity seasonal MA parameter may also imply that the seasonal structure is deterministic. To illustrate this idea, consider the following monthly process:

$$\gamma_t = \sum_{i=1}^{12} \beta_i D_t^i + a_t, \tag{5.71}$$

where γ_t is a purely seasonal variable, β_i is the monthly fluctuation in month i, and D_t^i is a seasonal dummy variable, which value is 1 when the period t corresponds to month i and zero otherwise. Obviously, model (5.71) implies that the seasonal fluctuation corresponding to a given month is always the same. Applying a seasonal difference to both sides of (5.71) we obtain:

$$(1-\mathrm{B}^{12})\gamma_t = (1-\mathrm{B}^{12})a_t, \tag{5.72}$$

so a seasonal MA root on the unit circle is also an indication of a deterministic seasonal structure in the time series. To illustrate these ideas, Figure 5.5 displays the seasonal components implied by models (5.69) and (5.70), extracted by the methods explained in Chapter 8. Note that the seasonal fluctuation is exactly the same in different years, meaning that seasonality for this series is deterministic. If the seasonal MA term in models (5.69) and (5.70) were invertible, seasonality would be stochastic and, accordingly, the seasonal fluctuation would change over the sample.

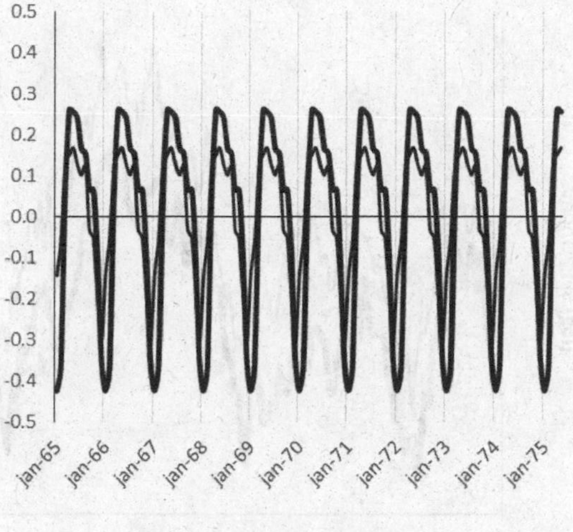

Figure 5.5 *Additive seasonal component of (log) Housing Starts (thick line) and Housing Sales (thin line), from January 1965 through May 1975.*

The code required to replicate this example can be found in the file `Example0543.m`, see Appendix D.

Example 5.4.4 A nonstandard VARMAX model with a cointegration constraint and calendar effects

As we said before, previous analyses often found the above series to be cointegrated, meaning that there is a long-term equilibrium between the housing starts and sales. The results in the previous example help us to visualize this equilibrium. Figure 5.6 shows the profile of both series (in logs) after adjusting for seasonality and calendar effects. Note that they follow each other closely which is a clear (if somewhat informal) evidence of cointegration.

On the basis of the previous analysis, and after a standard multivariate specification analysis, see Jenkins and Alavi [131], we fitted the following model to these series:

$$\begin{bmatrix} 1 & \underset{(.010)}{-1.028} \\ 0 & 1 \end{bmatrix} \begin{bmatrix} \log ST_t \\ \log SA_t \end{bmatrix} = \begin{bmatrix} \underset{(.006)}{.016} \\ 0 \end{bmatrix} L_t + \begin{bmatrix} \hat{N}_t^{ST} \\ \hat{N}_t^{SA} \end{bmatrix} \tag{5.73}$$

$$\begin{bmatrix} 1-\underset{(.085)}{.213}\mathrm{B} & 0 \\ 0 & 1-\mathrm{B} \end{bmatrix} \begin{bmatrix} 1-\mathrm{B}^{12} & 0 \\ 0 & 1-\mathrm{B}^{12} \end{bmatrix} \begin{bmatrix} \hat{N}_t^{ST} \\ \hat{N}_t^{SA} \end{bmatrix} =$$

$$\begin{bmatrix} 1 & 0 \\ \underset{(.082)}{0.237}\mathrm{B} & 1 \end{bmatrix} \begin{bmatrix} 1-\mathrm{B}^{12} & 0 \\ 0 & 1-\underset{(.039)}{.880}\mathrm{B}^{12} \end{bmatrix} \begin{bmatrix} \hat{a}_t^{ST} \\ \hat{a}_t^{SA} \end{bmatrix}$$

$$\hat{\Sigma}_{\boldsymbol{a}} = \begin{bmatrix} .007 & -.004 \\ -.004 & .006 \end{bmatrix}, \quad \boldsymbol{Q}(10) = \begin{bmatrix} 7.20 & 7.92 \\ 8.24 & 10.93 \end{bmatrix}, \tag{5.74}$$

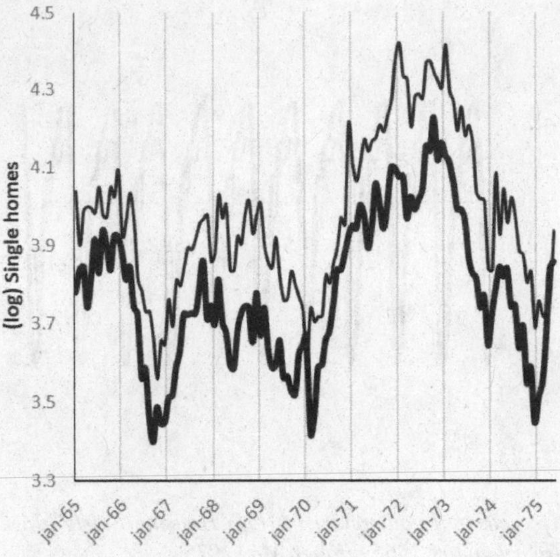

Figure 5.6 *Seasonality and calendar-adjusted (log) Housing Starts and Housing Sales series, from January 1965 through May 1975.*

where $Q(10)$ denotes the matrix of portmanteau statistics for the null that the model errors are white noise, computed with the first 10 sample auto- and cross-correlations.

Note that equations (5.73)–(5.74) define a simultaneous equation model with VARMAX errors [6]. This model has some interesting implications:

1. The autoregressive parameter in (5.74) provides a cointegration test. A value close enough to unity would imply that there was no cointegration, because the series would require the same orders of differencing as those in the previously fitted univariate models, thus rejecting cointegration. On the other hand, the estimate obtained (.213, with a .085 standard error) is small enough to provide positive evidence in favor of cointegration.
2. The first coefficient in Equation (5.73) is the cointegration parameter. Its estimate (-1.028 with a 0.010 standard error) means that there exists a long term equilibrium between both series. Deviations from this equilibrium are transient and given by the cointegrating relationship $\log ST_t - 1.028 \log SA_t$.
3. The estimated cointegrating parameter is close to unity. If its true value were -1 it would mean that, in the long term, the ratio between Housing Starts and Housing Sales is approximately stable.

The code required to replicate this example can be found in the file `Example0544.m`, see Appendix D.

[6] See Subsection 3.2.2 for the details about the SS formulation of these combined models.

Chapter 6

The likelihood of models with varying parameters

6.1 Regression with time-varying parameters 66
  6.1.1 SS formulation 66
  6.1.2 Maximum likelihood estimation 67
6.2 Periodic models 68
  6.2.1 All the seasons have the same dynamic structure 69
  6.2.2 The *s* models do not have the same dynamic structure 70
  6.2.3 Stationarity and invertibility 71
  6.2.4 Maximum likelihood estimation 72
6.3 The likelihood of models with GARCH errors 74
6.4 Examples 76
  6.4.1 A time-varying CAPM regression 76
  6.4.2 A periodic model for West German Consumption 78
  6.4.3 A model with vector-GARCH errors for two exchange rate series 79

In a statistical model, any value to be estimated must be either a constant, or a function of constant parameters. Thus, the reference to "varying parameters" in the title of this chapter is somewhat misleading. However, it is standard terminology, referring to models which allow for the variation over time of some feature which is considered fixed by most standard specifications.

In particular, we will discuss how to compute the Gaussian likelihood for three specifications of this kind. First, a time-varying parameter regression model, where the coefficients are allowed to change according to a given law of motion. Second, a periodic VARMAX model (PVARMAX), in which the parameters may change over different seasons. Third and last, a general dynamic model with conditionally heteroscedastic (GARCH) errors, where the traditional assumption of homoscedasticity is relaxed by parameterizing the conditional variance of the disturbances as a function of its own past.

6.1 Regression with time-varying parameters

The literature shows an increasing interest in estimating regressions with time-varying parameters. For example, Cooley and Prescott [56] and Rosenberg [188] study how different specification errors, such as structural changes, heteroscedasticity or incorrect functional forms, may result in a model with changing parameters. We can see a survey about varying coefficient regression models and new developments in Park et al. [177]. Economic theory suggests that many relationships may change over time. Along this line, Lucas [154] showed that econometric models may be inadequate to assess economic policy in the long term because they inherently assume parametric stability.

According to this interest, the application of time-varying regressions is becoming widespread in areas such as financial economics, see Fabozzi and Francis [89], Alexander and Benson [4] or Bera, Higgins, and Lee [22], macroeconomics, see Margaritis [162], Evans, Kim, and Oh [88], Stock and Watson [199], microeconomics, see Hackl [114], Chang and Hsing [50], Park and Zhao [178], as well as in other fields such as e.g. politics, epidemiology, or medical science, where the constant parameters assumption is not necessary or realistic because of the increasing availability of disaggregated data.

Varying coefficient models (not regression models) are natural extensions of standard parametric models with good interpretability and are becoming popular in data analysis, thus experiencing fast development on the methodological, theoretical, and applied side; see, for example, Fan and Zhang [93], Cai, Fan, and Yao [36], Fan *et al.* [91], Fan and Zhang [92], Kauermann and Tutz [135], Hastie and Tibshirani [122], or Nicholls and Quinn [167].

6.1.1 *SS formulation*

A general regression model with changing parameters is[1]:

$$z_t = \chi_t \beta_t \tag{6.1}$$

$$\beta_t = \boldsymbol{H} \boldsymbol{x}_t + \boldsymbol{D} \boldsymbol{u}_t + \boldsymbol{C} \boldsymbol{v}_t \tag{6.2}$$

$$\boldsymbol{x}_{t+1} = \boldsymbol{\Phi} \boldsymbol{x}_t + \boldsymbol{\Gamma} \boldsymbol{u}_t + \boldsymbol{E} \boldsymbol{w}_t, \tag{6.3}$$

where z_t is an observable endogenous variable, χ_t is a row $(1 \times k)$ of explanatory variables with $\chi_{1t} = 1$, for all t, β_t is a k-dimensional vector of unknown parameters, and $\boldsymbol{u}_t$ is a r-dimensional vector of variables affecting the coefficients, with $u_{1t} = 1$, for all t.

To write the model (6.1)–(6.3) in SS form, we need to substitute (6.2) in (6.1):

$$z_t = (\chi_t \boldsymbol{H}) \boldsymbol{x}_t + (\chi_t \boldsymbol{D}) \boldsymbol{u}_t + (\chi_t \boldsymbol{C}) \boldsymbol{v}_t, \tag{6.4}$$

or, equivalently:

$$z_t = \boldsymbol{H}_t \boldsymbol{x}_t + \boldsymbol{D}_t \boldsymbol{u}_t + \boldsymbol{C}_t \boldsymbol{v}_t, \tag{6.5}$$

[1]The model (6.1)–(6.3) generalizes the formulation in Swamy and Tavlas [205].

with $\boldsymbol{H}_t = \chi_t \boldsymbol{H}$, $\boldsymbol{D}_t = \chi_t \boldsymbol{D}$ and $\boldsymbol{C}_t = \chi_t \boldsymbol{C}$. Therefore, equations (6.5) and (6.3) define an SS model with a time-varying observation equation, which can be trivially generalized to the case of multiple outputs.

There are three specifications for the parameter change which have special interest: random walk, random coefficients, and mean-reverting parameters.

The *random walk model* assumes that the coefficients follow a random walk process:

$$z_t = \chi_t \boldsymbol{\beta}_t + \varepsilon_t \tag{6.6}$$

$$\boldsymbol{\beta}_{t+1} = \boldsymbol{\beta}_t + \boldsymbol{a}_t, \tag{6.7}$$

where ε_t is a white noise sequence that is uncorrelated with $\boldsymbol{a}_t$. The formulation of this model in SS form involves defining $\boldsymbol{\Phi} = \boldsymbol{I}$, $\boldsymbol{\Gamma} = 0$, $\boldsymbol{E} = \boldsymbol{I}$, $\boldsymbol{H}_t = \chi_t$, $\boldsymbol{D}_t = \boldsymbol{0}$, $\boldsymbol{C}_t = \boldsymbol{I}$, $\boldsymbol{Q} = \boldsymbol{\Sigma}_a$, $\boldsymbol{S} = 0$, and $\boldsymbol{R} = \sigma_\varepsilon^2$.

In the *random coefficients model* the transition matrix is null and the parameters are assumed to change randomly around a constant mean value $\bar{\boldsymbol{\beta}}$:

$$z_t = \chi_t \boldsymbol{\beta}_t + \varepsilon_t \tag{6.8}$$

$$\boldsymbol{\beta}_t = \bar{\boldsymbol{\beta}} + \boldsymbol{a}_t, \tag{6.9}$$

where $\boldsymbol{\Phi} = \boldsymbol{0}$, $\boldsymbol{\Gamma} = \boldsymbol{0}$, $\boldsymbol{E} = \boldsymbol{I}$, $\boldsymbol{H}_t = \chi_t$, $\boldsymbol{D}_t = \chi_t \bar{\boldsymbol{\beta}}$, $\boldsymbol{C}_t = \boldsymbol{I}$, $\boldsymbol{Q} = \boldsymbol{\Sigma}_a$, $\boldsymbol{S} = \boldsymbol{0}$ and $\boldsymbol{R} = \sigma_\varepsilon^2$.

Last, the parameters in the *mean-reverting model* follow a stationary AR(1) process given by:

$$z_t = \chi_t \boldsymbol{\beta}_t + \varepsilon_t \tag{6.10}$$

$$(\boldsymbol{\beta}_{t+1} - \bar{\boldsymbol{\beta}}) = \boldsymbol{F}(\boldsymbol{\beta}_t - \bar{\boldsymbol{\beta}}) + \boldsymbol{a}_t, \tag{6.11}$$

where $\boldsymbol{\Phi} = \boldsymbol{F}$, $\boldsymbol{\Gamma} = 0$, $\boldsymbol{E} = \boldsymbol{I}$, $\boldsymbol{H}_t = \chi_t$, $\boldsymbol{D}_t = \chi_t \bar{\boldsymbol{\beta}}$, $\boldsymbol{C}_t = \boldsymbol{I}$, $\boldsymbol{Q} = \boldsymbol{\Sigma}_a$, $\boldsymbol{S} = \boldsymbol{0}$, and $\boldsymbol{R} = \sigma_\varepsilon^2$.

6.1.2 *Maximum likelihood estimation*

Estimation of the models defined in Subsection 6.1.1 is generally addressed by the procedures described in Chapter 5 for the MEM formulation. In particular, the log-likelihood function to be evaluated is the one given in expression (5.46), its addends being the result of the following KF:

$$\hat{\boldsymbol{x}}_{t+1|t} = \boldsymbol{\Phi} \hat{\boldsymbol{x}}_{t|t-1} + \boldsymbol{\Gamma} \boldsymbol{u}_t + \boldsymbol{K}_t \tilde{\boldsymbol{z}}_t \tag{6.12}$$

$$\tilde{\boldsymbol{z}}_t = \boldsymbol{z}_t - \boldsymbol{H}_t \hat{\boldsymbol{x}}_{t|t-1} - \boldsymbol{D}_t \boldsymbol{u}_t \tag{6.13}$$

$$\boldsymbol{B}_t = \boldsymbol{H}_t \boldsymbol{P}_{t|t-1} \boldsymbol{H}_t^T + \boldsymbol{C}_t \boldsymbol{R} \boldsymbol{C}_t^T \tag{6.14}$$

$$\boldsymbol{K}_t = (\boldsymbol{\Phi} \boldsymbol{P}_{t|t-1} \boldsymbol{H}_t^T + \boldsymbol{E} \boldsymbol{S} \boldsymbol{C}_t^T) \boldsymbol{B}_t^{-1} \tag{6.15}$$

$$\boldsymbol{P}_{t+1|t} = \boldsymbol{\Phi} \boldsymbol{P}_{t|t-1} \boldsymbol{\Phi}^T + \boldsymbol{E} \boldsymbol{Q} \boldsymbol{E}^T - \boldsymbol{K}_t \boldsymbol{B}_t \boldsymbol{K}_t^T. \tag{6.16}$$

The model given by (6.3) and (6.5) can also be estimated using the log-likelihood (5.50) where the terms: (a) $\tilde{z}_t$ and B_t result from the $KF(\mathbf{0},\mathbf{0})$ given by (6.12)–(6.16), and (b) w_N and W_N in (5.50) are computed by the recursions:

$$w_t = w_{t-1} + \bar{\Phi}_{t-1} H_t^T B_t^{-1} \tilde{z}_t \tag{6.17}$$

$$W_t = W_{t-1} + \bar{\Phi}_{t-1} H_t^T B_t^{-1} H_t \bar{\Phi}_{t-1} \tag{6.18}$$

$$\bar{\Phi}_t = (\Phi_t - K_t H_t)\bar{\Phi}_{t-1}, \tag{6.19}$$

where w_t and W_t are initialized to zero, $\bar{\Phi}_0 = I$, and K_t is the $KF(\mathbf{0},\mathbf{0})$ gain.

The proper choice of initial conditions μ and P_1 in (5.50) depends on whether the model (6.3) and (6.5) is stationary or not, and on the nature (deterministic or stochastic) of its inputs, see Sections 5.3.1 and 5.3.2.

6.2 Periodic models

A periodic PVARMAX model extends the standard VARMAX framework by allowing the model parameters to change over different seasons. The PVARMAX model assumes that the observations in each of the seasons can be described using a different model. Such a property may be useful to describe some economic time series, since in several situations one may expect economic agents to behave differently in different seasons; see Ghysels and Osborn [107]. Most applications exclude the MA part of models. Periodic AR models have been used successfully in macroeconomic time series; see, for example, Novales and Flores [168], Franses [100] or Osborn and Smith [171].

As is known, it is said that the random vector z_t, for all $t = 1,2,...,N$, follows a process $PVARMAX_s(p_i,q_i,g_i)$, for all $i = 1,2,...,s$ where s is the number of sample periods required to complete a seasonal cycle, if:

$$F_t(\mathrm{B})z_t = G_t(\mathrm{B})u_t + \Theta_t(\mathrm{B})a_t, \tag{6.20}$$

where the parameters' matrices are:

$$F_t(\mathrm{B}) = I + F_{t,1}\mathrm{B} + F_{t,2}\mathrm{B}^2 + ... + F_{t,p_i}\mathrm{B}^{p_i} \tag{6.21}$$

$$G_t(\mathrm{B}) = I + G_{t,1}\mathrm{B} + G_{t,2}\mathrm{B}^2 + ... + G_{t,g_i}\mathrm{B}^{g_i} \tag{6.22}$$

$$\Theta_t(\mathrm{B}) = I + \Theta_{t,1}\mathrm{B} + \Theta_{t,2}\mathrm{B}^2 + ... + \Theta_{t,q_i}\mathrm{B}^{q_i} \tag{6.23}$$

$$a_t \sim \mathrm{IID}(\mathbf{0}, \Sigma_t). \tag{6.24}$$

and the coefficient matrices change according to a deterministic variation law with period s, such that:

$$F_t(\mathrm{B}) = F_{t+s}(\mathrm{B}); \quad G_t(\mathrm{B}) = G_{t+s}(\mathrm{B}); \quad \Theta_t(\mathrm{B}) = \Theta_{t+s}(\mathrm{B}), \tag{6.25}$$

and

$$\Sigma_t = \Sigma_{t+s}; \quad \text{for all } t = 1,2,...,N \tag{6.26}$$

so that the PVARMAX model can be seen to be a VARMAX model whose coefficients change periodically over time; see Lutkepohl [158].

Equations (6.20)–(6.26) can be written in SEM form as:

$$\boldsymbol{x}_{t+1} = \boldsymbol{\Phi}_t \boldsymbol{x}_t + \boldsymbol{\Gamma}_t \boldsymbol{u}_t + \boldsymbol{E}_t \boldsymbol{a}_t \tag{6.27}$$

$$\boldsymbol{z}_t = \boldsymbol{H}_t \boldsymbol{x}_t + \boldsymbol{D}_t \boldsymbol{u}_t + \boldsymbol{a}_t, \tag{6.28}$$

where the covariance matrix of the error vector $\boldsymbol{a}_t$ is $\boldsymbol{Q}_t$, and the dimensions of $\boldsymbol{\Phi}_t$, $\boldsymbol{\Gamma}_t$, $\boldsymbol{E}_t$, $\boldsymbol{H}_t$, and $\boldsymbol{D}_t$ are, respectively, $(n_{t+1} \times n_t)$, $(n_{t+1} \times r)$, $(n_{t+1} \times m)$, $(m \times n_t)$, and $(m \times r)$. Finally, the parameter variation laws are: $\boldsymbol{\Phi}_t = \boldsymbol{\Phi}_{t+s}$, $\boldsymbol{\Gamma}_t = \boldsymbol{\Gamma}_{t+s}$, $\boldsymbol{E}_t = \boldsymbol{E}_{t+s}$, $\boldsymbol{H}_t = \boldsymbol{H}_{t+s}$, $\boldsymbol{D}_t = \boldsymbol{D}_{t+s}$, and $\boldsymbol{Q}_t = \boldsymbol{Q}_{t+s}$, for all $t = 1, 2, \ldots, N$.

Note that the dimension of state vector $\boldsymbol{x}_{t+1}$ may change over different seasons and, therefore, the transition matrix $\boldsymbol{\Phi}$ may not be square in some cases; see e.g. Example 6.2.2 below.

To determine the relationship between the model (6.20)–(6.26) and matrices of (6.27)–(6.28), it is convenient to distinguish between the case when the dynamic structure is always the same and when it may be different in some seasons.

6.2.1 All the seasons have the same dynamic structure

When all the seasons have the same dynamic structure, it holds that $k_i = k$ for all $i = 1, 2, \ldots, s$ where $k_i = \max\{p_i, q_i, g_i\}$ and the matrices of (6.27)–(6.28) are:

$$\boldsymbol{\Phi}_t = \begin{bmatrix} -\boldsymbol{F}_{t+1,1} & \boldsymbol{I} & \boldsymbol{0} & \ldots & \boldsymbol{0} \\ -\boldsymbol{F}_{t+2,2} & \boldsymbol{0} & \boldsymbol{I} & \ldots & \boldsymbol{0} \\ \vdots & \vdots & \vdots & \ldots & \vdots \\ -\boldsymbol{F}_{t+k-2,k-1} & \boldsymbol{0} & \boldsymbol{0} & \ldots & \boldsymbol{I} \\ -\boldsymbol{F}_{t+k-1,k} & \boldsymbol{0} & \boldsymbol{0} & \ldots & \boldsymbol{0} \end{bmatrix}, \quad \boldsymbol{\Gamma}_t = \begin{bmatrix} \boldsymbol{G}_{t+1,1} - \boldsymbol{F}_{t+1,1}\boldsymbol{G}_{t,0} \\ \boldsymbol{G}_{t+2,2} - \boldsymbol{F}_{t+2,2}\boldsymbol{G}_{t,0} \\ \vdots \\ \boldsymbol{G}_{t+k-2,k-1} - \boldsymbol{F}_{t+k-2,k-1}\boldsymbol{G}_{t,0} \\ \boldsymbol{G}_{t+k-1,k} - \boldsymbol{F}_{t+k-1,k}\boldsymbol{G}_{t,0} \end{bmatrix}, \tag{6.29}$$

$$\boldsymbol{E}_t = \begin{bmatrix} \boldsymbol{\Theta}_{t+1,1} - \boldsymbol{F}_{t+1,1} \\ \boldsymbol{\Theta}_{t+2,2} - \boldsymbol{F}_{t+2,2} \\ \vdots \\ \boldsymbol{\Theta}_{t+k-2,k-1} - \boldsymbol{F}_{t+k-2,k-1} \\ \boldsymbol{\Theta}_{t+k-1,k} - \boldsymbol{F}_{t+k-1,k} \end{bmatrix}, \quad \boldsymbol{H}_t = \begin{bmatrix} \boldsymbol{I} & \boldsymbol{0} & \ldots & \boldsymbol{0} \end{bmatrix}, \quad \boldsymbol{D}_t = \boldsymbol{G}_{t,0}, \tag{6.30}$$

where $\boldsymbol{F}_{t+j,j+1} = \boldsymbol{F}_{t+j+s,j+1}$, $\boldsymbol{G}_{t+j,j+1} = \boldsymbol{G}_{t+j+s,j+1}$, $\boldsymbol{\Theta}_{t+j,j+1} = \boldsymbol{\Theta}_{t+j+s,j+1}$ for all $j = 0, 1, \ldots, k-1$.

Example 6.2.1

If $s = 3$ and the model for each season is a stationary AR(1) process, the standard representation of the periodic model $PAR_3(1)$ is:

$$\begin{aligned} z_{t+1} &= f_{1,1} z_t + a_{t+1} \\ z_{t+2} &= f_{2,1} z_{t+1} + a_{t+2} \\ z_{t+3} &= f_{3,1} z_{t+2} + a_{t+3}, \qquad \text{for } t = 0, 3, 6, 9, \ldots, \end{aligned} \tag{6.31}$$

where the SS formulation of each of the terms above is respectively given by:

$$z_{t+1} = x_{t+1} + a_{t+1} \tag{6.32}$$
$$x_{t+2} = f_{2,1}x_{t+1} + f_{2,1}a_{t+1}, \tag{6.33}$$

$$z_{t+2} = x_{t+2} + a_{t+2} \tag{6.34}$$
$$x_{t+3} = f_{3,1}x_{t+2} + f_{3,1}a_{t+2}, \tag{6.35}$$

$$z_{t+3} = x_{t+3} + a_{t+3} \tag{6.36}$$
$$x_{t+4} = f_{1,1}x_{t+3} + f_{1,1}a_{t+3}. \tag{6.37}$$

6.2.2 *The s models do not have the same dynamic structure*

When the s models do not have the same structure the standard conversion of the periodic VARMAX model to its corresponding SS representation may yield a non-minimal state vector. Casals, Sotoca and Jerez [44] proposed an SS representation of (6.20)–(6.26) which assures minimality. In this SS representation the number of rows of $\boldsymbol{\Phi_t}$ is $n_{t+1} = m \times \sum_{j=1}^{s}(k_j^+ + k_j^-)$, where k_j^+ is the integer part of k_j/s, and

$$k_j^- = \begin{cases} 1, & \text{if } k_j - k_j^+ s \geq j - t \text{ when } j > t, \\ 1, & \text{if } k_j - k_j^+ s \geq j + s - t \text{ when } j \leq t, \\ 0, & \text{otherwise.} \end{cases}$$

The conversion rule to achieve a minimum dimension representation is to remove redundant states, that is, those that are not affected by any observable ($\boldsymbol{u_t}$) or unobservable ($\boldsymbol{a_t}$) input. The way to build the matrices $\boldsymbol{\Phi_t}$, $\boldsymbol{\Gamma_t}$, and $\boldsymbol{E_t}$ is as follows. Let $k = \max\{k_i\}, i = 1, 2, \ldots, s$. Then, if $k_{i+j} \geq j$ for all $j = 1, 2, \ldots, k$:

- the $\boldsymbol{\Gamma}_t$ matrix is constructed by adding blocks of rows $[\boldsymbol{G}_{i+j,j} - \boldsymbol{F}_{i+j,j}\boldsymbol{G}_{i,0}]$;
- the $\boldsymbol{E}_t$ matrix is constructed by adding blocks of rows $[\boldsymbol{\Theta}_{i+j,j} - \boldsymbol{F}_{i+j,j}]$;
- the $\boldsymbol{\Phi}_t$ matrix is constructed by adding blocks of rows $[-\boldsymbol{F}_{i+j,j}\ \boldsymbol{0}\ \ldots\ \boldsymbol{0}]$.

If $k_{i+j} > j$ for all $j = 1, 2, \ldots, k$, columns $[\boldsymbol{0}\ \boldsymbol{0}\ \ldots\ \boldsymbol{I}]^T$ are added to the matrix $\boldsymbol{\Phi}_t$.

Example 6.2.2

Consider a model where $s = 3$, $k_1 = 6$, $k_2 = 1$ and $k_3 = 1$. According the conventional conversion rule, the non minimum dimension SS formulation is:

$$\boldsymbol{\Phi}_1 = \begin{bmatrix} -f_{2,1} & 1 & 0 & 0 & 0 & 0 \\ -f_{3,2} & 0 & 1 & 0 & 0 & 0 \\ -f_{1,3} & 0 & 0 & 1 & 0 & 0 \\ -f_{2,4} & 0 & 0 & 0 & 1 & 0 \\ -f_{3,5} & 0 & 0 & 0 & 0 & 1 \\ -f_{1,6} & 0 & 0 & 0 & 0 & 0 \end{bmatrix}, \quad \boldsymbol{\Phi}_2 = \begin{bmatrix} -f_{3,1} & 1 & 0 & 0 & 0 & 0 \\ -f_{1,2} & 0 & 1 & 0 & 0 & 0 \\ -f_{2,3} & 0 & 0 & 1 & 0 & 0 \\ -f_{3,4} & 0 & 0 & 0 & 1 & 0 \\ -f_{1,5} & 0 & 0 & 0 & 0 & 1 \\ -f_{2,6} & 0 & 0 & 0 & 0 & 0 \end{bmatrix} \tag{6.38}$$

$$\mathbf{\Phi}_3 = \begin{bmatrix} -f_{1,1} & 1 & 0 & 0 & 0 & 0 \\ -f_{2,2} & 0 & 1 & 0 & 0 & 0 \\ -f_{3,3} & 0 & 0 & 1 & 0 & 0 \\ -f_{1,4} & 0 & 0 & 0 & 1 & 0 \\ -f_{2,5} & 0 & 0 & 0 & 0 & 1 \\ -f_{3,6} & 0 & 0 & 0 & 0 & 0 \end{bmatrix}. \tag{6.39}$$

However, the application of the rule above removes redundant states providing the following minimum dimension matrices:

$$\mathbf{\Phi}_1 = \begin{bmatrix} -f_{2,1} & 0 \\ -f_{1,3} & 1 \\ -f_{1,6} & 0 \end{bmatrix}, \quad \mathbf{\Phi}_2 = \begin{bmatrix} -f_{3,1} & 0 & 0 \\ -f_{1,2} & 1 & 0 \\ -f_{1,5} & 0 & 1 \end{bmatrix}, \tag{6.40}$$

$$\mathbf{\Phi}_3 = \begin{bmatrix} -f_{1,1} & 1 & 0 \\ -f_{1,4} & 0 & 1 \end{bmatrix}, \tag{6.41}$$

where the dimension of the state vector changes from being 6 (equal to the maximum dynamic, $k_i = 6$), to be varying with the season, but never exceeding 3.

6.2.3 *Stationarity and invertibility*

The stationarity and invertibility conditions of a periodic process often are characterized by its equivalent representation VARMA with constant parameters; see, for example, Franses and Paap [101] or Boswijk and Franses [27]. Proceeding in the same way, to characterize the stationarity and invertibility conditions of a PVARMAX process, it is necessary to write the state equation (6.27) in the equivalent form:

$$\boldsymbol{x}_{t+s} = (\mathbf{\Phi}_{t+s-1} \times \mathbf{\Phi}_{t+s-2} \times \cdots \times \mathbf{\Phi}_t)\boldsymbol{x}_t + \sum_{i=1}^{s} \prod_{j=1}^{s-i} (\mathbf{\Phi}_{t+s-j}\mathbf{\Gamma}_{t+i-1}\boldsymbol{u}_{t+i-1} + \mathbf{\Phi}_{t+s-1}\boldsymbol{E}_{t+i-j}\boldsymbol{a}_{t+i-1}), \tag{6.42}$$

whose matrices do not change every s periods due to the law of variation $\mathbf{\Phi}_t = \mathbf{\Phi}_{t+s}$, $\mathbf{\Gamma}_t = \mathbf{\Gamma}_{t+s}$, and $\boldsymbol{E}_t = \boldsymbol{E}_{t+s}$, $t = 1, 2, \ldots, N$. Once described, the model in (6.28) and (6.42), is said to be *cycle-stationary* if all eigenvalues of $\mathbf{\Phi}^* = \prod_{j=1}^{s} \mathbf{\Phi}_{s-j+1}$ are within the unit circle.

Similarly, to characterize the invertibility conditions of a PVARMAX model one should use its equivalent VARX(∞) representation. Thus the model can be written as:

$$\boldsymbol{x}_{t+1} = \bar{\mathbf{\Phi}}^* \boldsymbol{x}_{t+1-s} + \bar{\mathbf{\Gamma}}^* \begin{bmatrix} \boldsymbol{u}_t \\ \boldsymbol{u}_{t-1} \\ \vdots \\ \boldsymbol{u}_{t+1-s} \end{bmatrix} + \bar{\boldsymbol{E}}^* \begin{bmatrix} \boldsymbol{z}_t \\ \boldsymbol{z}_{t-1} \\ \vdots \\ \boldsymbol{z}_{t+1-s} \end{bmatrix} \tag{6.43}$$

$$z_{t+1} = H_{t+1}\bar{\Phi}^* x_{t+1-s} + H_{t+1}\bar{\Gamma}^* \begin{bmatrix} u_t \\ u_{t-1} \\ \vdots \\ u_{t+1-s} \end{bmatrix} + H_{t+1}\bar{E}^* \begin{bmatrix} z_t \\ z_{t-1} \\ \vdots \\ z_{t+1-s} \end{bmatrix} + D_t u_t + a_{t+1}, \tag{6.44}$$

where

$$\bar{\Phi}^* = (\Phi_t - E_t H_t) \times (\Phi_{t-1} - E_{t-1} H_{t-1}) \times \cdots \times (\Phi_{t-1+s} - E_{t-1+s} H_{t-1+s}) \tag{6.45}$$

$$\bar{E}^* = [E_t, (\Phi_t - E_t H_t) E_{t-1}, (\Phi_t - E_t H_t)(\Phi_{t-1} - E_{t-1} H_{t-1} E_{t-2}), \ldots, \prod_{i=0}^{s-2} (\Phi_{t-i} + E_{t-i} H_{t-i}) E_{t+1-s}] \tag{6.46}$$

$$\bar{\Gamma}^* = [\bar{\Gamma}_t, (\Phi_t - E_t H_t)\bar{\Gamma}_{t-1}, (\Phi_t - E_t H_t)(\Phi_{t-1} - E_{t-1} H_{t-1})\bar{\Gamma}_{t-2}, \ldots, \prod_{i=0}^{s-2} (\Phi_{t-i} - E_{t-i} H_{t-i})\bar{\Gamma}_{t+1-s}], \tag{6.47}$$

with $\bar{\Gamma}_t = \Gamma_t - E_t D_t$. Since (6.44) can be interpreted as a VARX(∞) process, substituting recursively x_{t+1-s} according to (6.43), it is straightforward to see that, for the past values of z_t to have a decreasing effect on the current value (invertibility) it is necessary that the eigenvalues in $\bar{\Phi}^*$ lie within the unit circle. Bentarzi and Hallin [21] obtained the same necessary and sufficient condition for invertibility.

6.2.4 *Maximum likelihood estimation*

The Gaussian likelihood of model (6.27)–(6.28) is given by (5.32). Its evaluation can be accomplished in three phases by computing: (a) $\tilde{z}_t$ and B_t, (b) W_N and w_N, and (c) using the initial conditions μ and P_1. As usual, the innovations and their covariances are calculated by propagating a $KF(\mathbf{0},\mathbf{0})$:

$$\hat{x}_{t+1|t} = \Phi_t \hat{x}_{t|t-1} + \Gamma_t u_t + K_t \tilde{z}_t \tag{6.48}$$

$$\tilde{z}_t = z_t - H_t \hat{x}_{t|t-1} - D_t u_t \tag{6.49}$$

$$B_t = Q_t \tag{6.50}$$

$$K_t = E_t \tag{6.51}$$

$$P_{t+1|t} = \mathbf{0}, \tag{6.52}$$

where we have taken into account that $P = \mathbf{0}$ is the solution of the Riccati equation of the KF for the SEM (6.27)–(6.28), even when some matrices are time-varying. Moreover, the vector w_N and the matrix W_N are computed by once again propagating (6.17)–(6.19) with the same initial conditions.

The choice of initial conditions μ and P_1 should take into account whether the model (6.27)–(6.28) is stationary or not, and if it has deterministic and/or stochastic inputs (see Sections 5.2 and 5.3), according to the following procedure.

1. *Stationary PVARMAX models.* If the model is stationary and has no exogenous variables, the proper initialization of the KF is $\boldsymbol{\mu} = \mathbf{0}$, with $\boldsymbol{P_1}$ chosen to be the solution of the Lyapunov equation corresponding to (6.42). The likelihood function in this case would be:

$$\ell(\boldsymbol{Z};\boldsymbol{\theta}) = \log|\boldsymbol{P_1}| + \sum_{t=1}^{N} \log|\boldsymbol{Q_t}| + \sum_{t=1}^{N} \tilde{\boldsymbol{z}}_t^T \boldsymbol{Q_t}^{-1} \tilde{\boldsymbol{z}}_t + \log\left|\boldsymbol{P_1}^{-1} + \boldsymbol{W_N}\right| - \boldsymbol{w}_N^T(\boldsymbol{P}_1^{-1} + \boldsymbol{W_N})^{-1}\boldsymbol{w_N}, \quad (6.53)$$

where $\boldsymbol{w_N}$ and $\boldsymbol{W_N}$ are calculated from $\boldsymbol{w_t} = \boldsymbol{w}_{t-1} + \bar{\boldsymbol{\Phi}}_{t-1}\boldsymbol{H}_t^T\boldsymbol{Q}_t^{-1}\tilde{\boldsymbol{z}}_t$, and $\boldsymbol{W_t} = \boldsymbol{W}_{t-1} + \bar{\boldsymbol{\Phi}}_{t-1}\boldsymbol{H}_t^T\boldsymbol{Q}_t^{-1}\boldsymbol{H}_t\bar{\boldsymbol{\Phi}}_{t-1}$, being $\boldsymbol{\Phi_t} = (\boldsymbol{\Phi_t} - \boldsymbol{E_t}\boldsymbol{H_t})\bar{\boldsymbol{\Phi}}_{t-1}$, with $\bar{\boldsymbol{\Phi}}_1 = \boldsymbol{I}$. Now, denoting by $\boldsymbol{M}$ the Cholesky factor of $\boldsymbol{P_1}$, we have that $\boldsymbol{P_1} = \boldsymbol{M}\boldsymbol{M}^T$, $|\boldsymbol{P_1}| = |\boldsymbol{M}\boldsymbol{M}^T| = |\boldsymbol{M}|^2$, and $\log|\boldsymbol{P_1}| + \log|\boldsymbol{P}_1^{-1} + \boldsymbol{W_N}| = \log|\boldsymbol{I} + \boldsymbol{M}^T\boldsymbol{W_N}\boldsymbol{M}|$. Therefore, (6.53) can be written as:

$$\ell(\boldsymbol{Z};\boldsymbol{\theta}) = \log\left|\boldsymbol{I} + \boldsymbol{M}^T\boldsymbol{W_N}\boldsymbol{M}\right| + \sum_{t=1}^{N} \log|\boldsymbol{Q_t}| + \sum_{t=1}^{N} \tilde{\boldsymbol{z}}_t^T \boldsymbol{Q_t}^{-1} \tilde{\boldsymbol{z}}_t - \boldsymbol{w}_N^T\boldsymbol{M}(\boldsymbol{I} + \boldsymbol{M}^T\boldsymbol{W_N}\boldsymbol{M})^{-1}\boldsymbol{M}^T\boldsymbol{w_N}. \quad (6.54)$$

If $\boldsymbol{L}$ is the Cholesky factor of $\boldsymbol{I} + \boldsymbol{M}^T\boldsymbol{W_N}\boldsymbol{M}$ such that $(\boldsymbol{I} + \boldsymbol{M}^T\boldsymbol{W_N}\boldsymbol{M})^{-1} = (\boldsymbol{L}\boldsymbol{L}^T)^{-1}$, it is easy to obtain recursively a $\boldsymbol{\lambda}$ vector that satisfies the triangular system of equations $\boldsymbol{L}\boldsymbol{\lambda} = \boldsymbol{M}^T\boldsymbol{w_N}$, and consequently, (6.54) can be written in simplified form as:

$$\ell(\boldsymbol{Z};\boldsymbol{\theta}) = \log|\boldsymbol{L}|^2 + \sum_{t=1}^{N} \log|\boldsymbol{Q_t}| + \sum_{t=1}^{N} \tilde{\boldsymbol{z}}_t^T \boldsymbol{Q_t}^{-1} \tilde{\boldsymbol{z}}_t - \boldsymbol{\lambda}^T\boldsymbol{\lambda}, \quad (6.55)$$

where if N is the total number of observations in the sample and N_i the number of observations in the i-th season, so that $\sum_{i=1}^{s} N_i = N$. Note that the term $\sum_{t=1}^{N} \log|\boldsymbol{Q_t}|$ in (6.55) can therefore be written as $\sum_{i=1}^{s} N_i \log|\boldsymbol{Q_i}|$.

2. *Nonstationary PVARMAX models.* If the model is nonstationary and has no exogenous variables, the Kalman filter must be initialized with $\boldsymbol{\mu} = \mathbf{0}$ and $\boldsymbol{P}_1^{-1} = \mathbf{0}$, which is associated with infinite initial uncertainty. In this case, we can evaluate the minimally conditioned likelihood defined in Subsection 5.2.2.

3. *Partially Nonstationary PVARMAX models.* If the model is partially nonstationary and has no exogenous variables, the eigenvalues of $\prod_{j=1}^{s} \boldsymbol{\Phi}_{s-j+1}$ will be either on or inside the unit circle. In this situation, the appropriate initial conditions are $\boldsymbol{\mu} = \mathbf{0}$, with $\boldsymbol{P}_1^{-1}$ set to a finite non-null matrix; see De Jong and Chu-Chun-Lin [69].

The presence of exogenous variables in the model and its deterministic or stochastic nature will affect the initial conditions for the KF; see Sections 5.2 and 5.3.

6.3 The likelihood of models with GARCH errors

The idea that a model may have conditional heteroscedastic errors, originally proposed by Engle [86], motivated the development of a large family of ARCH/GARCH-type models. In such models, the state vector may provide information about the future variances and covariances as a function of the past history.

A model with GARCH errors has two main elements: the *model for the mean*, which is a standard econometric model for the endogenous variables, and the *model for the variance*, which is a dynamic equation describing the law governing changes in variances and covariances[2].

We will consider that the model for the mean is (or can be written as) the SEM:

$$\boldsymbol{x}_{t+1} = \boldsymbol{\Phi x}_t + \boldsymbol{\Gamma u}_t + \boldsymbol{E\varepsilon}_t \tag{6.56}$$

$$\boldsymbol{z}_t = \boldsymbol{H x}_t + \boldsymbol{D u}_t + \boldsymbol{\varepsilon}_t, \tag{6.57}$$

where the distribution of the error $\boldsymbol{\varepsilon}_t$, conditional on the information available up to time $t-1$, is $NID(\boldsymbol{0}, \boldsymbol{\Sigma}_t)$[3], where the error covariance matrix satisfies:

$$\begin{aligned}\text{vech}(\boldsymbol{\Sigma}_t) = \text{vech}(\bar{\boldsymbol{\Sigma}}) + \sum_{i=1}^{p_\varepsilon} \boldsymbol{F}_i^\varepsilon \text{vech}(\varepsilon_{t-i}\varepsilon_{t-i}^T) + \sum_{i=0}^{q_\varepsilon} \boldsymbol{G}_i^\varepsilon \boldsymbol{u}_t^\varepsilon \\ + \sum_{i=1}^{s_\varepsilon} \boldsymbol{\Xi}_i^\varepsilon \text{vech}(\boldsymbol{\Sigma}_{t-i}),\end{aligned} \tag{6.58}$$

and vech() denotes the half-vectorization operator, which vectorizes the upper triangle and main diagonal of its (matrix) argument.

The vector $\boldsymbol{u}_t^\varepsilon$ allows for exogenous variables explaining changes in the volatility of the process. These variables are often dummies associated with an event, or with some calendar effect, e.g., a holiday. Therefore the conditional variance matrix $\boldsymbol{\Sigma}_t$ has a complex dependence on past information.

Assuming that the initial state follows a normal distribution, it is easy to prove that the process $\boldsymbol{z}_t$ is conditionally normally distributed, being its first and second-

[2]Models with GARCH errors have been very successful, especially in the financial area, because high frequency time series in finance (returns, transaction volumes, etc.) often display periods of high volatility followed by more quiescent phases. This fact, which is known as *volatility clustering*, suggests that there is an autocorrelation structure for the volatility which can be captured by GARCH-type specifications. The theoretical and empirical literature in this field is too numerous to provide even a quick overview. For a survey, see Bollerslev, Engle and Nelson [25].

[3]In many cases and, in particular, when working with high-frequency financial data, the assumption of normality is not realistic. Despite this, one can still obtain consistent estimates for the model parameters by optimizing the Gaussian log-likelihood; see also Appendix B. These are sometimes referred to as pseudo-likelihood estimates.

order moments the output of the following KF:

$$\tilde{\boldsymbol{z}}_t = \boldsymbol{z}_t - \boldsymbol{H}_t\hat{\boldsymbol{x}}_{t|t-1} - \boldsymbol{D}_t\boldsymbol{u}_t \tag{6.59}$$

$$\hat{\boldsymbol{x}}_{t+1|t} = \boldsymbol{\Phi}_t\hat{\boldsymbol{x}}_{t|t-1} + \boldsymbol{\Gamma}_t\boldsymbol{u}_t + \boldsymbol{K}_t\tilde{\boldsymbol{z}}_t \tag{6.60}$$

$$\boldsymbol{K}_t = (\boldsymbol{\Phi}\boldsymbol{P}_{t|t-1}\boldsymbol{H}^T + \boldsymbol{E}\boldsymbol{\Sigma}_t)\boldsymbol{B}_t^{-1} \tag{6.61}$$

$$\boldsymbol{P}_{t+1|t} = \boldsymbol{\Phi}\boldsymbol{P}_{t|t-1}\boldsymbol{\Phi}^T + \boldsymbol{E}\boldsymbol{\Sigma}_t\boldsymbol{E}^T - \boldsymbol{K}_t\boldsymbol{B}_t\boldsymbol{K}_t^T \tag{6.62}$$

$$\boldsymbol{B}_t = \boldsymbol{H}\boldsymbol{P}_{t|t-1}\boldsymbol{H}^T + \boldsymbol{\Sigma}_t. \tag{6.63}$$

Therefore, the propagation of (6.59)–(6.63) provides the variables required to evaluate the (minus) log-likelihood of the innovations $\tilde{\boldsymbol{z}}_t$ (5.46), taking into account the nature of $\boldsymbol{\Sigma}_t$.

As $\mathrm{E}[\mathrm{vech}(\boldsymbol{\varepsilon}_t\boldsymbol{\varepsilon}_t^T)\,|\,\Omega_{t-1}] = \mathrm{vech}(\boldsymbol{\Sigma}_t)$, we can define a vector, $\boldsymbol{\xi}_t = \mathrm{vech}(\boldsymbol{\varepsilon}_t\boldsymbol{\varepsilon}_t^T) - \mathrm{vech}(\boldsymbol{\Sigma}_t)$, such that $\mathrm{E}(\boldsymbol{\xi}_t\,|\,\Omega_{t-1}) = \boldsymbol{0}$. Now, adding $\boldsymbol{\xi}_t$ to both sides of (6.58), we obtain:

$$\begin{aligned}\mathrm{vech}(\boldsymbol{\varepsilon}_t\boldsymbol{\varepsilon}_t^T) &= \mathrm{vech}(\bar{\boldsymbol{\Sigma}}) + \sum_{i=1}^{p_\varepsilon}\boldsymbol{F}_i^\varepsilon\,\mathrm{vech}(\boldsymbol{\varepsilon}_{t-i}\boldsymbol{\varepsilon}_{t-i}^T) + \sum_{i=0}^{q_\varepsilon}\boldsymbol{G}_i^\varepsilon\boldsymbol{u}_t^\varepsilon\\ &+ \sum_{i=1}^{s_\varepsilon}\boldsymbol{\Xi}_i^\varepsilon\,\mathrm{vech}(\boldsymbol{\varepsilon}_t\boldsymbol{\varepsilon}_t^T) + \boldsymbol{\xi}_t - \sum_{i=1}^{s_\varepsilon}\boldsymbol{\Xi}_i^\varepsilon\boldsymbol{\xi}_{t-i},\end{aligned} \tag{6.64}$$

or the equivalent VARMAX representation:

$$\begin{aligned}\mathrm{vech}(\boldsymbol{\varepsilon}_t\boldsymbol{\varepsilon}_t^T) &= \mathrm{vech}(\bar{\boldsymbol{\Sigma}}) + \sum_{i=1}^{\bar{p}_\varepsilon}\bar{\boldsymbol{F}}_i^\varepsilon\,\mathrm{vech}(\boldsymbol{\varepsilon}_{t-i}\boldsymbol{\varepsilon}_{t-i}^T)\\ &\qquad + \sum_{i=0}^{q_\varepsilon}\boldsymbol{G}_i^\varepsilon\boldsymbol{u}_t^\varepsilon + \boldsymbol{\xi}_t - \sum_{i=1}^{s_\varepsilon}\boldsymbol{\Xi}_i^\varepsilon\boldsymbol{\xi}_{t-i},\end{aligned} \tag{6.65}$$

where $\bar{p}_\varepsilon = \max(p_\varepsilon, q_\varepsilon)$, and $\bar{\boldsymbol{F}}_i^\varepsilon = \boldsymbol{F}_i^\varepsilon + \boldsymbol{\Xi}_i^\varepsilon$, with $\boldsymbol{F}_i^\varepsilon = \boldsymbol{0}$ and $\boldsymbol{\Xi}_i^\varepsilon = \boldsymbol{0}$ for all $i > q_\varepsilon$. The VARMAX formulation (6.65) has an SS representation given by:

$$\boldsymbol{x}_{t+1}^\varepsilon = \boldsymbol{\Phi}^\varepsilon\boldsymbol{x}_t^\varepsilon + \boldsymbol{\Gamma}^\varepsilon\boldsymbol{u}_t^\varepsilon + \boldsymbol{E}^\varepsilon\boldsymbol{\xi}_t \tag{6.66}$$

$$\mathrm{vech}(\boldsymbol{\varepsilon}_t\boldsymbol{\varepsilon}_t^T) = \boldsymbol{H}^\varepsilon\boldsymbol{x}_t^\varepsilon + \boldsymbol{D}^\varepsilon\boldsymbol{u}_t^\varepsilon + \boldsymbol{\xi}_t, \tag{6.67}$$

and, as the vector $\mathrm{vech}(\boldsymbol{\varepsilon}_t\boldsymbol{\varepsilon}_t^T)$ in equation (6.67) is not observable, it is replaced by:

$$\mathrm{vech}(\tilde{\boldsymbol{z}}_t\tilde{\boldsymbol{z}}_t^T) = \boldsymbol{H}^\varepsilon\boldsymbol{x}_t^\varepsilon + \boldsymbol{D}^\varepsilon\boldsymbol{u}_t^\varepsilon + \boldsymbol{\xi}_t, \tag{6.68}$$

where $\tilde{\boldsymbol{z}}_t = \boldsymbol{z}_t - \hat{\boldsymbol{z}}_{t/t-1}$, and its covariance matrix is given by (6.63).

Note that, in a representation such as (6.56)–(6.57), $\bar{\boldsymbol{P}} = \boldsymbol{0}$ is the solution of the algebraic Riccati equation:

$$\begin{aligned}\bar{\boldsymbol{P}} &= \boldsymbol{\Phi}\bar{\boldsymbol{P}}\boldsymbol{\Phi}^T + \boldsymbol{E}\boldsymbol{\Sigma}_t\boldsymbol{E}^T\\ &\qquad - (\boldsymbol{\Phi}\bar{\boldsymbol{P}}\boldsymbol{H}^T + \boldsymbol{E}\boldsymbol{\Sigma}_t)(\boldsymbol{H}\bar{\boldsymbol{P}}\boldsymbol{H}^T + \boldsymbol{\Sigma}_t)^{-1}(\boldsymbol{\Phi}\bar{\boldsymbol{P}}\boldsymbol{H}^T + \boldsymbol{E}\boldsymbol{\Sigma}_t)^T,\end{aligned} \tag{6.69}$$

associated with equation (6.62); see Appendix A.

Since $\text{vech}(\boldsymbol{\Sigma}_t) = \text{E}[\text{vech}(\boldsymbol{\varepsilon}_t \boldsymbol{\varepsilon}_t^T)\,|\boldsymbol{\Omega}_{t-1}]$, a KF applied to the model given by (6.66) and (6.68) provides the minimum variance linear estimator of $\text{vech}(\boldsymbol{\varepsilon}_t \boldsymbol{\varepsilon}_t^T)$ conditional on the information available up to time $t-1$, regardless of the error distribution $\boldsymbol{\xi}_t$; see Anderson and Moore [6]. This property allows us to define $\text{vech}(\boldsymbol{\Sigma}_t) = \boldsymbol{H}^{\varepsilon} \hat{\boldsymbol{x}}^{\varepsilon}_{t|t-1} + \boldsymbol{D}^{\varepsilon} \boldsymbol{u}^{\varepsilon}_t$, where:

$$\hat{\boldsymbol{x}}^{\varepsilon}_{t+1|t} = (\boldsymbol{\Phi}^{\varepsilon} - \boldsymbol{K}^{\varepsilon}_t \boldsymbol{H}^{\varepsilon})\hat{\boldsymbol{x}}^{\varepsilon}_{t|t-1} + \boldsymbol{\Gamma}^{\varepsilon} \boldsymbol{u}^{\varepsilon}_t + \boldsymbol{K}^{\varepsilon}_t \text{vech}(\tilde{\boldsymbol{z}}_t \tilde{\boldsymbol{z}}_t^T) \tag{6.70}$$

$$\boldsymbol{K}^{\varepsilon}_t = [\boldsymbol{\Phi}^{\varepsilon} \boldsymbol{P}^{\varepsilon}_{t|t-1} (\boldsymbol{H}^{\varepsilon})^T + \boldsymbol{E}^{\varepsilon} \boldsymbol{\Sigma}^{\varepsilon}](\boldsymbol{B}^{\varepsilon}_t)^{-1} \tag{6.71}$$

$$\boldsymbol{P}^{\varepsilon}_{t+1|t} = \boldsymbol{\Phi}^{\varepsilon} \boldsymbol{P}^{\varepsilon}_{t|t-1} (\boldsymbol{\Phi}^{\varepsilon})^T + \boldsymbol{E}^{\varepsilon} \boldsymbol{B}^{\varepsilon}_t (\boldsymbol{E}^{\varepsilon})^T - \boldsymbol{K}^{\varepsilon}_t \boldsymbol{B}^{\varepsilon}_t (\boldsymbol{K}^{\varepsilon}_t)^T \tag{6.72}$$

$$\boldsymbol{B}^{\varepsilon}_t = \boldsymbol{H}^{\varepsilon} \boldsymbol{P}^{\varepsilon}_{t|t-1} (\boldsymbol{H}^{\varepsilon})^T + \boldsymbol{\Sigma}^{\varepsilon}. \tag{6.73}$$

Assuming that $\boldsymbol{x}^{\varepsilon}_1$ is a null vector, the filter (6.70)–(6.73) collapses to the expression:

$$\hat{\boldsymbol{x}}^{\varepsilon}_{t+1|t} = (\boldsymbol{\Phi}^{\varepsilon} - \boldsymbol{E}^{\varepsilon} \boldsymbol{H}^{\varepsilon})\hat{\boldsymbol{x}}^{\varepsilon}_{t|t-1} + \boldsymbol{\Gamma}^{\varepsilon} \boldsymbol{u}^{\varepsilon}_t + \boldsymbol{E}^{\varepsilon} \text{vech}(\tilde{\boldsymbol{z}}_t \tilde{\boldsymbol{z}}_t^T). \tag{6.74}$$

Note that this assumption implies that $\boldsymbol{P}^{\varepsilon}_1 = \boldsymbol{0}$ is the solution to the algebraic Riccati equation associated with (6.71), and therefore $\boldsymbol{P}^{\varepsilon}_{t|t-1} = \boldsymbol{0}$ and $\boldsymbol{K}^{\varepsilon}_t = \boldsymbol{E}^{\varepsilon}$ for all t.

If the eigenvalues of $\boldsymbol{\Phi}^{\varepsilon}$ are inside the unit circle, model (6.56)–(6.57) implies that the unconditional variance is finite. On the other hand, if the model has some unit roots it would reduce to an IGARCH; see Engle and Bollerslev [84].

The conditional variance model (6.58) does not assure that $\boldsymbol{\Sigma}_t$ will be positive semi-definite. In the univariate case is easy to constrain the parameters to avoid negative variances. For example, in a model where the conditional variance is:

$$\sigma_t^2 = \bar{\sigma}^2 + f^{\varepsilon} \varepsilon_{t-1}^2 + \xi^{\varepsilon} \sigma_{t-1}^2, \tag{6.75}$$

the conditions assuring non-negativity of the variance are $\bar{\sigma}^2 > 0$, $f^{\varepsilon} \geq 0$, and $\xi^{\varepsilon} \geq 0$. In the multivariate case however, the constraints required to ensure that $\boldsymbol{\Sigma}_t$ is positive-semidefinite are not trivial. Because of this, Engle and Kroner [85] proposed an alternative formulation known as BEKK, whose functional form assures this condition without additional constraints. The BEKK model can also be interpreted as a constrained form of (6.58).

6.4 Examples

6.4.1 A time-varying CAPM regression

A well known econometric model in finance is the Capital-Asset Pricing Model (CAPM), given by a linear regression between the return of the i-th asset or portfolio, r_t^i and that of a market index r_t^m:

$$r_t^i = \alpha^i + \beta^i r_t^m + \varepsilon_t^i, \tag{6.76}$$

where (a) the slope parameter (β^i) measures the change in the asset profitability due to changes in the market return, (b) the constant term (α^i) is an extra return not related to fluctuations in the market, and (c) the error term ε_t^i captures specific changes in r_t^i not explained by the market performance.

The β^i parameter is useful in many ways. For example, it is often used to determine an optimal hedging strategy for the market risk associated with the i-th asset, using synthetic assets related to the market return r_t^m. On the other hand, its value is useful in characterizing the performance of the asset in comparison with that of the market. For example, the returns of an asset with $\beta^i > 1$ typically amplify the market fluctuations, so it would make a good investment in "bull" markets. On the other hand, an asset with $0 < \beta^i < 1$ mitigates the market fluctuations, so it would be a sensible defensive investment in "bear" markets.

Despite its usefulness, there is a broad consensus in the literature about the instability of the β^i coefficient, so it is relevant to develop methods to test and measure its changes over time. We tested this hypothesis with a sample of closing prices of General Electric (GE) shares and the Standard & Poor's 500 (SP500) Market index, observed during 2,362 consecutive market days from November 1, 1993 to April 3, 2003. Building on this dataset, we first computed the daily return of the GE shares and the Market index, and then subtracted the simultaneous return of a US Treasury Bill divided by 360 to obtain the corresponding excess return series, $erGE_t$ and $erSP500_t^m$. Last, we formulated a time-varying coefficient version of the CAPM model (6.76), assuming that the "alpha" and "beta" parameters follow a random-walk process as in (6.6)–(6.7):

$$erGE_t = \alpha_t^{GE} + \beta_t^{GE} erSP500_t^m + \varepsilon_t^{GE} \tag{6.77}$$

$$\alpha_{t+1}^{GE} = \alpha_t^{GE} + a_t^{\alpha} \tag{6.78}$$

$$\beta_{t+1}^{GE} = \beta_t^{GE} + a_t^{\beta}\,. \tag{6.79}$$

The model was fitted via ML estimation, resulting in a log-likelihood value of 3929.41. The estimates and standard deviations are displayed in Table 6.1.

Table 6.1 *Estimation results for model (6.77)–(6.79).*

Parameter	Estimate	Standard Error
standard deviation of a_t^{α}	0.000	0.001
standard deviation of a_t^{β}	0.015	0.006
standard deviation of ε_t^{GE}	1.267	0.019

The parameter sequences implied by this model can be easily estimated by fixed-interval smoothing; see Chapter 4. Note that the estimate for the variance of the "alpha" parameter is very close to zero, so α_t^{GE} is approximately constant, with a fixed-interval smoothed estimate of $\hat{\alpha}_t^{GE} = .043$ and a standard error of .026. The "beta" parameter, on the other hand, has nonzero variance, and so it is indeed time-varying. Figure 6.1 displays the corresponding sequence of smoothed estimates, $\hat{\beta}_t^{GE}$, as well as approximate 95% confidence bounds.

The code required to replicate this example can be found in the file `Example0641.m`; see Appendix D.

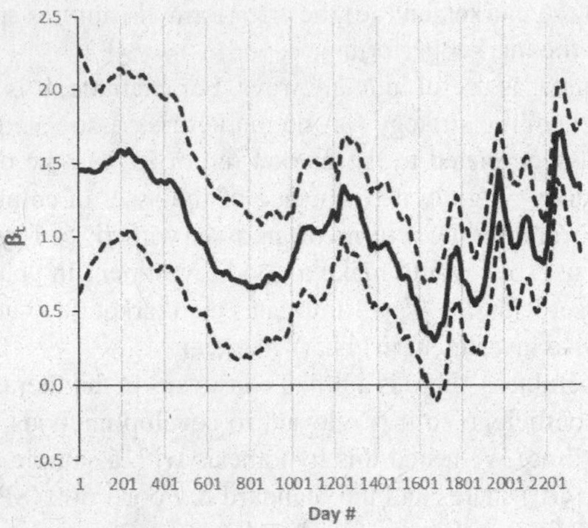

Figure 6.1 *Fixed-interval smoothed estimates for β_t (solid line) in the time-varying parameter CAPM regression with $\pm 2\sigma$ limits (dashed line).*

6.4.2 *A periodic model for West German Consumption*

In this example we will model the quarterly unadjusted West German Consumption data for the years 1960–1987, see Table E.4 in Appendix E of Lutkepohl [156]. The profile of this series, see Figure 6.2, displays a strong seasonal pattern and Lutkepohl [156], after testing for parametric variation, concludes that there could be a periodic structure, perhaps affected by structural shifts during the sample period used. We will ignore such potential problems because our goal here is to provide an illustrative example for the periodic models given in Section 6.2.

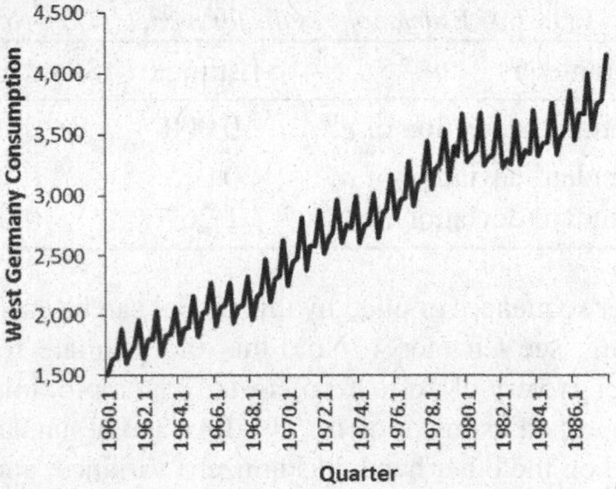

Figure 6.2 *Quarterly unadjusted West German Consumption. The data is in billions of Deutsch Marks.*

Table 6.2 *Estimation results for the model (6.80). The values on convergence of the (log) likelihood function and the information criteria are, respectively,* $\ell^*(\hat{\theta})$*=-313.766; AIC = -5.715, SBC = -5.515*

t	$\hat{\phi}_1$	$\hat{\phi}_2$	d_t	$\hat{\theta}_1$	$\hat{\sigma}^2_{a,t}$
t = 1, 5, 9, ...	.56 (.16)	.48 (.17)	1	–	.011 (.002)
t = 2, 6, 10, ...	–	–	1	.58 (.17)	.019 (.003)
t = 3, 7, 11, ...	–	–	0	–	.010 (.001)
t = 4, 8, 12, ...	–	–	1	.27 (.26)	.011 (.001)

A standard ARIMA specification analysis for each season provided the following general structure:

$$(1-\phi_{1,t}B-\phi_{2,t}B^2)(1-B)^{d_t}\log C_t = a_t, \qquad a_t \sim \text{IID}(0,\sigma^2_{a,t}), \qquad (6.80)$$

where C_t denotes the West German Consumption at quarter t. The estimation results for model (6.80) are summarized in Table 6.2.

On the other hand, C_t can be represented by the following non-periodic ARIMA model:

$$\begin{aligned}(1-B)(1-B^4)\log C_t &= (1-\underset{(.080)}{.259}B)(1-\underset{(.092)}{.548}B^4)\hat{a}_t, \qquad \sigma^2_a = \underset{(.001)}{.014} \\ \ell^*(\hat{\theta}) &= -307.017, \qquad AIC = -5.687, \qquad SBC = -5.608,\end{aligned} \qquad (6.81)$$

where the value of likelihood at the optimum is -307.017, while the corresponding value for the periodic model is -313.766.

The code required to replicate this example can be found in the file `Example0642.m`; see Appendix D.

6.4.3 *A model with vector-GARCH errors for two exchange rate series*

This example focuses on estimating the short-run correlation between the spot bid exchange rates of Deutsch Mark (dm) and Japanese Yen (jy) against the US Dollar, observed in the London Market during the 695 week period extending from January 1985 to April 1998. The data has been logged, differenced, and scaled by a factor of 100, to obtain the corresponding log percent yields. Excess returns were then computed by subtracting the sample mean.

Univariate analyses suggest that these (log) excess returns: (a) are stationary, (b) may have a short autoregressive structure in the mean, and (c) some conditional heteroscedasticity in the variance. We will model these features using VAR(1) model for the mean and a diagonal GARCH(1,1) model for the variance. The pseudo-ML estimation results for this model are:

$$\begin{bmatrix} 1-\underset{(.032)}{.244}B & 0 \\ 0 & 1-\underset{(.034)}{.300}B \end{bmatrix} \begin{bmatrix} dm_t \\ jy_t \end{bmatrix} = \begin{bmatrix} \hat{\varepsilon}_t^{dm} \\ \hat{\varepsilon}_t^{jy} \end{bmatrix} \quad (6.82)$$

$$\begin{bmatrix} (\hat{\varepsilon}_t^{dm})^2 \\ \hat{\varepsilon}_t^{dm} \cdot \hat{\varepsilon}_t^{jy} \\ (\hat{\varepsilon}_t^{jy})^2 \end{bmatrix} = \begin{bmatrix} \underset{(.282)}{1.810} \\ \underset{(.155)}{1.145} \\ \underset{(.171)}{1.666} \end{bmatrix} + \begin{bmatrix} 1-\underset{(.012)}{.973}B & 0 & 0 \\ 0 & 1-\underset{(.015)}{.942}B & 0 \\ 0 & 0 & 1-\underset{(.035)}{.861}B \end{bmatrix}^{-1}$$
$$\cdot \begin{bmatrix} 1-\underset{(.021)}{.916}B & 0 & 0 \\ 0 & 1-\underset{(.024)}{.873}B & 0 \\ 0 & 0 & 1-\underset{(.047)}{.739}B \end{bmatrix} \begin{bmatrix} \hat{\xi}_t^{dm} \\ \hat{\xi}_t^{dm,jy} \\ \hat{\xi}_t^{jy} \end{bmatrix}, \quad (6.83)$$

where dm_t and jy_t denote, respectively, the excess returns of the Deutsch Mark and the Japanese Yen. Note that Equation (6.83) is a vector GARCH model in VARMAX form; see equation (6.65).

Finally, Figure 6.3 shows the resulting conditional correlations. Note that they fluctuate around its unconditional mean value of 0.64, with highs and lows of 0.9 and 0.3, respectively. A high conditional correlation value means that the co-movement of both series is strong in this period. On the contrary, low correlations indicate periods when both currencies fluctuate independently.

The code required to replicate this example can be found in the file `Example0643.m`; see Appendix D.

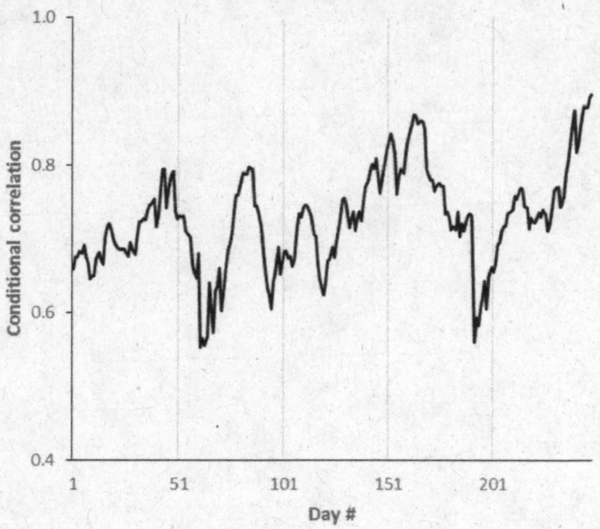

Figure 6.3 *Conditional correlation between Deutsch Mark and Japanese Yen returns.*

Chapter 7

Subspace methods

7.1 Theoretical foundations 83
7.1.1 Subspace structure and notation 84
7.1.2 Assumptions, projections and model reduction 86
7.1.3 Estimating the system matrices 87
7.2 System order estimation 89
7.2.1 Preliminary data analysis methods 89
7.2.2 Model comparison methods 91
7.3 Constrained estimation 92
7.3.1 State sequence structure 92
7.3.2 Subspace-based likelihood 93
7.4 Multiplicative seasonal models 94
7.5 Examples 95
7.5.1 Univariate models 95
7.5.2 A multivariate model for the interest rates 99

This chapter presents the basic theory of a family of system identification procedures known generically as "subspace methods." Their aim is to estimate the matrices of an SS model given a series of input and output measurements.

7.1 Theoretical foundations

Subspace methods are a set of algorithms based on systems theory, linear algebra, and statistics, which estimate the parameters in an SS model, i.e., the matrices in (2.1)–(2.2) or (2.19)–(2.20), given a series of input ($\boldsymbol{u}_t$) and output ($\boldsymbol{z}_t$) measures. They are based on fast and efficient linear algebra tools, such as the QR decompositions or the Singular Value (SVD), see Appendix A.1 and A.4, respectively. As a consequence, subspace methods are not iterative and, therefore, are very appropriate when dealing with large time series datasets.

The basic idea of these methods consists of estimating the sequence of system states from $\boldsymbol{u}_t$ and $\boldsymbol{z}_t$, without further information on the model dynamics. On the basis of these estimates, the system matrices can be estimated consistently by linear Least Squares (hereafter, LS), so that the model residuals can be obtained in order to ultimately estimate the error covariance matrices.

7.1.1 *Subspace structure and notation*

Assume that the system inputs, $\boldsymbol{u}_t \in \mathbb{R}^k$, and outputs, $\boldsymbol{z}_t \in \mathbb{R}^m$, are related by an SS model in SEM form (2.19)–(2.20)[1]. By substituting (2.20) into (2.19) in $\boldsymbol{a}_t$ and solving the resulting recursion, we obtain:

$$\boldsymbol{x}_t = (\boldsymbol{\Phi} - \boldsymbol{K}\boldsymbol{H})^{t-1}\boldsymbol{x}_1 + \sum_{j=1}^{t-1}(\boldsymbol{\Phi} - \boldsymbol{K}\boldsymbol{H})^{t-1-j}\boldsymbol{K}\boldsymbol{z}_j + \sum_{j=1}^{t-1}(\boldsymbol{\Phi} - \boldsymbol{K}\boldsymbol{H})^{t-1-j}(\boldsymbol{\Gamma} - \boldsymbol{K}\boldsymbol{D})\boldsymbol{u}_j, \quad (7.1)$$

so that the states at time t depend on the initial state vector, $\boldsymbol{x}_1$, as well as the past values of the input and output.

On the other hand, substituting the recursive solution of (2.19) into the observation equation (2.20), yields:

$$\boldsymbol{z}_t = \boldsymbol{H}\boldsymbol{\Phi}^{t-1}\boldsymbol{x}_1 + \boldsymbol{H}\sum_{j=1}^{t-1}\boldsymbol{\Phi}^{t-1-j}\boldsymbol{\Gamma}\boldsymbol{u}_j + \boldsymbol{D}\boldsymbol{u}_t + \boldsymbol{H}\sum_{j=1}^{t-1}\boldsymbol{\Phi}^{t-1-j}\boldsymbol{K}\boldsymbol{a}_j + \boldsymbol{a}_t, \quad (7.2)$$

which implies that the endogenous variable, $\boldsymbol{z}_t$, depends on the initial state, as well as current and past values of the innovations, $\boldsymbol{a}_t$, and inputs, $\boldsymbol{u}_t$.

Equations (7.1) and (7.2) can be written in matrix form as:

$$\boldsymbol{X}_f = \boldsymbol{L}_x^i \boldsymbol{X}_p + \boldsymbol{L}_z \boldsymbol{Z}_p + \boldsymbol{L}_u \boldsymbol{U}_p, \quad (7.3)$$

and

$$\boldsymbol{Z}_f = \boldsymbol{O}_i \boldsymbol{X}_f + \boldsymbol{T}_{d,i}\boldsymbol{U}_f + \boldsymbol{T}_{s,i}\boldsymbol{A}_f, \quad (7.4)$$

respectively, where i is a user-defined index which represents the number of past block rows, p. For simplicity we will assume henceforth that the dimension of past and future blocks is equal to i, so that $p = i$ and $f = 2i$. This dimensioning implies that the $\boldsymbol{L}$ matrices in (7.3) are defined as:

$$\begin{aligned} \boldsymbol{L}_x &:= (\boldsymbol{\Phi} - \boldsymbol{K}\boldsymbol{H})_{n\times n}, \\ \boldsymbol{L}_z &:= \left[(\boldsymbol{\Phi} - \boldsymbol{K}\boldsymbol{H})^{i-1}\boldsymbol{K} \quad \ldots \quad (\boldsymbol{\Phi} - \boldsymbol{K}\boldsymbol{H})\boldsymbol{K} \quad \boldsymbol{K}\right]_{n\times im}, \end{aligned}$$

[1] The basic ideas of subspace methods can also be implemented using an MEM form. For example, the N4SID algorithm (see Van Overschee and De Moor [215]) builds on model (2.1)–(2.2), particularly with $\boldsymbol{E}$ and $\boldsymbol{C}$ equal to the identity.

and

$$L_u := \left[(\Phi - KH)^{i-1}(\Gamma - KD) \quad \dots \quad (\Phi - KH)(\Gamma - KD) \quad (\Gamma - KD) \right]_{n \times ik}.$$

Thus the matrices L_x, L_z, and L_u have a number of rows equal to the system order, n, while the number of columns is n, im, and ik, respectively.

The system order is typically unknown in practice, but the literature provides several methods to estimate it consistently; see e.g. Garcia–Hiernaux, Jerez, and Casals [106] or Bauer [16]. We will describe some of those procedures later. In this section, and for simplicity, we will assume that n is known or has been previously determined.

In what follows, we will define additional matrices in equations (7.3) and (7.4), following the scheme presented by Favoreel, De Moor and Van Overschee [94]:

1. The output and input block Hankel matrices, whose dimensions are determined by the integers p and f, are such that:

$$Z_p := \begin{bmatrix} z_1 & z_2 & \dots & z_{T-p-f+1} \\ z_2 & z_3 & \dots & z_{T-p-f+2} \\ \vdots & \vdots & & \vdots \\ z_p & z_{p+1} & \dots & z_{T-f} \end{bmatrix}; \; Z_f := \begin{bmatrix} z_{p+1} & z_{p+2} & \dots & z_{T-f+1} \\ z_{p+2} & z_{p+3} & \dots & z_{T-f+2} \\ \vdots & \vdots & & \vdots \\ z_{p+f} & z_{p+f+1} & \dots & z_T \end{bmatrix}. \tag{7.5}$$

   On the other hand, A_p, A_f, U_p, and U_f are defined, respectively, as Z_p and Z_f, but with their own corresponding observations instead of z_t. Therefore, Z_p, Z_f, A_p and A_f are $im \times T - 2i + 1$ matrices, while U_p and U_f are $ik \times T - 2i + 1$.

2. State sequences are crucial in the derivation of subspace methods. The past and future state sequences are defined as $X_p := [x_1 \; x_2 \; \dots \; x_{T-p-f+1}]$ and $X_f := [x_{p+1} \; x_{p+2} \; \dots \; x_{T-f+1}]$. Since $p = i$ and $f = 2i$, both matrices are $n \times T - 2i + 1$.

3. The Extended Observability matrix:

$$O_i := \left[H^T \quad (H\Phi)^T \quad (H\Phi^2)^T \quad \dots \quad (H\Phi^{i-1})^T \right]^T_{im \times n}. \tag{7.6}$$

4. The deterministic lower block triangular Toeplitz matrix, T_d, defined as:

$$T_{d,i} := \begin{bmatrix} D_m & 0 & 0 & \dots & 0 \\ H\Gamma & D_m & 0 & \dots & 0 \\ H\Phi\Gamma & H\Gamma & D_m & \dots & 0 \\ \vdots & \vdots & \vdots & & \vdots \\ H\Phi^{i-2}\Gamma & H\Phi^{i-3}\Gamma & H\Phi^{i-4}\Gamma & \dots & D_m \end{bmatrix}_{im}. \tag{7.7}$$

5. The stochastic lower block triangular Toeplitz matrix, T_s, defined as:

$$T_{s,i} := \begin{bmatrix} I_m & 0 & 0 & \dots & 0 \\ HK & I_m & 0 & \dots & 0 \\ H\Phi K & HK & I_m & \dots & 0 \\ \vdots & \vdots & \vdots & & \vdots \\ H\Phi^{i-2}K & H\Phi^{i-3}K & H\Phi^{i-4}K & \dots & I_m \end{bmatrix}_{im}. \tag{7.8}$$

7.1.2 Assumptions, projections and model reduction

To set up the foundation of the subspace methods, we need the following assumptions:

A1. The system is strictly minimum-phase. That is, the eigenvalues of $\boldsymbol{L}_x$ are inside the unit circle.

A2. The system is minimal, which implies that the pair $(\boldsymbol{\Phi}, \boldsymbol{H})$ is observable and $(\boldsymbol{\Phi}, [\boldsymbol{\Gamma}, \boldsymbol{K}])$ is controllable; see Subsection 2.1.3.

A3. *Open-loop* data assumption: the innovations, $\boldsymbol{a}_t$, and the input, $\boldsymbol{u}_t$, are uncorrelated.

A4. The innovations, $\boldsymbol{a}_t$, follow a stationary, zero-mean, white noise process with $\mathrm{E}(\boldsymbol{a}_t \boldsymbol{a}_t^T) = \boldsymbol{Q}$.

A5. The input $\boldsymbol{u}_t$ is quasi-stationary in the sense of Ljung [152].

From A1 and for sufficiently large i, $\boldsymbol{L}_x^i \approx 0$. Consequently, substituting equation (7.3) in (7.4) gives:

$$\boldsymbol{Z}_f = \boldsymbol{O}_i\big(\boldsymbol{L}_z \boldsymbol{Z}_p + \boldsymbol{L}_u \boldsymbol{U}_p\big) + \boldsymbol{T}_{d,i}\boldsymbol{U}_f + \boldsymbol{T}_{s,i}\boldsymbol{A}_f, \tag{7.9}$$

which can be written in regression form as:

$$\boldsymbol{Z}_f = \boldsymbol{\Theta}_i \boldsymbol{V}_p + \boldsymbol{T}_{d,i}\boldsymbol{U}_f + \boldsymbol{T}_{s,i}\boldsymbol{A}_f, \tag{7.10}$$

where $\boldsymbol{\Theta}_i := \boldsymbol{O}_i \boldsymbol{L}_i$, $\boldsymbol{L}_i := [\boldsymbol{L}_z \quad \boldsymbol{L}_u]$, and $\boldsymbol{V}_p := [\boldsymbol{Z}_p^T \quad \boldsymbol{U}_p^T]^T$.

Equation (7.10) is the basis of subspace methods. It is a regression model which can be consistently estimated by LS because the columns in $\boldsymbol{T}_{s,i}\boldsymbol{A}_f$ are uncorrelated with the regressors, given by columns in $\boldsymbol{V}_p$ and $\boldsymbol{U}_f$.[2]

Note also that the matrix $\boldsymbol{\Theta}_i$ is the product of the observability and controllability matrices of the process, and has a reduced rank, n, equal to the system order, which is less than the matrix dimensions $im \times (im + ik)$. Because of this, estimation of (7.10) requires reduced-rank LS; see Appendix A.9. The estimation problem can be written as:

$$\min_{\{\boldsymbol{\Theta}_i, \boldsymbol{T}_{d,i}\}} \left\| \boldsymbol{W}_1\big(\boldsymbol{Z}_f - \boldsymbol{\Theta}_i \boldsymbol{V}_p - \boldsymbol{T}_{d,i}\boldsymbol{U}_f\big) \right\|_F^2, \qquad \text{such that } \operatorname{rank}(\boldsymbol{\Theta}_i) = n, \tag{7.11}$$

where $\|\cdot\|_F$ is the Frobenius norm. This problem is solved by first eliminating the term that includes $\boldsymbol{U}_f$. To this end, we post-multiply (7.10) by $\boldsymbol{\Pi}_{U_f}^{\perp} = \boldsymbol{I} - \boldsymbol{U}_f^T(\boldsymbol{U}_f\boldsymbol{U}_f^T)^{-1}\boldsymbol{U}_f$, which is the projection on the orthogonal complement of $\boldsymbol{U}_f$. Then we can run a LS regression on the result, which yields:

$$\hat{\boldsymbol{\Theta}}_i = \boldsymbol{Z}_f \boldsymbol{\Pi}_{U_f}^{\perp} \boldsymbol{V}_p^T (\boldsymbol{V}_p \boldsymbol{\Pi}_{U_f}^{\perp} \boldsymbol{V}_p^T)^{-1}, \tag{7.12}$$

where we used the idempotency of $\boldsymbol{\Pi}_{U_f}^{\perp}$. An SVD decomposition, see Appendix A.4, is now required on $\hat{\boldsymbol{\Theta}}_i$ to deal with the rank reduction. Van Overschee and De Moor

---

[2] This is assured by assumptions A3 and A4.

[216] proved that several subspace methods are equivalent to performing:

$$\boldsymbol{W}_1 \hat{\boldsymbol{\Theta}}_i \boldsymbol{W}_2 \overset{svd}{=} \boldsymbol{U}_n \boldsymbol{S}_n \boldsymbol{V}_n^T + \hat{\boldsymbol{E}}_n, \tag{7.13}$$

where $\hat{\boldsymbol{E}}_n$ denotes the approximation error arising from the use of a system order n, and $\boldsymbol{W}_1$, $\boldsymbol{W}_2$ are two nonsingular weighting matrices, whose composition varies according to the specific algorithm being employed; see Table 7.1. Furthermore, $\boldsymbol{W}_2$ should not reduce the rank for $\hat{\boldsymbol{\Theta}}_i \boldsymbol{W}_2$, and must be uncorrelated with $\boldsymbol{A}_f$.

Table 7.1 *Definition of the weight matrices* $\mathbf{W_1}$ *and* $\mathbf{W_2}$ *in equation (7.13). Note that when the CVA approach is used, then the singular values obtained from the SVD coincide with the canonical correlation coefficients between* $\boldsymbol{Z}_f$ *and* $\boldsymbol{V}_p$*.*

Weighting	CVA [147]	MOESP [217]	N4SID [215]
$\boldsymbol{W}_1$	$(\boldsymbol{Z}_f \boldsymbol{\Pi}_{\boldsymbol{U}_f}^{\perp} \boldsymbol{Z}_f^T)^{-\frac{1}{2}}$	$\boldsymbol{I}$	$\boldsymbol{I}$
$\boldsymbol{W}_2$	$(\boldsymbol{V}_p \boldsymbol{\Pi}_{\boldsymbol{U}_f}^{\perp} \boldsymbol{V}_p^T)^{\frac{1}{2}}$	$(\boldsymbol{V}_p \boldsymbol{\Pi}_{\boldsymbol{U}_f}^{\perp} \boldsymbol{V}_p^T)^{\frac{1}{2}}$	$(\boldsymbol{V}_p \boldsymbol{V}_p^T)^{\frac{1}{2}}$

7.1.3 *Estimating the system matrices*

Once the matrices $\boldsymbol{\Theta}_i$ and $\boldsymbol{T}_{d,i}$ have been estimated by solving problem (7.11), the matrices in (2.19)–(2.20) can be computed via two different approaches. The first one uses the extended observability matrix, while the second uses the state sequence.

Let us first focus on the methods based on the extended observability matrix. From (7.13), we can estimate the extended observability matrix, defined in (7.6), as $\hat{\boldsymbol{O}}_i = \boldsymbol{W}_1^{-1} \boldsymbol{U}_n \boldsymbol{S}_n^{1/2}$, where we have used the definition $\boldsymbol{\Theta}_i := \boldsymbol{O}_i [\boldsymbol{L}_z \quad \boldsymbol{L}_u]$ presented in equation (7.10), as well as the result in (7.13). Note that if a factorization differing from (7.13) is used, the resulting estimates will be observationally equivalent.

Following Kung [145], Arun and Kung [12], and Verhaegen [217], we can obtain the matrix $\boldsymbol{\Phi}$ in (2.19)–(2.20) as:

$$\hat{\boldsymbol{\Phi}} = \underline{\hat{\boldsymbol{O}}_i}^{\dagger} \overline{\hat{\boldsymbol{O}}_i}, \tag{7.14}$$

where $\underline{\hat{\boldsymbol{O}}_i}$ and $\overline{\hat{\boldsymbol{O}}_i}$ denote the result of removing the last and first m rows, respectively, from $\hat{\boldsymbol{O}}_i$, and where $\boldsymbol{A}^{\dagger}$ denotes the Moore–Penrose pseudo-inverse of matrix $\boldsymbol{A}$; see Appendix A.4, Property 2. The estimate of $\boldsymbol{H}$ is given by the first m rows of $\hat{\boldsymbol{O}}_i$.

On the other hand, we can compute the matrices $\boldsymbol{\Gamma}$ and $\boldsymbol{D}$ using the matrix $\hat{\boldsymbol{T}}_{d,i}$ resulting from the solution of problem (7.11). In fact, from equation (7.10), we have:

$$\boldsymbol{O}_i^{\perp} \boldsymbol{Z}_f \boldsymbol{U}_f^{\dagger} = \boldsymbol{O}_i^{\perp} \boldsymbol{T}_{d,i}, \tag{7.15}$$

where $\boldsymbol{O}_i^{\perp} = \boldsymbol{I} - \boldsymbol{O}_i \boldsymbol{O}_i^{\dagger}$ is a reduced rank matrix that satisfies $\boldsymbol{O}_i^{\perp} \boldsymbol{O}_i = \boldsymbol{0}$, and assumption A3 assures that $\boldsymbol{T}_{s,i} \boldsymbol{A}_f \boldsymbol{U}_f^{\dagger} = \boldsymbol{0}$.

It is straightforward to prove that $Z_f U_f^\dagger$ coincides with $\hat{T}_{d,i}$. Then, taking into account the structure of $T_{d,i}$ in (7.7), the procedure for obtaining $\hat{D}$ and $\hat{\Gamma}$ consists in solving a linear system where Φ, H, and $O_i^\perp$ are replaced by their corresponding estimates[3]. The problem can also be written as:

$$\arg\min_{\{\Gamma, D\}} \left\| \hat{O}_i^\perp \left(\hat{T}_{d,i} - T_{d,i}(\hat{\Phi}, \Gamma, \hat{H}, D) \right) \right\|_F^2. \tag{7.16}$$

Finally, the matrix K can be determined from an estimate of the extended controllability matrix, L_i, introduced in (7.10). From the SVD (7.13) we get:

$$\hat{L}_i = S_n^{1/2} V_n^T W_2^{-1}. \tag{7.17}$$

Similar to the estimation of H, we can get $\hat{K}$ from $[\hat{L}_z \quad \hat{L}_u] = \hat{L}_i$ and its corresponding structure defined in (7.3). Therefore, all the system matrices can be estimated using the observability matrix approach.

We will now focus on the methods based on the estimation of the state sequence such as, e.g., those presented in Larimore [147] or Van Overschee and De Moor [215]. These algorithms estimate all model parameters by solving a simple set of LS problems.

From the estimate of the controllability matrix in (7.17), we obtain the state sequence as:

$$\hat{X}_f \simeq \hat{L}_i V_p, \tag{7.18}$$

and similarly, $\hat{X}_{f^+} \simeq \hat{L}_i V_{p^+}$, where V_{p^+} is like V_p, defined in (7.10), but adding a '+1' to all the subscripts[4]. Once the state sequence is approximated, the system matrix estimates $(\hat{\Phi}, \hat{\Gamma}, \hat{H}, \hat{D}, \hat{K})$ are obtained by LS as:

$$[\hat{H} \quad \hat{D}] = Z_{f_1} \begin{bmatrix} \hat{X}_f^T & U_{f_1}^T \end{bmatrix} \left[\begin{bmatrix} \hat{X}_f \\ U_{f_1} \end{bmatrix} \begin{bmatrix} \hat{X}_f^T & U_{f_1}^T \end{bmatrix} \right]^{-1}, \tag{7.19}$$

where Z_{f_1} and U_{f_1} denote the first row of Z_f and U_f, respectively, in (7.5). From these estimates we then compute the residuals $\hat{A}_{f_1} = Z_{f_1} - \hat{H}\hat{X}_f - \hat{D}U_{f_1}$, finally arriving at the expression:

$$[\hat{\Phi} \quad \hat{\Gamma} \quad \hat{K}] = \hat{X}_{f^+} \begin{bmatrix} \hat{X}_f^T & U_{f_1}^T & \hat{A}_{f_1}^T \end{bmatrix} \left[\begin{bmatrix} \hat{X}_f \\ U_{f_1} \\ \hat{A}_{f_1} \end{bmatrix} \begin{bmatrix} \hat{X}_f^T & U_{f_1}^T & \hat{A}_{f_1}^T \end{bmatrix} \right]^{-1}. \tag{7.20}$$

Equations (7.19)–(7.20) therefore demonstrate how the parameter matrices in the SEM can also be estimated from the state sequence approach.

---

[3]We follow the convention of topping a matrix with a hat, $\hat{A}$, to denote an estimate of that matrix, A.
[4]The $\simeq$ sign recalls that the approximation $L_x^i \approx 0$ for large i is used in the two expressions.

7.2 System order estimation

In Section 7.1 we assumed that the system order, the minimum number of dynamic components required to represent the data dynamics, was known. However, n is unknown in almost all empirical analysis, and specifying its value is critical in applied data modeling. Accordingly, there is an extensive literature about this subject, which can be divided into two broad categories: (i) preliminary data analysis methods, and (ii) procedures based on final estimates.

The first group includes very different proposals. Some of them check whether the model dimension is too high using the condition number of the information matrix; see Stoica and Soderstrom [203]. Other authors, such as Cooper and Wood [57], Tsay [212] or Tiao and Tsay [209], estimate the system order using canonical correlations. In a recent paper, Toscano and Reisen [211] revise and compare these methods. Aoki [11] applies the same technique to state space models. Closely related to this approach, Peternell [180] and Bauer [16] present some subspace-based criteria to estimate the system order, while Sorelius [197] test the rank of a particular sample covariance matrix; see also Woodside [228] with the same aim. In a different context, Bujosa, Garcia–Ferrer and Young [33] apply spectral techniques to obtain valuable information about resonance peaks and phase shifts that aids in identifying the system order required to represent the dynamics.

On the other hand, the methods based on final estimates include the procedures that reduce the system order by Wald-type tests, and those that typically rely on information criteria. This last methodology consists of comparing several alternative models by balancing their "goodness-of-fit" with a penalty depending on the sample size and a measure of model complexity. The pioneering criterion in this vein is due to Akaike [2], whose seminal work motivated other proposals such as those of Schwarz [191] or Hannan and Quinn [117] .

Finally, it is worth noting that determining the system order in an SS model is intrinsically more parsimonious than, for instance, the popular VAR approach, which represents a fair alternative to the subspace methods in terms of balancing model simplicity with computational cost. To clarify this remark, consider a m-vector of time series and the decision to increase the order of a VAR(p) model to a VAR($p+1$). In a LS VAR framework, this would imply increasing the dimension of the state vector by adding m new states, while in a SEM representation one could use the above methods to test all the possible choices in the sequence $1, 2, \ldots, m$ of additional states, without imposing any parameter restrictions.

For the sake of simplicity, in the remaining Section 7.2 we will assume the SEM (2.19)–(2.20) without deterministic inputs $\boldsymbol{u}_t$. In this case, the model dynamics is determined exclusively from the relation between $\boldsymbol{z}_t$ with its own past.

7.2.1 *Preliminary data analysis methods*

In this section we present three procedures specifically adapted or designed for subspace methods.

1. First, one may determine the system order by testing the null hypothesis that $\sigma_{n+1} = \sigma_{n+2} = \ldots = \sigma_{im} = 0$, where σ_{n+1} denotes the $(n+1)$-th *canonical correlation* between Z_p and Z_f; see Appendix A.5 for additional details on canonical correlations. These two matrices were introduced in the previous section and defined in (7.5). To do so, we suggest the statistic

$$D(n) = -(T-2i+1) \sum_{k=n+1}^{im} \log\left(1 - \frac{\hat{\sigma}_k^2}{\hat{d}_k}\right), \tag{7.21}$$

with $\hat{d}_k = 1 + 2\sum_{l=1}^{i} \hat{\rho}_{pk}(l)\hat{\rho}_{fk}(l)$, and where $\hat{\rho}_{fk}(l)$ and $\hat{\rho}_{pk}(l)$ are the sample l-order autocorrelations of the canonical variables. The term $\hat{d}_k$ is related to the asymptotic variance of the estimated canonical correlation coefficients. Under the null, $D(n)$ follows a χ^2 distribution with $2(im-n)$ degrees of freedom; see Tsay [212] for further details.

2. Second, Bauer [16] proposes the *SVC* (Singular Values Criterion) which is based on the analysis of the largest neglected singular value in the SVD step; see (7.13). This criterion is defined as:

$$SVC(n) = \hat{\sigma}_{n+1}^2 + C(T)d(n), \tag{7.22}$$

where $\hat{\sigma}_{n+1}$ is the $(n+1)$-th singular value of $\boldsymbol{W}_1\hat{\boldsymbol{\Theta}}_i\boldsymbol{W}_2$, and $d(n) = 2nm$ denotes the number of parameters in the original SEM model. Note that we also denote the singular values by σ_j, for all $j = 1, 2, \ldots, im$, since, when CVA methods are used, the weighting matrices $\boldsymbol{W}_1$ and $\boldsymbol{W}_2$ are such that the singular values of $\boldsymbol{W}_1\hat{\boldsymbol{\Theta}}_i\boldsymbol{W}_2$ coincide with the canonical correlations between Z_p and Z_f; see Table 7.1. In this case, the method closely approximates that of Tsay [212]. The estimated order is obtained as the argument that minimizes (7.22). Bauer [16] shows in a simulation exercise that *SVC* with $C(T) = \log T/T$ outperforms several information criteria when the sample is relatively large, but recognizes that: (i) the choice of $\boldsymbol{W}_1$ has a large impact on the performance of *SVC* in finite samples, and (ii) the penalty function is somewhat arbitrary.

3. Third, to overcome the latter drawbacks, Garcia–Hiernaux, Jerez, and Casals [106] propose two refinements. They first substitute $\boldsymbol{W}_1 = (Z_f Z_f^T)^{-\frac{1}{2}}$ by the matrix $\tilde{\boldsymbol{W}}_1 = (\tilde{Z}_f \tilde{Z}_f')^{-\frac{1}{2}}$, where $\tilde{Z}_f$ contains the prediction errors of the output future block, e.g., the residuals of previously regressing Z_f onto Z_p. $\boldsymbol{W}_1$ is chosen for its asymptotic properties, not its finite sample behavior. $\tilde{\boldsymbol{W}}_1$ is suggested for two reasons: (i) a lessening in the effects of heteroscedasticity (a consequence of the fact that uncertainty grows when the prediction horizon increases) as well as a reduction in the correlation among prediction errors, and (ii) the estimation results obtained with $\tilde{\boldsymbol{W}}_1$ and $\boldsymbol{W}_1$ are asymptotically equivalent[5], so that the asymptotic properties of *SVC* hold.

[5]The parameters estimated with both matrices have the same asymptotic distributions; see e.g. Bauer and Ljung [17].

The second procedure consists of using a refined penalty function, denoted by $H(T,i)$, which depends not only on the sample size, but also on the number of block rows in the output block Hankel matrix. From a geometric point of view, σ_j can be interpreted as the cosine of the angles between subspaces. In this way, σ_1 corresponds to the minimum angle between all pairs of vectors that belong to each subspace. Obviously, increasing the dimension of the subspace reduces the angle. The assessment of the canonical correlations should necessarily take this fact into account; a basic argument in favor of the definition of this penalty function so as to improve its power in finite samples. Intuitively, a larger subspace dimension implies, *caeteris paribus*, higher singular values, since adding new explanatory variables increases the coefficient of determination in a regression model. With this in mind, the "n identification criterion" (*NIDC*), is defined as:

$$NIDC(n) = \tilde{\sigma}_{n+1}^2 + H(T,i)d(n), \tag{7.23}$$

where $\tilde{\sigma}_{n+1}^2$ is calculated with the weighting matrix $\tilde{\boldsymbol{W}}_1 = (\tilde{\boldsymbol{Z}}_f \tilde{\boldsymbol{Z}}_f')^{-\frac{1}{2}}$, and $d(n) = 2nm$ denotes the number of parameters, for $n = 0, 1, ..., i-1$. Garcia–Hiernaux, Jerez, and Casals [106] show that the penalty function optimized through Monte Carlo simulations, $H(T,i) = e^{-2}T^{-9/10}i^{3/2}$, ensures strong consistency of the system order estimated by minimizing $NIDC(n)$.

7.2.2 Model comparison methods

Other useful tools to determine the system order are the so-called information criteria, such as *AIC*, see Akaike [2], *SBC* (also known as *BIC*), see Schwarz [191], or *HQ*, see Hannan and Quinn [117], among others. These criteria are encompassed by the expression:

$$IC(n) = \frac{-2\log l(n)}{T} + C(T)d(n), \tag{7.24}$$

where, $l(n)$ and $d(n)$ denote, respectively, the maximized log-likelihood value, and the number of parameters corresponding to a n-order system. $C(T)$ is again a penalty function that depends on the criterion; for instance, in *AIC*, $C(T) = 2/T$, in *SBC*, $C(T) = \log T/T$ and in *HQ*, $C(T) = 2\log\log T/T$. These procedures are well known and widely employed; for recent references see for example Stoica, Yan, and Jian [202] or Bengtsson and Cavanaugh [20]. Their computational cost could be relevant for high-dimensional outputs, where the extra cost of one gradient step might be large (although it is considered acceptable in many cases).

Using the procedures introduced in this section, Garcia–Hiernaux, Jerez, and Casals [106] define an automatic model-building criterion based on the mode of the system order chosen by the previous *a priori* and *a posteriori* methods. A simulation analysis confirms that the mode-based criterion is more robust than any specific procedure alone. Its performance from a forecasting perspective is studied in Garcia–Hiernaux, Jerez, and Casals[103].

7.3 Constrained estimation

As the pros and cons of LS and ML are complementary, many works focus on devising methods that provide the best of both worlds. In particular, Spliid [198], Koreisha and Pukkila [143, 144], Flores and Serrano [97] and Dufour and Pelletier [81] extend LS and GLS to VARMAX modeling. Also Lutkepohl and Poskitt [159] propose an LS-based method to specify and estimate a canonical VARMA structure and Francq and Zakoian [99] propose weak GARCH representations which can be estimated by two-stage LS. All these methods share two substantial limitations. First, when there are exclusion constraints they require iterating, which partially offsets their computational advantage. Second, they are typically devised for specific parameterizations, such as VARMAX, and their implementation in other frameworks is not trivial in general.

In this section we present a fast and stable algorithm to estimate time series models written in their equivalent SS form. This procedure is useful, either to obtain adequate initial conditions for iterative ML, or to provide final estimates when ML is considered inadequate or too expensive. On the other hand, its SS foundation provides a wide scope, as it can be applied to any linear fixed-coefficients model with an equivalent SS form; see Chapter 9 and references therein for further details.

The algorithm is based on the subspace methods introduced in Section 7.1. Any exclusion or fixed-value constraint is treated in the same way. The procedure, termed SUBML from now on, provides a Gaussian ML solution to the subspace regressions. Iteration is required as there are constraints relating the parameters in the standard and SS representation of the model. Monte Carlo experiments show that SUBML is numerically more stable, computationally faster, and provides estimates close to those of ML in finite samples; see Garcia–Hiernaux, Jerez, and Casals [104].

The basic problem consists in estimating $\boldsymbol{\beta}$, which is the vector of parameters in the original model (2.19)–(2.20). The system matrices $\boldsymbol{\Phi}, \boldsymbol{\Gamma}, \boldsymbol{K}, \boldsymbol{H}, \boldsymbol{D}$ and $\boldsymbol{Q}$ will be, in general, nonlinear functions of $\boldsymbol{\beta}$ to be estimated by iterating on $\boldsymbol{\beta}$. To this end, we will first discuss how to compute the state sequences defined in the previous section.

7.3.1 State sequence structure

Our starting point is the regression model (7.4). This expression implies that the expected value of future outputs conditional on $\boldsymbol{U}$ (which contains the past and future information of the observed inputs) and on $\boldsymbol{Z}_p$ (which contains the past values of the output), is:

$$\mathrm{E}(\boldsymbol{Z}_f|\boldsymbol{U},\boldsymbol{Z}_p) = \boldsymbol{O}_i\mathrm{E}(\boldsymbol{X}_f|\boldsymbol{U},\boldsymbol{Z}_p) + \boldsymbol{T}_{d,i}\mathrm{E}(\boldsymbol{U}_f|\boldsymbol{U},\boldsymbol{Z}_p) + \boldsymbol{T}_{s,i}\mathrm{E}(\boldsymbol{A}_f|\boldsymbol{U},\boldsymbol{Z}_p). \quad (7.25)$$

Note that $\boldsymbol{U}_f$ is included in $\boldsymbol{U}$. On the other hand, assumption A3 guarantees that the innovations are uncorrelated with the inputs and, by construction, $\mathrm{E}(\boldsymbol{z}_{t-j}\boldsymbol{a}_t^T) = \boldsymbol{0}$, for all $j > 0$. Thus, (7.25) simplifies to:

$$\mathrm{E}(\boldsymbol{Z}_f|\boldsymbol{U},\boldsymbol{Z}_p) = \boldsymbol{O}_i\mathrm{E}(\boldsymbol{X}_f|\boldsymbol{U},\boldsymbol{Z}_p) + \boldsymbol{T}_{d,i}\boldsymbol{U}_f. \quad (7.26)$$

By rearranging the terms in this equation, we obtain:

$$\mathrm{E}(\boldsymbol{X}_f|\boldsymbol{U},\boldsymbol{Z}_p) = \boldsymbol{O}_i^{\dagger}\left[\mathrm{E}(\boldsymbol{Z}_f|\boldsymbol{U},\boldsymbol{Z}_p) - \boldsymbol{T}_{d,i}\boldsymbol{U}_f\right]. \tag{7.27}$$

As $\boldsymbol{Z}_p$ and $\boldsymbol{U}$ are observable, from (7.27) we can estimate the states using orthogonal projections:

$$\hat{\boldsymbol{X}}_f = \boldsymbol{O}_i^{\dagger}(\boldsymbol{Z}_f\boldsymbol{\Pi}_{U,Z_p} - \boldsymbol{T}_{d,i}\boldsymbol{U}_f), \tag{7.28}$$

where $\boldsymbol{\Pi}_{U,Z_p}$ is the orthogonal projection of the stacked matrix $[\boldsymbol{Z}_p^T \quad \boldsymbol{U}^T]^T$, and $\hat{\boldsymbol{X}}_f$ is an estimate of the state sequence conditional on the past of the output ($\boldsymbol{Z}_p$) as well as the entire input ($\boldsymbol{U}$). Parameters are included in matrices $\boldsymbol{O}_i^{\dagger}$ and $\boldsymbol{T}_{d,i}$, which are nonlinear functions of $\boldsymbol{\beta}$.

7.3.2 *Subspace-based likelihood*

Equation (7.4) implies that the error term can be written as:

$$\boldsymbol{T}_{s,i}\boldsymbol{A}_f = \boldsymbol{Z}_f - \boldsymbol{O}_i\boldsymbol{X}_f - \boldsymbol{T}_{i,d}\boldsymbol{U}_f, \tag{7.29}$$

where the states, $\boldsymbol{X}_f$, will be replaced with the expression given by equation (7.28).

The error term $\boldsymbol{T}_{s,i}\boldsymbol{A}_f$ has a particular structure, with one-step-ahead errors in the first row, two-step-ahead errors in the second row, and so on. On this basis, the Gaussian log-likelihood can then be written as,

$$\log l(\boldsymbol{\beta}) = -\frac{im}{2}\log(2\pi) - \frac{i}{2}\log(|\boldsymbol{\Sigma}|) - \frac{1}{2}\mathrm{tr}(\boldsymbol{A}_f^T\boldsymbol{T}_{s,i}^T\boldsymbol{\Sigma}^{-1}\boldsymbol{T}_{s,i}\boldsymbol{A}_f), \tag{7.30}$$

where $\mathrm{tr}(\cdot)$ is the trace operator and $\boldsymbol{\Sigma}$ is the covariance matrix of the prediction errors. The noise term is obviously autocorrelated, and thus $\boldsymbol{\Sigma}$ has the following structure:

$$\boldsymbol{\Sigma} = \boldsymbol{O}_i\boldsymbol{P}_i^*\boldsymbol{O}_i^T + \boldsymbol{T}_{s,i}(\boldsymbol{I}_i \otimes \boldsymbol{B}_i^*)\boldsymbol{T}_{s,i}^T, \tag{7.31}$$

where $\boldsymbol{P}_i^*$ is the covariance matrix of the estimation error of the Kalman states and $\boldsymbol{B}_i^*$ is the covariance of the Kalman innovations. Note that the first addend on the right-hand side of (7.31) refers to the error covariance of the states, while the second addend corresponds to the future output error variance, conditional on the estimated states. On account of the structure of $\boldsymbol{T}_{s,i}\boldsymbol{A}_f$, its covariance matrix is:

$$\boldsymbol{\Sigma} := \begin{cases} \Sigma_{jk} \equiv \mathrm{PE}_j \text{ covariance matrix}, & \text{when } j = k \\ \Sigma_{jk} \equiv \mathrm{cov}(\mathrm{PE}_j, \mathrm{PE}_k), & \text{when } j \neq k \end{cases} \quad j,k = 1,2,\ldots,i, \tag{7.32}$$

where PE_j denotes the j-step-ahead prediction errors. Computing expression (7.31) requires propagating the Kalman filter covariance equations i times:

$$\begin{aligned} \hat{\boldsymbol{K}}_i^* &= (\hat{\boldsymbol{\Phi}}\hat{\boldsymbol{P}}_i^*\hat{\boldsymbol{H}}^T + \hat{\boldsymbol{E}}\hat{\boldsymbol{S}}\hat{\boldsymbol{C}}^T)\hat{\boldsymbol{B}}_i^{*-1}, \\ \hat{\boldsymbol{P}}_{i+1}^* &= \hat{\boldsymbol{\Phi}}\hat{\boldsymbol{P}}_i^*\hat{\boldsymbol{\Phi}}^T + \hat{\boldsymbol{E}}\hat{\boldsymbol{Q}}\hat{\boldsymbol{E}}^T - \hat{\boldsymbol{K}}_i^*\hat{\boldsymbol{B}}_i^*\hat{\boldsymbol{K}}_i^{*T}, \\ \hat{\boldsymbol{B}}_i^* &= \hat{\boldsymbol{H}}\hat{\boldsymbol{P}}_i^*\hat{\boldsymbol{H}}^T + \hat{\boldsymbol{C}}\hat{\boldsymbol{R}}\hat{\boldsymbol{C}}^T, \end{aligned} \tag{7.33}$$

in order to obtain an estimate of $\boldsymbol{P}_i^*$ and $\boldsymbol{B}_i^*$. Note that $\hat{\boldsymbol{K}}_i^*$ is the Kalman filter gain introduced in Chapter 4.

This subspace-based maximum likelihood (SUBML) procedure requires the SS model to be in the SEM form (2.19)–(2.20). However, it is easy to extend it to the MEM (2.1)–(2.2) by just transforming the MEM structure in each iteration to the equivalent SEM form (using the method described in Subsection 2.2.3). Note that the likelihood value corresponding to two equivalent models is the same, and thus in practice it is indifferent whether the log-likelihood value is computed on the basis of the original MEM form or its equivalent SEM.

7.4 Multiplicative seasonal models

In models for seasonal data, the state vector typically has to undergo a substantial increase in size in order to capture the seasonal dynamic structure. This multiplies the computational cost of most estimation algorithms applied to these models. Additionally, seasonality also increases the dimension of the Block Hankel matrices (7.5) where data is organized, which is a second source of higher computational load.

We can cope with these drawbacks by decomposing a seasonal dynamic model (without deterministic inputs) into the following SEM subsystems:

$$\begin{aligned} \boldsymbol{x}_{s,t+s} &= \boldsymbol{\Phi}_s \boldsymbol{x}_{s,t} + \boldsymbol{E}_s \boldsymbol{r}_t \\ \boldsymbol{z}_t &= \boldsymbol{H}_s \boldsymbol{x}_{s,t} + \boldsymbol{r}_t, \end{aligned} \tag{7.34}$$

and,

$$\begin{aligned} \boldsymbol{x}_{t+1} &= \boldsymbol{\Phi} \boldsymbol{x}_t + \boldsymbol{E} \boldsymbol{\varepsilon}_t \\ \boldsymbol{r}_t &= \boldsymbol{H} \boldsymbol{x}_t + \boldsymbol{\varepsilon}_t. \end{aligned} \tag{7.35}$$

which can be combined into a single "cascade" formulation:

$$\begin{aligned} \begin{bmatrix} \boldsymbol{x}_{s,t+s} \\ \boldsymbol{x}_{t+1} \end{bmatrix} &= \begin{bmatrix} \boldsymbol{\Phi}_s & \boldsymbol{E}_s \boldsymbol{H} \\ \boldsymbol{0} & \boldsymbol{\Phi} \end{bmatrix} \begin{bmatrix} \boldsymbol{x}_{s,t} \\ \boldsymbol{x}_t \end{bmatrix} + \begin{bmatrix} \boldsymbol{E}_s \\ \boldsymbol{E} \end{bmatrix} \boldsymbol{\varepsilon}_t \\ \boldsymbol{z}_t &= \begin{bmatrix} \boldsymbol{H}_s & \boldsymbol{H} \end{bmatrix} \begin{bmatrix} \boldsymbol{x}_{s,t} \\ \boldsymbol{x}_t \end{bmatrix} + \boldsymbol{\varepsilon}_t, \end{aligned} \tag{7.36}$$

where the state vector propagates simultaneously in two frequencies: 1 and s. From subsystem (7.35), it can be shown that $\mathrm{E}(\boldsymbol{r}_{t+s}\boldsymbol{r}_t^T) = \boldsymbol{H}\boldsymbol{\Phi}^{s-1}\mathrm{E}(\boldsymbol{x}_{t+1}\boldsymbol{r}_t^T)$. Then, when s is much larger than the order of the subsystem (7.35), we have $\mathrm{E}(\boldsymbol{r}_{t+s}\boldsymbol{r}_t^T) \simeq 0$, for all $i = s, 2s, 3s, ...$, a result which holds with equality when the regular structure corresponds to a moving average of order q, and $q < s$. This property allows one to consider the input $\boldsymbol{r}_t$ in the subsystem (7.34) as a white noise process in the frequency s and, therefore, uncorrelated for all t with the state vector $\boldsymbol{x}_{s,t}$.

To estimate the subsystem (7.34) using the subspace methods presented in Section 7.1, we define the s-frequency Block Hankel matrices:

$$Z_p^s := \begin{bmatrix} z_1 & z_2 & \cdots & z_{T-s(p+f)} \\ z_{s+1} & z_{s+2} & \cdots & z_{T-s(p+f-1)} \\ \vdots & \vdots & & \vdots \\ z_{ps+1} & z_{ps+2} & \cdots & z_{T-sf} \end{bmatrix}, \tag{7.37}$$

$$Z_f^s := \begin{bmatrix} z_{s(p+1)+1} & z_{s(p+1)+2} & \cdots & z_{T-s(f-1)} \\ z_{s(p+2)+1} & z_{s(p+2)+2} & \cdots & z_{T-s(f-2)} \\ \vdots & \vdots & & \vdots \\ z_{s(p+f)+1} & z_{s(p+f)+2} & \cdots & z_T \end{bmatrix}, \tag{7.38}$$

analogous to the *regular* Block Hankel matrices introduced in (7.5).

Therefore, when working with seasonal data, the estimation procedure would be:

1. Estimate the seasonal subsystem (7.34) using the above seasonal Block Hankel matrices, (7.37)–(7.38).
2. Compute the residuals, $\hat{r}_t$, from the previous subsystem.
3. Estimate the matrices in the regular subsystem (7.35), by means of $\hat{r}_t$ organized in the Block Hankel matrices (7.5).

Note that the same strategy can be applied to estimate the system matrices using the subspace methods presented in Section 7.1, or the SUBML algorithm described in Section 7.3.

7.5 Examples

In this section we present some examples using the algorithms previously described. The first block illustrates the performance of these methods with univariate models. The second block is devoted to the identification of a multivariate model.

7.5.1 Univariate models

We first detail the steps to identify a univariate seasonal model using the well-known airline passengers data. Later, we will show the results obtained from applying the same methodology to nine time series commonly found in the literature.

The airline passengers time series was used by Box, Jenkins and Reinsel [30] to illustrate how to model seasonal data. These and other authors specify and estimate a double IMA structure. Figure 7.1 (top) depicts this data.

We start from the Box–Cox [28] transformed (logarithms of the) original series, and apply one regular and one seasonal difference as is usually done in the literature, $z_t = \nabla\nabla_{12}\log(y_t)$, where $\nabla_d = (1 - \mathrm{B}^d)$. The identification (Box–Jenkins sense) of the number of regular and seasonal units roots is beyond the scope of this book. However, the reader can refer to Garcia–Hiernaux [102] and Garcia–Hiernaux, Jerez, and Casals [105] to see how to cope with this problem from a subspace point of view. Table 7.2 displays the results of applying E^4 function `nid`, see appendices C and D, to z_t in order to identify the system order via the procedures presented in Section 7.2.

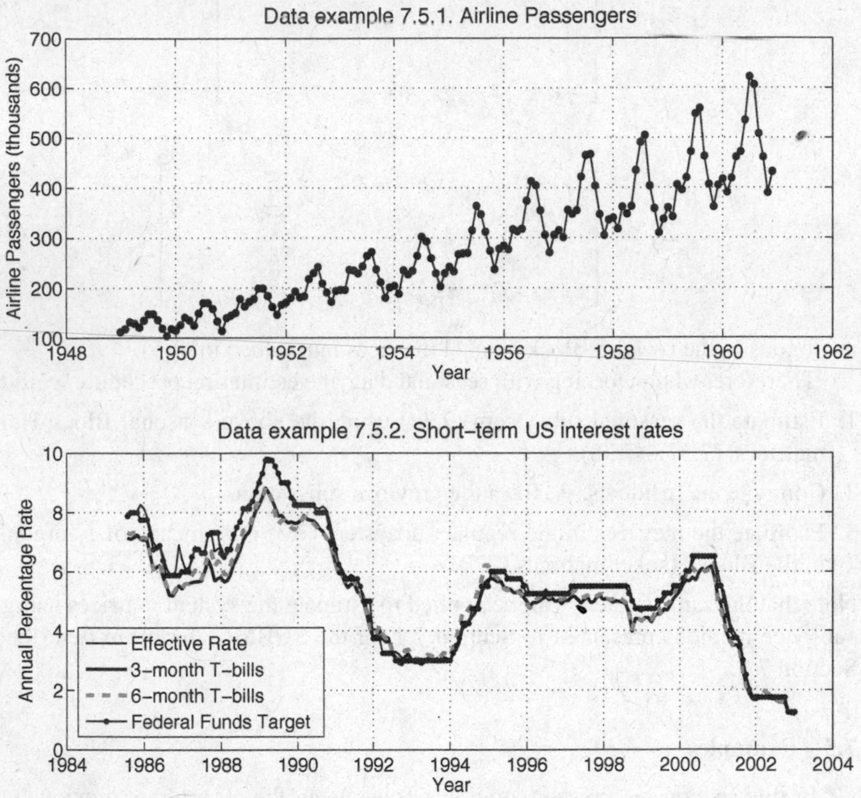

Figure 7.1 *Data in examples 7.5.1 and 7.5.2. Top: Airline passengers time series from Box, Jenkins and Reinsel [30]. Bottom: US short-term interest rates.*

All the procedures find a regular system order equal to one, i.e., $\hat{n}_r = 1$. To estimate the system order of the seasonal subsystem we also use `mid`, but with seasonal block Hankel matrices where the monthly frequency is set to $s = 12$. The output is shown in Table 7.3. All the procedures again find the same system order for the seasonal subsystem, i.e., $\hat{n}_s = 1$.

With these system orders, we then apply the subspace methods in cascade, see (7.34)–(7.35), obtaining the following balanced model:

$$(1 + \underset{(.21)}{.05}\,\mathrm{B})(1 - \underset{(.17)}{.08}\,\mathrm{B}^{12})z_t = (1 - \underset{(.20)}{.35}\,\mathrm{B})(1 - \underset{(.16)}{.58}\,\mathrm{B}^{12})\hat{a}_{1t}, \qquad \hat{\sigma}_{a_1} = 3.68\%, \quad (7.39)$$

where standard errors for the parameters are given in parentheses below the corresponding estimates, and $\hat{\sigma}_{a_1}$ denotes the standard deviation of the residuals.

Table 7.2 *Regular system order estimation for z_t.*

n	AIC	SBC	HQ	SVC	NIDC	P-value χ^2
	Results for different methods and values of n					
0	−3.319	−3.297	−3.310	0.196	0.244	–
1	−3.459	−3.393	−3.432	0.152	0.128	0.000
2	−3.433	−3.323	−3.388	0.208	0.150	0.066
3	−3.405	−3.252	−3.343	0.237	0.144	0.205
	Chosen regular system order					
$\hat{n}_r$	1	1	1	1	1	1

Table 7.3 *Seasonal system order estimation for z_t with seasonal structure ($s = 12$) Block Hankel matrices.*

n	AIC	SBC	HQ	SVC	NIDC	P-value χ^2
	Results for different methods and values of n					
0	−3.319	−3.300	−3.310	0.280	0.389	–
1	−3.566	−3.500	−3.540	0.108	0.079	0.013
2	−3.543	−3.433	−3.498	0.168	0.107	0.341
3	−3.513	−3.359	−3.450	0.235	0.143	0.341
	Chosen seasonal system order					
$\hat{n}_s$	1	1	1	1	1	1

If the user wishes to prune the model of non-significant parameters and obtain, with a low computational cost, a more precise estimation (or good initial conditions for an ML estimation), it may be useful to employ the SUBML algorithm. Doing so provides the following results:

$$z_t = (1 - .39\text{B})(1 - .52\text{B}^{12})\hat{a}_{2t}, \qquad \hat{\sigma}_{a_2} = 3.68\%. \tag{7.40}$$

Regardless of when a decision is made to stop, (7.39), (7.40), or even after a full-blown ML estimation, it is recommended that a diagnostic exercise be carried out. To illustrate, we apply `nid` once again to the residuals from model (7.40), for both the $s = 1$ and $s = 12$ frequencies. Tables 7.4 and 7.5 reveal that there is no additional regular or seasonal dynamical structure, because the system order estimated by all the criteria (except NIDC in Table 7.4) is zero.

The residuals $\hat{a}_{2t}$ do not display any evidence of misspecification, either in the regular or seasonal components of the model. The model identified by the subspace algorithm corresponds to the structure typically found in the literature. ML estimation using the procedures detailed in Chapter 5 yields:

$$z_t = (1 - \underset{(.08)}{.40}\text{B})(1 - \underset{(.08)}{.56}\text{B}^{12})\hat{a}_t, \qquad \hat{\sigma}_a = 3.67\%. \tag{7.41}$$

Table 7.4 *Regular system order estimation for* $\hat{a}_{2t}$.

n	AIC	SBC	HQ	SVC	NIDC	P-value χ^2
	Results for different methods and values of n					
0	−3.766	−3.744	−3.757	0.108	0.120	–
1	−3.757	−3.691	−3.730	0.134	0.107	0.365
2	−3.729	−3.619	−3.684	0.175	0.114	0.234
3	−3.711	−3.558	−3.649	0.229	0.136	0.434
	Chosen regular system order					
$\hat{n}_r$	0	0	0	0	1	0

Table 7.5 *Seasonal system order estimation for* $\hat{a}_{2t}$ *with the seasonal Block Hankel matrix and* $s = 12$.

n	AIC	SBC	HQ	SVC	NIDC	P-value χ^2
	Results for different methods and values of n					
0	−3.766	−3.744	−3.757	0.043	0.044	–
1	−3.736	−3.670	−3.709	0.113	0.083	0.292
2	−3.724	−3.614	−3.679	0.161	0.099	0.660
3	−3.693	−3.539	−3.631	0.233	0.140	0.762
	Chosen seasonal system order					
$\hat{n}_s$	0	0	0	0	0	0

As expected, the ML estimates are very close to those obtained from SUBML. Obviously, in univariate systems the advantage of using SUBML is not as large as in multivariate systems, but it is interesting to see how the mechanical application of these methods simplifies the identification process (specification and estimation), and, in particular, how they allow one to perform the univariate analysis of massive amounts of time series with a relatively small computational and human cost.

The code required to replicate this example can be found in the file `Example0751.m`; see Appendix D.

We now present the results of applying the same analysis to nine univariate time series. Data analyzed is briefly described in Table 7.6, while Tables 7.7, 7.8, and 7.9 compare the results reported by the literature with those obtained by sequentially applying a subspace method (SM), the SUBML algorithm, and ML estimation. The only significant difference seen is in the series H, where the results produced by SUBML are close to those of Wei [221]. However, when estimating the model by ML, the moving average parameter converges to unity. As the sample is very short ($T = 32$), this could be an example of what is called in the literature the "Pile-up Effect"[6]. Regarding this matter, Garcia–Hiernaux, Jerez, and Casals [104] shows that: (i) the algorithm SUBML described in Section 7.3 is less sensitive than ML to the pile-up effect, and (ii) discarding this effect, estimates of the moving average parameters are very similar for SUBML and ML.

Hence, the sequential use of SM, SUBML, and ML offers flexible alternatives for model estimation, as it allows one to attain final results that are consistently similar, depending on the modeling goals and the resources available. For example, even though we may sacrifice using SM in lieu of ML because of computational considerations, we can have some assurance that the results will not be too different.

7.5.2 A multivariate model for the interest rates

This example shows the performance of the previously discussed methods in a multivariate framework. Consider now four short-term interest rates of the US economy. Specifically, we consider the Federal Funds Target (FT_t), the Federal Funds effective rate (ER_t), and the 3-month and 6-month T-bills in the secondary market ($T3m_t$ and $T6m_t$, respectively), in monthly averages of yields for the period August 1985 to January 2003. The four series are depicted in Figure 7.1 (bottom).

We assume ER_t, $T3m_t$, and $T6m_t$ to be I(1) and CI(1,1) cointegrated with FT_t. Although the unit root, cointegrating rank, and cointegrating parameter estimation can be performed using subspace-based methods, see Bauer and Wagner [18], Garcia–Hiernaux [102], Garcia–Hiernaux, Jerez, and Casals [105], this is not the purpose of this book and therefore it is not shown here. We will therefore analyze a stationary multivariate process representing the spread of some short-term interest rates defined as[7]:

$$\boldsymbol{Y}_{1t} = \begin{bmatrix} \nabla & 0 & 0 & 0 \\ --1.01 & 1 & 0 & 0 \\ --0.92 & 0 & 1 & 0 \\ --0.93 & 0 & 0 & 1 \end{bmatrix} \begin{bmatrix} FT_t \\ ER_t \\ T3m_t \\ T6m_t \end{bmatrix}. \tag{7.42}$$

When studying the dynamics of the multivariate process $\boldsymbol{Y}_{1t}$, most of the procedures yield an estimated system order of 4; see Table 7.10. Therefore, we proceed to estimate, via subspace methods, a model for $\boldsymbol{Y}_{1t}$ with $\hat{n} = 4$ (the mode of the chosen system order), which implies an unrestricted VARMA(1,1) structure. In this first estimation we found ten coefficients that are not significantly different from zero. Those coefficients were then removed from the model, which was then re-estimated using the SUBML algorithm. Table 7.11 summarizes the results produced by these two steps. The values of the parameters vary slightly, getting closer (as we will see later) to the ML estimates. This second step yields adequate initial conditions for an ML estimation. Note that the relevance of adequate initial conditions grows with

[6] This effect means that the distribution of ML estimates for moving average coefficients is multimodal, with one of the modes on the invertibility boundary; see Ansley and Newbold [10], Davis and Song [65], and Paige, Trindade, and Wickramasinghe [174].

[7] See Garcia–Hiernaux [102] for details about the estimated values of the cointegration matrix.

Table 7.6 *Datasets used to compare the SM, SUBML and ML methods in Tables 7.7, 7.8, and 7.9.*

Time series ID	# obs.	Reference and description
A	45	[35]: Daily Average Number of Truck Manufacturing Defects.
B*	150	[221]: Simulated Quarterly Time Series
C	35	[221]: Monthly Unemployed Females between 16 and 19 in the USA, 01/1961-12/1985.
D*	200	[227]: Simulated Time Series # 2.
E	226	[30]: Series C - Chemical Process Temperature, readings every minute.
F	197	[30]: Series A - Chemical Process Concentration, readings every two hours.
G	178	[170]: Wisconsin Monthly Employment Time Series, 01/1961-10/1975.
H	32	[221]: Quarterly USA Beer Production, 01/1975-04/1982.
I*	36	[160]: Simulated Series.

* These are simulated series. However, while in Tables 7.7 and 7.8, series B and D present the data generating processes, series I presents the specification and estimation performed by the author.

Table 7.7 *Univariate models. Comparison of models estimated for the data described in Table 7.6 - Part 1. Standard errors are in parenthesis.* $\hat{\sigma}_a^2$ *denotes the residual variance. SM is a subspace methods estimation with weights corresponding to the CVA algorithm (see Table 7.1) and using an extended observability matrix approach (see Section 7.1).*

Series	Obs	Specification and estimation by the literature	SM
A	45	$(1-\underset{(.13)}{.43}\,B)(Z_t-\underset{(.07)}{1.79})=a_t$ $\hat{\sigma}_a^2=.21$	$(1-\underset{(.30)}{.61}\,B)(Z_t-\underset{(.13)}{3.94})=(1-\underset{(.31)}{.24}\,B)a_t$ $\hat{\sigma}_a^2=.21$
B	150	$\nabla\nabla_4 Z_t=(1-.80B)(1-.60B^4)a_t$ $\hat{\sigma}_a^2=1.00$	$(1+\underset{(.13)}{.22}\,B)(1+\underset{(.13)}{.21}\,B^4)\nabla\nabla_4 Z_t=(1-\underset{(.13)}{.56}\,B)(1-\underset{(.12)}{.49}\,B^4)a_t$ $\hat{\sigma}_a^2=.97$
C	35	$\nabla Z_t=(1-\underset{(.05)}{.51}B)a_t$ $\hat{\sigma}_a^2=1397.27$	$(1-\underset{(.11)}{.00}B)\nabla Z_t=(1-\underset{(.11)}{.51}B)a_t$ $\hat{\sigma}_a^2=1380.35$
D	200	$(1-.70B+.49B^2)Z_t=a_t$ $\hat{\sigma}_a^2=.60$	$(1-\underset{(.11)}{.84}\,B+\underset{(.10)}{.51}\,B^2)Z_t=(1-\underset{(.11)}{.16}\,B+\underset{(.11)}{.05}\,B^2)a_t$ $\hat{\sigma}_a^2=.66$
E	226	$(1-\underset{(.04)}{.82}\,B)\nabla Z_t=a_t$ $\hat{\sigma}_a^2=.02$	$(1-\underset{(.06)}{.82}\,B)\nabla Z_t=(1-\underset{(.08)}{.02}\,B)a_t$ $\hat{\sigma}_a^2=.02$
F	197	$(1-\underset{(.04)}{.92}\,B)Z_t=1.45+(1-\underset{(.08)}{.58}\,B)a_t$ $\hat{\sigma}_a^2=.10$	$(1-\underset{(.09)}{.88}\,B)Z_t=\underset{(.02)}{2.08}+(1-\underset{(.10)}{.50}\,B)a_t$ $\hat{\sigma}_a^2=.10$
G	178	$(1-\underset{(.05)}{.79}\,B)\nabla_{12}Z_t=(1-\underset{(.05)}{.76}\,B^{12})a_t$ $\hat{\sigma}_a^2=1.24$	$(1-\underset{(.10)}{.56}\,B)(1-\underset{(.18)}{.05}\,B^{12})\nabla_{12}Z_t=-\underset{(.10)}{2.11}+(1-\underset{(.11)}{.18}\,B)(1-\underset{(.17)}{.35}\,B^{12})a_t$ $\hat{\sigma}_a^2=1.54$
H	32	$\nabla_4 Z_t=\underset{(.09)}{1.49}+(1-\underset{(.16)}{.87}\,B^4)a_t$ $\hat{\sigma}_a^2=2.39$	$(1-\underset{(.26)}{.14}\,B^4)\nabla_4 Z_t=\underset{(.30)}{1.21}+(1-\underset{(.23)}{.71}\,B^4)a_t$ $\hat{\sigma}_a^2=2.03$
I	36	$Z_t=\underset{(4.84)}{51.03}+a_t$ $\hat{\sigma}_a^2=818.81$	$Z_t=\underset{(4.77)}{47.31}+a_t$ $\hat{\sigma}_a^2=790.54$

Table 7.8 *Univariate models. Comparison of models estimated for the data described in Table 7.6 - Part 2. Standard errors are in parenthesis.* $\hat{\sigma}_a^2$ *denotes the residual variance. SUBML values were obtained with the algorithm described in Section 7.3, by pruning the non-significant parameters found in Table 7.7.*

Series	Obs	Specification and estimation by the literature	SM + SUBML
A	45	$(1-\underset{(.13)}{.43}\mathrm{B})(Z_t-\underset{(.07)}{1.79})=a_t$ $\hat{\sigma}_a^2=.21$	$(1-.35\mathrm{B})(Z_t-1.87)=a_t$ $\hat{\sigma}_a^2=.21$
B	150	$\nabla\nabla_4 Z_t=(1-.80\mathrm{B})(1-.60\mathrm{B}^4)a_t$ $\hat{\sigma}_a^2=1.00$	$\nabla\nabla_4 Z_t=(1-.74\mathrm{B})(1-.66\mathrm{B}^4)a_t$ $\hat{\sigma}_a^2=.88$
C	35	$\nabla Z_t=(1-\underset{(.05)}{.51}B)+a_t$ $\hat{\sigma}_a^2=1397.27$	$\nabla Z_t=(1-.50\mathrm{B})+a_t$ $\hat{\sigma}_a^2=1393.54$
D	200	$(1-.70\mathrm{B}+.49\mathrm{B}^2)Z_t=a_t$ $\hat{\sigma}_a^2=.60$	$(1-.71\mathrm{B}+.42\mathrm{B}^2)Z_t=a_t$ $\hat{\sigma}_a^2=.66$
E	226	$(1-\underset{(.04)}{.82}\mathrm{B})\nabla Z_t=a_t$ $\hat{\sigma}_a^2=.02$	$(1-.81\mathrm{B})\nabla Z_t=a_t$ $\hat{\sigma}_a^2=.02$
F	197	$(1-\underset{(.04)}{.92}\mathrm{B})Z_t=1.45+(1-\underset{(.08)}{.58}\mathrm{B})a_t$ $\hat{\sigma}_a^2=.10$	$(1-.88\mathrm{B})Z_t=2.07+(1-.51B)a_t$ $\hat{\sigma}_a^2=.10$
G	178	$(1-\underset{(.05)}{.79}\mathrm{B})\nabla_{12}Z_t=(1-\underset{(.05)}{.76}\mathrm{B}^{12})a_t$ $\hat{\sigma}_a^2=1.24$	$(1-.68\mathrm{B})\nabla_{12}Z_t=.07+(1-.56\mathrm{B}^{12})a_t$ $\hat{\sigma}_a^2=1.37$
H	32	$\nabla_4 Z_t=\underset{(.09)}{1.49}+(1-\underset{(.16)}{.87}\mathrm{B}^4)a_t$ $\hat{\sigma}_a^2=2.39$	$\nabla_4 Z_t=1.36+(1-.59\mathrm{B}^4)a_t$ $\hat{\sigma}_a^2=2.15$
I	36	$Z_t=\underset{(4.84)}{51.03}+a_t$ $\hat{\sigma}_a^2=818.81$	$Z_t=51.03+a_t$ $\hat{\sigma}_a^2=796.68$

Table 7.9 *Univariate models. Comparison of models estimated for the data described in Table 7.6 - Part 3. Standard errors are in parenthesis.* $\hat{\sigma}_a^2$ *denotes the residual variance. ML estimates were initiated with the parameters obtained by SUBML shown in Table 7.8.*

Series	Obs	Specification and estimation by the literature	SM + SUBML + ML
A	45	$(1-\underset{(.13)}{.43}B)(Z_t-\underset{(.07)}{1.79})=a_t$ $\hat{\sigma}_a^2=.21$	$(1-\underset{(.15)}{.42}B)(Z_t-\underset{(.07)}{1.81})=a_t$ $\hat{\sigma}_a^2=.21$
B	150	$\nabla\nabla_4 Z_t=(1-.80B)(1-.60B^4)a_t$ $\hat{\sigma}_a^2=1.00$	$\nabla\nabla_4 Z_t=(1-\underset{(.06)}{.75}B)(1-\underset{(.07)}{.63}B^4)a_t$ $\hat{\sigma}_a^2=.91$
C	35	$\nabla Z_t=(1-\underset{(.05)}{.51}B)a_t$ $\hat{\sigma}_a^2=1397.27$	$\nabla Z_t=(1-\underset{(.05)}{.51}B)a_t$ $\hat{\sigma}_a^2=1393.12$
D	200	$(1-.70B+.49B^2)Z_t=a_t$ $\hat{\sigma}_a^2=.60$	$(1-\underset{(.06)}{.72}B+\underset{(.06)}{.41}B^2)Z_t=a_t$ $\hat{\sigma}_a^2=.66$
E	226	$(1-\underset{(.04)}{.82}B)\nabla Z_t=a_t$ $\hat{\sigma}_a^2=.02$	$(1-\underset{(.04)}{.82}B)\nabla Z_t=a_t$ $\hat{\sigma}_a^2=.02$
F	197	$(1-\underset{(.04)}{.92}B)Z_t=1.45+(1-\underset{(.08)}{.58}B)a_t$ $\hat{\sigma}_a^2=.10$	$(1-\underset{(.04)}{.92}B)Z_t=\underset{(.67)}{1.45}+(1-\underset{(.08)}{.60}B)a_t$ $\hat{\sigma}_a^2=.10$
G	178	$(1-\underset{(.05)}{.79}B)\nabla_{12}Z_t=(1-\underset{(.05)}{.76}B^{12})a_t$ $\hat{\sigma}_a^2=1.24$	$(1-\underset{(.05)}{.77}B)\nabla_{12}Z_t=(1-\underset{(.07)}{.61}B^{12})a_t$ $\hat{\sigma}_a^2=1.31$
H	32	$\nabla_4 Z_t=\underset{(.09)}{1.49}+(1-\underset{(.16)}{.87}B^4)a_t$ $\hat{\sigma}_a^2=2.39$	$\nabla_4 Z_t=\underset{(.13)}{1.55}+\nabla_4 a_t$ $\hat{\sigma}_a^2=1.64$
I	36	$Z_t=\underset{(4.84)}{51.03}+a_t$ $\hat{\sigma}_a^2=818.81$	$Z_t=\underset{(4.84)}{51.03}+a_t$ $\hat{\sigma}_a^2=818.86$

the complexity of the model. In the third step, the calculation of the exact information matrix and the ML estimates allows the user to conduct more accurate tests, and thereby remove some parameters whose lack of significance was not previously detected.

Table 7.10 *System order estimation for* $\mathbf{Y}_{1t}$.

			Results for different methods and values of n			
n	AIC	SBC	HQ	SVC	NIDC	P-value χ^2
2	−4.036	−3.621	−3.868	1.076	2.305	0.000
3	−4.636	−4.093	−4.416	1.117	1.474	0.000
4	−4.877	−4.205	−4.605	1.230	1.310	0.000
5	−4.811	−4.011	−4.488	1.379	1.316	0.304
			Chosen system order			
$\hat{n}$	4	4	4	2	4	4

In the ensuing diagnostic analysis, three new parameters are specified and the final model results in a restricted VARMA (3,1). The sequence of estimations is shown in Table 7.11. The covariance matrix of the residuals is found to be:

$$\hat{\boldsymbol{\Sigma}}_{\hat{\mathbf{a}}_1}(\%) = \begin{bmatrix} \underset{(.33)}{3.33} & - & - & - \\ \underset{(.22)}{-1.59} & \underset{(.23)}{2.30} & - & - \\ \underset{(.23)}{-1.33} & \underset{(.20)}{1.35} & \underset{(.28)}{2.84} & - \\ \underset{(.22)}{-.83} & \underset{(.19)}{0.98} & \underset{(.26)}{2.43} & \underset{(.28)}{2.87} \end{bmatrix},$$

where $\hat{\boldsymbol{a}}_{1t}$ are the residuals obtained from the ML estimates. The Ljung–Box [151] statistics are given by:

$$\boldsymbol{Q}_{(7)}(\%) = \begin{bmatrix} 5.8 & 10.3 & 5.4 & 8.9 \\ 8.6 & 7.7 & 4.3 & 9.9 \\ 10.6 & 17.0 & 11.1 & 12.6 \\ 6.5 & 15.5 & 8.3 & 10.1 \end{bmatrix}.$$

This example illustrates the ability to identify a multivariate process using the procedures presented in this chapter. The initial specification deviates slightly from that of the final model, likely due to the (barely significant) persistence of the second and third order autoregressive components in the $\hat{\boldsymbol{\Phi}}(1,1)$ element. However, note that application of the first two steps with SM and SUBML already yields a reasonable approximation to the final model; these two steps being obtained at low computational and human cost.

The code required to replicate this example can be found in the file `Example0752.m`; see Appendix D.

Table 7.11 *Sequence of identified models for the process* $\mathbf{Y}_{1t}$. *The figures in parentheses are the standard errors of the parameters and the symbol "-" denotes that the corresponding parameter has been constrained to its current value, so it does not have a standard error.*

Method	$\hat{\Phi}(B)$	$\hat{\Theta}(B)$
SM	$\begin{bmatrix} 1-\underset{(.08)}{.49}B & \underset{(.18)}{.19}B & \underset{(.15)}{.45}B & -\underset{(.12)}{.35}B \\ \underset{(.04)}{.20}B & 1-\underset{(.10)}{.29}B & \underset{(.08)}{.18}B & \underset{(.06)}{.11}B \\ \underset{(.06)}{.32}B & \underset{(.14)}{.13}B & 1-\underset{(.11)}{.80}B & -\underset{(.09)}{.10}B \\ \underset{(.08)}{.23}B & \underset{(.18)}{.0}B & -\underset{(.15)}{.02}B & 1-\underset{(.12)}{.94}B \end{bmatrix}$	$\begin{bmatrix} 1+\underset{(.08)}{.04}B & \underset{(.15)}{.36}B & \underset{(.16)}{1.36}B & -\underset{(.13)}{.53}B \\ \underset{(.06)}{.20}B & 1+\underset{(.10)}{.09}B & -\underset{(.10)}{.32}B & \underset{(.09)}{.20}B \\ \underset{(.07)}{.40}B & -\underset{(.12)}{.06}B & 1-\underset{(.12)}{.29}B & -\underset{(.11)}{.06}B \\ \underset{(.09)}{.35}B & -\underset{(.16)}{.30}B & -\underset{(.17)}{.02}B & 1-\underset{(.14)}{.15}B \end{bmatrix}$
SUBML	$\begin{bmatrix} 1-.50B & 0 & .40B & -.30B \\ .12B & 1-.54B & -.03B & 0 \\ .21B & 0 & 1-.85B & 0 \\ .18B & 0 & 0 & 1-.86B \end{bmatrix}$	$\begin{bmatrix} 1 & .20B & 1.10B & -.28B \\ .21B & 1 & -.15B & -.01B \\ .34B & 0 & 1-.18B & 0 \\ .38B & 0 & 0 & 1 \end{bmatrix}$
ML	$\begin{bmatrix} 1-\underset{(.06)}{.46}B-\underset{(.05)}{.04}B^2-\underset{(.05)}{.19}B^3 & \underset{(-)}{0} & \underset{(.06)}{.19}B & -\underset{(.05)}{.10}B \\ \underset{(.05)}{.16}B & 1-\underset{(.08)}{.66}B & \underset{(-)}{0} & \underset{(-)}{0} \\ \underset{(.05)}{.19}B & \underset{(-)}{0} & 1-\underset{(.03)}{.87}B & \underset{(-)}{0} \\ \underset{(.06)}{.23}B & \underset{(-)}{0} & \underset{(-)}{0} & 1-\underset{(.03)}{.89}B \end{bmatrix}$	$\begin{bmatrix} 1 & \underset{(.08)}{.30}B & \underset{(.09)}{.67}B & \underset{(-)}{0} \\ \underset{(.08)}{.28}B & 1-\underset{(.11)}{.23}B & \underset{(-)}{0} & \underset{(-)}{0} \\ \underset{(.07)}{.28}B & \underset{(-)}{0} & 1-\underset{(.06)}{.26}B & \underset{(-)}{0} \\ \underset{(.08)}{.43}B & \underset{(-)}{0} & \underset{(-)}{0} & 1-\underset{(.06)}{.19}B \end{bmatrix}$

Chapter 8

Signal extraction

8.1 Input and error-related components 108
8.1.1 The deterministic and stochastic subsystems 108
8.1.2 Enforcing minimality 109
8.2 Estimation of the deterministic components 110
8.2.1 Estimating the total effect of the inputs 111
8.2.2 Estimating the individual effect of each input 113
8.3 Decomposition of the stochastic component 114
8.3.1 Characterization of the structural components 114
8.3.2 Estimation of the structural components 116
8.4 Structure of the method 116
8.5 Examples 116
8.5.1 Comparing different methods with simulated data 116
8.5.2 Common features in wheat prices 120
8.5.3 The effect of advertising on sales 123

The term "signal extraction" refers to the definition and estimation of meaningful signals that are buried in a vector of time series. In the time series literature, it often refers to the decomposition of a single time series (z_t) into the sum of "structural components" named as "trend" (t_t), "cycle" (c_t), "seasonal" (s_t) and "irregular" (ε_t):

$$z_t = t_t + c_t + s_t + \varepsilon_t. \tag{8.1}$$

In this chapter we describe a signal extraction method for a vector of time series, $\mathbf{z}_t$, which is the output of a SEM (2.19)–(2.20). Building on an SS model as the starting point has a clear advantage in terms of generality, as it includes as particular cases common specifications; see Chapter 2.

On the other hand, the methodology described here has some convenient features. In particular it:

1. ... allows for both, single and multiple time series,

2. ... considers a component not included in (8.1) which is driven by exogenous variables
3. ... is "revision-free," meaning that the estimates for the components do not change when the sample size increases, as happens with most alternative procedures including those of Burman [34], Hillmer and Tiao [125], or Harvey [118]
4. ... does nor require specifying *a priori* models for the components.

8.1 Input and error-related components

Let z_t be an $m \times 1$ random vector including all the series to be decomposed and assume that it is realized by the minimal SEM model:

$$\boldsymbol{x}_{t+1} = \boldsymbol{\Phi}\,\boldsymbol{x}_t + \boldsymbol{\Gamma}\,\boldsymbol{u}_t + \boldsymbol{K}\,\boldsymbol{a}_t \tag{8.2}$$

$$\boldsymbol{z}_t = \boldsymbol{H}\,\boldsymbol{x}_t + \boldsymbol{D}\,\boldsymbol{u}_t + \boldsymbol{a}_t, \tag{8.3}$$

which coincides with the formulation (2.19)–(2.20) that was already defined in Chapter 2. We will further consider that the model (8.2)–(8.3) has been transformed so that the matrix $\boldsymbol{\Phi}$ is block-diagonal, see Subsection 3.1.3 and Appendix A.

8.1.1 *The deterministic and stochastic subsystems*

Following the procedure described in Subsection 3.1.1, we can decompose model (8.2)–(8.3) into the corresponding deterministic and stochastic subsystems:

$$\boldsymbol{x}^d_{t+1} = \boldsymbol{\Phi}\,\boldsymbol{x}^d_t + \boldsymbol{\Gamma}\,\boldsymbol{u}_t \tag{8.4}$$

$$\boldsymbol{z}^d_t = \boldsymbol{H}\,\boldsymbol{x}^d_t + \boldsymbol{D}\,\boldsymbol{u}_t \tag{8.5}$$

$$\boldsymbol{x}^s_{t+1} = \boldsymbol{\Phi}\,\boldsymbol{x}^s_t + \boldsymbol{K}\,\boldsymbol{a}_t \tag{8.6}$$

$$\boldsymbol{z}^s_t = \boldsymbol{H}\,\boldsymbol{x}^s_t + \boldsymbol{a}_t. \tag{8.7}$$

Note that this decomposition preserves the block-diagonal structure of the system as the transition matrices in (8.4) and (8.6) coincide with that in (8.2).

Building on the deterministic-stochastic decomposition, Section 8.2 discusses the calculation of the deterministic component $\boldsymbol{z}^d_t$, which is the output of (8.4)–(8.5).

Estimating the deterministic component is useful in many situations. For example, if the inputs are control variables, their effect on the output provides a measure of the influence of past controls. On the other hand, if the inputs are leading indicators, then their individual influence shows how each one affects the forecasts for $\boldsymbol{z}_t$. Last, if the variables in $\boldsymbol{u}_t$ are intervention terms (Box and Tiao [31]) or calendar variables (Cleveland and Devlin [54]), such a decomposition provides estimates for the individual effect of each input, and, if required, the means to eliminate them.

After determining the deterministic component, the stochastic term is trivially given by: $\boldsymbol{z}^s_t = \boldsymbol{z}_t - \boldsymbol{z}^d_t$. In Section 8.3 we will show how to further decompose this component into the trend, cycle, seasonal, and irregular components implied by model (8.6)–(8.7).

Putting all these ideas together, the decomposition under consideration can be written as:

$$z_t = z_t^d + z_t^s, \tag{8.8}$$

$$z_t^s = t_t + c_t + s_t + \varepsilon_t, \tag{8.9}$$

where t_t, c_t, s_t and ε_t are $m \times 1$ vectors of trend, cycle, seasonal and irregular components, respectively.

8.1.2 *Enforcing minimality*

It is easy to show that, if the SEM model (8.2)–(8.3) is minimal, then the deterministic and stochastic components, z_t^d and z_t^s, are observable. On the other hand, decomposing the SEM into its deterministic and stochastic subsystems replicates the state equation because expressions (8.4) and (8.6) can be jointly written as:

$$\begin{bmatrix} x_{t+1}^d \\ x_{t+1}^s \end{bmatrix} = \begin{bmatrix} \Phi & 0 \\ 0 & \Phi \end{bmatrix} \begin{bmatrix} x_t^d \\ x_t^s \end{bmatrix} + \begin{bmatrix} \Gamma \\ 0 \end{bmatrix} u_t + \begin{bmatrix} 0 \\ K \end{bmatrix} a_t, \tag{8.10}$$

and adding (8.5) and (8.7) yields, bearing in mind that $z_t = z_t^d + z_t^s$, the following observation equation:

$$z_t = \begin{bmatrix} H & H \end{bmatrix} \begin{bmatrix} x_t^d \\ x_t^s \end{bmatrix} + D u_t + a_t. \tag{8.11}$$

Note that this model realizes z_t using $2n$ states, while (8.2)–(8.3) only requires n. Therefore (8.10)–(8.11) is a non-minimal system whose output is the sum of two components with identical dynamics. In most cases these components cannot be distinguished. However, such a distinction is possible in this case because the input of the deterministic subsystem, u_t, is exactly known. As we will see in Section 8.2, to estimate z_t^d we will only need adequate initial conditions for the deterministic subsystem.

On the other hand, minimality of both subsystems is not assured, even if their outputs are observable. This happens because controllability of (8.2)–(8.3) does not assure controllability of (8.4)–(8.5) and (8.6)–(8.7)[1].

Uncontrollable modes are redundant and wreak havoc on standard signal extraction algorithms. Fortunately, it is easy to detect and eliminate them by applying the Staircase algorithm (Rosenbrock [190]) to each subsystem. The following example illustrates the previous discussion and the elimination of uncontrollable modes.

Example 8.1.1 Elimination of uncontrollable modes

Consider the SISO transfer function:

$$z_t = \frac{.5}{1-.6\mathrm{B}} u_t + \frac{1}{1-\mathrm{B}} a_t, \tag{8.12}$$

[1]For example, if part of the model dynamics is uniquely associated with the inputs, as happens for example in a transfer function, the states describing these dynamics are not excited in the stochastic subsystem, and, as a consequence, this subsystem is uncontrollable.

which can be written as a block-diagonal SEM (see Subsection 3.1.3 and Appendix A) as:

$$\begin{bmatrix} x_{t+1}^1 \\ x_{t+1}^2 \end{bmatrix} = \begin{bmatrix} 1 & 0 \\ 0 & .6 \end{bmatrix} \begin{bmatrix} x_t^1 \\ x_t^2 \end{bmatrix} + \begin{bmatrix} 0 \\ -.103 \end{bmatrix} u_t + \begin{bmatrix} 1.166 \\ 0 \end{bmatrix} a_t, \tag{8.13}$$

$$z_t = \begin{bmatrix} .858 & -2.916 \end{bmatrix} \begin{bmatrix} x_t^1 \\ x_t^2 \end{bmatrix} + .5u_t + a_t. \tag{8.14}$$

According to (8.4)–(8.5) and (8.6)–(8.7), the deterministic and stochastic subsystems are given respectively by:

$$\begin{bmatrix} x_{t+1}^{d1} \\ x_{t+1}^{d2} \end{bmatrix} = \begin{bmatrix} 1 & 0 \\ 0 & .6 \end{bmatrix} \begin{bmatrix} x_t^{d1} \\ x_t^{d2} \end{bmatrix} + \begin{bmatrix} 0 \\ -.103 \end{bmatrix} u_t \tag{8.15}$$

$$z_t^d = \begin{bmatrix} .858 & -2.916 \end{bmatrix} \begin{bmatrix} x_t^{d1} \\ x_t^{d2} \end{bmatrix} + .5u_t \tag{8.16}$$

$$\begin{bmatrix} x_{t+1}^{s1} \\ x_{t+1}^{s2} \end{bmatrix} = \begin{bmatrix} 1 & 0 \\ 0 & .6 \end{bmatrix} \begin{bmatrix} x_t^{s1} \\ x_t^{s2} \end{bmatrix} + \begin{bmatrix} 1.166 \\ 0 \end{bmatrix} a_t \tag{8.17}$$

$$z_t^s = \begin{bmatrix} .858 & -2.916 \end{bmatrix} \begin{bmatrix} x_t^{s1} \\ x_t^{s2} \end{bmatrix} + a_t. \tag{8.18}$$

Note that the first state in (8.15) and the second state in (8.17) will never be excited, as they do not depend on the values of the input and the error, respectively. Therefore, they are uncontrollable and the overall system is not minimal. Applying the Staircase algorithm to (8.15)–(8.16) and (8.17)–(8.18), yields two minimal equivalent representations:

$$x_{t+1}^d = .6x_t^d - .103u_t \tag{8.19}$$

$$z_t^d = -2.916x_t^d + .5u_t, \tag{8.20}$$

and

$$x_{t+1}^s = x_t^s + 1.166a_t \tag{8.21}$$

$$z_t^s = .858x_t^s + a_t. \tag{8.22}$$

8.2 Estimation of the deterministic components

Assuming that the matrices for the minimal deterministic and stochastic subsystems have been computed, we will now address the problem of estimating the sequences $\boldsymbol{z}_t^s$ and $\boldsymbol{z}_t^d$, for a given sample $\{\boldsymbol{z}_t\,,\,\boldsymbol{u}_t\}_{t=1}^N$. To this end, note that the solution for $\boldsymbol{x}_t^d$ in (8.4) is given by:

$$\boldsymbol{x}_t^d = \boldsymbol{\Phi}^{t-1}\boldsymbol{x}_1^d + \sum_{i=1}^{t-1} \boldsymbol{\Phi}^{t-1-i}\boldsymbol{\Gamma}\boldsymbol{u}_i. \tag{8.23}$$

If we had an adequate initial condition for the deterministic subsystem $\hat{\boldsymbol{x}}_{1|N}^d$, computation of the decomposition would then reduce to:

1. propagating (8.23) to obtain the sequence $\left\{\hat{\boldsymbol{x}}^{\boldsymbol{d}}_{t|N}\right\}_{t=2}^{N}$;
2. estimating the output of the deterministic subsystem which, by (8.5), would be

$$\hat{\boldsymbol{z}}^{\boldsymbol{d}}_{t|N} = \boldsymbol{H}\,\hat{\boldsymbol{x}}^{\boldsymbol{d}}_{t|N} + \boldsymbol{D}\boldsymbol{u}_t; \tag{8.24}$$

3. estimating the output of the stochastic subsystem using (8.8)

$$\hat{\boldsymbol{z}}^{\boldsymbol{s}}_{t|N} = \boldsymbol{z}_t - \hat{\boldsymbol{z}}^{\boldsymbol{d}}_{t|N}. \tag{8.25}$$

De Jong [66] shows that one way to isolate the effect of the initial state $\boldsymbol{x}_1$ on the innovations is computing the sequence:

$$\tilde{\boldsymbol{z}}_t = \boldsymbol{H}\bar{\boldsymbol{\Phi}}_{t-1}\boldsymbol{x}_1 + \tilde{\boldsymbol{z}}^*_t, \tag{8.26}$$

where $\tilde{\boldsymbol{z}}_t$ are the innovations resulting from an arbitrary KF$(\mathbf{0},\mathbf{0})$, $\tilde{\boldsymbol{z}}^*_t$ are the KF innovations that would result if the true initial conditions were null, and $\bar{\boldsymbol{\Phi}}_t$ is the matrix sequence resulting from $\bar{\boldsymbol{\Phi}}_t = (\boldsymbol{\Phi} - \boldsymbol{K}_t\boldsymbol{H})\,\bar{\boldsymbol{\Phi}}_{t-1}$ with $\bar{\boldsymbol{\Phi}}_1 = \boldsymbol{I}$.

Expression (8.26) can be written in matrix form as:

$$\tilde{\boldsymbol{Z}} = \boldsymbol{X}\boldsymbol{x}_1 + \tilde{\boldsymbol{Z}}^*, \tag{8.27}$$

where $\tilde{\boldsymbol{Z}}$ and $\tilde{\boldsymbol{Z}}^*$ are $(m \times N) \times 1$ vectors containing the innovations $\tilde{\boldsymbol{z}}_t$ and $\tilde{\boldsymbol{z}}^*_t$, respectively, and $\boldsymbol{X}$ is a matrix whose row-blocks contain the values $\boldsymbol{H}\bar{\boldsymbol{\Phi}}_{t-1}$.

Taking into account that $\boldsymbol{x}_1 = \boldsymbol{x}^{\boldsymbol{d}}_1 + \boldsymbol{x}^{\boldsymbol{s}}_1$, it would be natural to write (8.27) as $\tilde{\boldsymbol{Z}} = \boldsymbol{X}\left(\boldsymbol{x}^{\boldsymbol{d}}_1 + \boldsymbol{x}^{\boldsymbol{s}}_1\right) + \tilde{\boldsymbol{Z}}^* = \boldsymbol{X}\boldsymbol{x}^{\boldsymbol{d}}_1 + \boldsymbol{X}\boldsymbol{x}^{\boldsymbol{s}}_1 + \tilde{\boldsymbol{Z}}^*$. However, if we only take into account the controllable modes of the subsystems (see Section 8.1) the matrices affecting $\boldsymbol{x}^{\boldsymbol{d}}_1$ and $\boldsymbol{x}^{\boldsymbol{s}}_1$ in this expression would be different. Accordingly, expression (8.27) can be written as:

$$\tilde{\boldsymbol{Z}} = \boldsymbol{X}^{\boldsymbol{d}}\boldsymbol{x}^{\boldsymbol{d}}_1 + \boldsymbol{X}^{\boldsymbol{s}}\boldsymbol{x}^{\boldsymbol{s}}_1 + \tilde{\boldsymbol{Z}}^*, \tag{8.28}$$

and the problem of estimating the deterministic component reduces to determining the initial conditions $\boldsymbol{x}^{\boldsymbol{d}}_1$, as we have seen above.

8.2.1 *Estimating the total effect of the inputs*

The initial conditions for the deterministic subsystem can be determined either by assuming that they are diffuse, or by computing the Generalized Least Squares (GLS) estimate of a non-zero initial state, $\boldsymbol{x}^{\boldsymbol{d}}_1$, with a null covariance. As we will see, both of them are equivalent.

Diffuse initialization implies assuming that $\boldsymbol{x}^{\boldsymbol{d}}_1 \sim (\bar{\boldsymbol{x}}^{\boldsymbol{d}}_1, \boldsymbol{P}^{\boldsymbol{d}}_1)$, with $(\boldsymbol{P}^{\boldsymbol{d}}_1)^{-1} = \mathbf{0}$. In this case (De Jong [66]) the initial conditions are:

$$\hat{\boldsymbol{x}}^{\boldsymbol{d}}_{1|N} = \left[(\boldsymbol{P}^{\boldsymbol{d}}_1)^{-1} + (\boldsymbol{X}^{\boldsymbol{d}})^T\bar{\boldsymbol{B}}^{-1}\boldsymbol{X}^{\boldsymbol{d}}\right]^{-1}\left[(\boldsymbol{P}^{\boldsymbol{d}}_1)^{-1}\bar{\boldsymbol{x}}^{\boldsymbol{d}}_1 + (\boldsymbol{X}^{\boldsymbol{d}})^T\bar{\boldsymbol{B}}^{-1}\tilde{\boldsymbol{Z}}\right], \tag{8.29}$$

where $\bar{B} = X^s P_1^s (X^s)^T + B$, and $B = \text{cov}(\tilde{Z}^*) = \text{diag}(B_1, B_2, \ldots, B_m)$, with $\{B_t\}_{t=1,2,\ldots,n}$ the sequence of covariances resulting from a KF(0,0). Taking into account that $\left(P_1^d\right)^{-1} = 0$, expression (8.29) simplifies to:

$$\hat{x}_{1|N}^d = \left[(X^d)^T \bar{B}^{-1} X^d\right]^{-1} \left[(X^d)^T \bar{B}^{-1} \tilde{Z}\right], \tag{8.30}$$

which is the GLS estimate of the non-null initial state, x_1^d; see De Jong [66] and Casals, Jerez, and Sotoca [40].

Note that (8.30) depends on $\bar{B}^{-1} = [X^s P_1^s (X^s)^T + B]^{-1}$, and thus on the covariance P_1^s. If the stochastic subsystem has unit roots, this matrix will have non-finite elements. Following De Jong and Chu-Chun-Lin [69], this inconvenience can be overcome by applying the matrix inversion lemma to $\bar{B}^{-1}$, which yields:

$$\bar{B}^{-1} = B^{-1} - B^{-1} X^s R (X^s)^T B^{-1}, \tag{8.31}$$

where $R = \left[(P_1^s)^{-1} + (X^s)^T B^{-1} X^s\right]^{-1}$ and, substituting (8.31) in (8.30) yields:

$$\hat{x}_{1|N}^d = \left\{(X^d)^T B^{-1} X^d - (X^d)^T B^{-1} X^s R (X^s)^T B^{-1} X^d\right\}^{-1} \cdot \left\{(X^d)^T B^{-1} \tilde{Z} - (X^d)^T B^{-1} X^s R (X^s)^T B^{-1} \tilde{Z}\right\}. \tag{8.32}$$

The initial condition (8.32) allows for both stationary and unit roots because it depends, not on P_1^s, but on its inverse $(P_1^s)^{-1}$. When the system has both types of roots, the matrix $(P_1^s)^{-1}$ converges to a finite value with as many null eigenvalues as nonstationary states; see Chapter 5.

Depending on the system dynamics, expression (8.32) can adopt different forms.

Case 1. The stochastic subsystem is static, so that $X^s = 0$, and (8.32) therefore simplifies to:

$$\hat{x}_{1|N}^d = \left[(X^d)^T B^{-1} X^d\right]^{-1} (X^d)^T B^{-1} \tilde{Z}. \tag{8.33}$$

Case 2. The deterministic and stochastic subsystems share the same dynamics, so that $X^d = X^s$, and (8.32) thus collapses to (8.33).

Case 3. The deterministic and stochastic subsystems share at least one nonstationary mode. This situation creates a severe identifiability problem because the matrix

$$(X^d)^T B^{-1} X^d - (X^d)^T B^{-1} X^s R (X^s)^T B^{-1} X^d, \tag{8.34}$$

whose inverse is required to compute (8.32), is singular. The only option in this case consists of using the pseudoinverse of (8.34) in order to compute the initial conditions.

8.2.2 *Estimating the individual effect of each input*

The procedure previously described is useful when one is interested in the total effect of all the exogenous variables. On the other hand, in models with several inputs one may want to estimate the individual effect of each one. This separation is easier if we write the deterministic subsystem (8.4)–(8.5) in its equivalent Luenberger Canonical Form (hereafter LCF; see Subsection 9.2.2), denoted by:

$$\bar{\boldsymbol{x}}_{t+1}^{d} = \bar{\boldsymbol{\Phi}}\,\bar{\boldsymbol{x}}_{t}^{d} + \bar{\boldsymbol{\Gamma}}\,\boldsymbol{u}_{t} \tag{8.35}$$

$$\boldsymbol{z}_{t}^{d} = \bar{\boldsymbol{H}}\,\bar{\boldsymbol{x}}_{t}{}^{d} + \boldsymbol{D}\,\boldsymbol{u}_{t}. \tag{8.36}$$

Model (8.35)–(8.36) can be obtained by a similar transformation of (8.4)–(8.5), given by $\bar{\boldsymbol{x}}_{t}^{d} = \boldsymbol{T}^{-1}\boldsymbol{x}_{t}^{d}$, see Subsection 2.1.2. Accordingly, the initial state of the LCF system corresponding to the vector $\hat{\boldsymbol{x}}_{1|N}^{d}$ defined in (8.32) would be:

$$\hat{\bar{\boldsymbol{x}}}_{1|N}^{d} = \boldsymbol{T}^{-1}\hat{\boldsymbol{x}}_{1|N}^{d}. \tag{8.37}$$

The controllability matrix of the LCF separates the dynamics associated with the individual inputs, therefore permitting one to discern whether a given input affects (or not) each system state. To see this, note that propagating the LCF according to (8.23) yields:

$$\bar{\boldsymbol{x}}_{t}^{d} = \bar{\boldsymbol{\Phi}}^{t-1}\bar{\boldsymbol{x}}_{1}^{d} + \begin{bmatrix} \bar{\boldsymbol{\Phi}}^{t-2}\bar{\boldsymbol{\Gamma}} & \bar{\boldsymbol{\Phi}}^{t-3}\bar{\boldsymbol{\Gamma}} & \dots & \bar{\boldsymbol{\Gamma}} \end{bmatrix} \begin{bmatrix} \boldsymbol{u}_1 \\ \boldsymbol{u}_2 \\ \vdots \\ \boldsymbol{u}_{t-1} \end{bmatrix}, \tag{8.38}$$

where $\boldsymbol{C} = \begin{bmatrix} \bar{\boldsymbol{\Phi}}^{t-2}\bar{\boldsymbol{\Gamma}} & \bar{\boldsymbol{\Phi}}^{t-3}\bar{\boldsymbol{\Gamma}} & \dots & \bar{\boldsymbol{\Gamma}} \end{bmatrix}$ is the controllability matrix of the LCF system whose zero/nonzero structure characterizes which inputs affect any given state. Specifically, if the i-th state is not excited by the j-th input, then the elements $c(i,j)$, $c(i,j+r)$, $c(i,j+2r)$,..., $c(i,j+(n-1)r)$ should be null, where r is the number of inputs.

Under these conditions, estimating the individual effect of each input requires the following two-step procedure.

Step 1: Decompose the initial condition (8.37) into r vectors, such that each one is exclusively related to one of the r system inputs.

Step 2: Propagate the r systems given by $\bar{\boldsymbol{\Phi}}$, $\bar{\boldsymbol{H}}$, and the j-th columns of $\bar{\boldsymbol{\Gamma}}$ and $\bar{\boldsymbol{D}}$, for $j = 1,\dots,r$, starting from the corresponding initial conditions.

Ideally, the initial vectors $\hat{\bar{\boldsymbol{x}}}_{1,j}^{d}$ will be orthogonal, and such that $\hat{\bar{\boldsymbol{x}}}_{1|N}^{d} = \sum_{j=1}^{r}\hat{\bar{\boldsymbol{x}}}_{1,j}^{d}$. This orthogonality can be achieved by distributing the initial state of the LCF system according to the structure of the controllability matrix $\boldsymbol{C}$, i.e., such that for $j = 1,\dots,r$ and $i = 1,\dots,n$, we set:

$$\bar{x}_{1,j}^{d}(i) = \begin{cases} 0, & \text{if } c(i,j+kr) = 0, \text{ for } k = 0,1,\dots,n-1, \\ \hat{\bar{x}}_{1|N}^{d}(i), & \text{otherwise,} \end{cases} \tag{8.39}$$

where $\hat{\hat{\boldsymbol{x}}}^{\boldsymbol{d}}_{1|N}(i)$ is the i-th element of $\hat{\hat{\boldsymbol{x}}}^{\boldsymbol{d}}_{1|N}$. Finally, the vectors $\tilde{\boldsymbol{x}}^{\boldsymbol{d}}_{1,j}$ resulting from (8.39), should be transformed according to:

$$\hat{\hat{\boldsymbol{x}}}^{d}_{1,j} = \boldsymbol{\Pi}_j \, \tilde{\boldsymbol{x}}^{\boldsymbol{d}}_{1,j}, \qquad j = 1, \ldots, r, \tag{8.40}$$

where $\boldsymbol{\Pi}_j$ is the orthogonal projector onto the null space of the matrix $[\; \hat{\hat{\boldsymbol{x}}}^{\boldsymbol{d}}_{1,1} \;\; \hat{\hat{\boldsymbol{x}}}^{\boldsymbol{d}}_{1,2} \;\; \ldots \;\; \hat{\hat{\boldsymbol{x}}}^{\boldsymbol{d}}_{1,j-1} \;\; \hat{\hat{\boldsymbol{x}}}^{\boldsymbol{d}}_{1,j+1} \;\; \ldots \;\; \hat{\hat{\boldsymbol{x}}}^{\boldsymbol{d}}_{1,r} \;]$. The initial conditions (8.40) can then be interpreted as, either the part of the LCF initial state that is orthogonal to the rest of the inputs, or as the marginal contribution of each input toward the initial condition of the system.

In many models, such as transfer functions, each mode is affected only by a single input, so that the orthogonal projector $\boldsymbol{\Pi}_j$ is the identity, and the condition $\hat{\hat{\boldsymbol{x}}}^{\boldsymbol{d}}_{1|N} = \sum_{j=1}^{r} \hat{\hat{\boldsymbol{x}}}^{\boldsymbol{d}}_{1,j}$ is thus assured. On the other hand, if a given input affects more than one mode, then the decomposition does not assure a complete split of (8.37), so that in general there will be a remainder, $\hat{\boldsymbol{x}}^{\boldsymbol{d}}_{1|N} - \sum_{j=1}^{r} \hat{\boldsymbol{x}}^{\boldsymbol{d}}_{1,j} \neq \boldsymbol{0}$. This remainder can be propagated with the pair $(\bar{\boldsymbol{\Phi}}, \bar{\boldsymbol{H}})$, and the resulting component interpreted as the common effect of the system inputs.

8.3 Decomposition of the stochastic component

8.3.1 *Characterization of the structural components*

After computing the deterministic-stochastic decomposition (8.8), one may want to estimate the generalized structural components in (8.9). To this end, we first need to characterize their properties.

The literature often describes the properties of the structural components in the frequency domain and, following Burman [34]: (a) the trend is characterized by a peak at the null frequency, (b) the irregular component displays no significant peak at any frequency, and (c) the seasonal component has peaks at the basic seasonal frequency and its multiples. To complete the decomposition, (d) any reminding component should be assigned to the cyclic term.

To implement these ideas, we build on the block-diagonal form for the stochastic components (8.6)–(8.7), where the eigenvalues of the transition matrix characterize the properties of the states in both time and frequency domains; see e.g., West [223].

In particular, assume that $\lambda_{j,k} = a_j \pm b_j i$ is a pair of complex conjugate eigenvalues of the transition matrix $\boldsymbol{\Phi}$ in state equation (8.6), associated with the j-th and k-th states. In this case, both states generate a peak in the pseudo-spectrum in the frequency $f_{j,k} = (2\pi)^{-1} \arctan(b_j/a_j)$, where $f_{j,k}$ is in cycles per unit time.

This general result has some particular cases that are worth discussing:

1. If the real part of $\lambda_{j,k}$ is zero, then $b_j/a_j = \infty$, implying that $f_{j,k} = 1/4$ if $b_j > 0$, or $f_{j,k} = 3/4$ if $b_j < 0$
2. If any eigenvalue is a real number, then $b_j/a_j = 0$, implying that $f_{j,k} = 0$ if $a_j > 0$, or $f_{j,k} = 1/2$ if $a_j < 0$
3. If $b_j = 0$ and a_j tends to one, then the spectral power tends to infinity at $f_{j,k} = 0$

If the output series are seasonal, it is easy to characterize the set of seasonal frequencies, defined as $\boldsymbol{F}_s = \{f_i = k_i/s; k_j = 1, 2, \ldots, [s/2]\}$, where s denotes the seasonal period (number of observations per seasonal cycle), and $[s/2] = s/2$ if s is even, or $[s/2] = (s-1)/2$ if s is odd.

Building on these ideas, the states in the block-diagonal stochastic subsystem (8.6)–(8.7) can be naturally assigned to the structural components with the rules summarized in Table 8.1.

Table 8.1 *Association between state variables and structural components according to the eigenvalues of the transition matrix and their spectral properties.*

<table>
<tr><th colspan="2">Eigenvalues of Φ</th><th>Spectral peaks</th><th>Component</th></tr>
<tr><td rowspan="4">Real</td><td>$\lambda_j = 1$</td><td>$f_j = 0$</td><td>Trend</td></tr>
<tr><td rowspan="3">$\lambda_{j,k} = a_j$,
$-1 < a_j < 1$</td><td>$f_j = 0$, if $0 < a_j < 1$</td><td>Cycle</td></tr>
<tr><td colspan="2">If $a_j = 0$, then x_j is either a redundant state, which has no effect on the structural components, or a lagged uncorrelated error, which does not create a spectral peak in any frequency. In this case, it should be included in the cycle</td></tr>
<tr><td>$f_j = 1/2$, if $-1 < a_j < 0$</td><td>Seasonal if $f_{j,k} \in \boldsymbol{F}_s$, cycle otherwise</td></tr>
<tr><td>Complex</td><td>$\lambda_{j,k} = a_j \pm b_j i$</td><td>$f_{j,k} = (2\pi)^{-1} \arctan(b_j/a_j)$</td><td>Seasonal if $f_{j,k} \in \boldsymbol{F}_s$, cycle otherwise</td></tr>
</table>

According to this classification, we can reorganize the block-diagonal model (8.6)–(8.7) as:

$$\begin{bmatrix} \bar{\boldsymbol{x}}_{t+1}^t \\ \bar{\boldsymbol{x}}_{t+1}^c \\ \bar{\boldsymbol{x}}_{t+1}^s \end{bmatrix} = \begin{bmatrix} \bar{\boldsymbol{\Phi}}^t & \boldsymbol{0} & \boldsymbol{0} \\ \boldsymbol{0} & \bar{\boldsymbol{\Phi}}^c & \boldsymbol{0} \\ \boldsymbol{0} & \boldsymbol{0} & \bar{\boldsymbol{\Phi}}^s \end{bmatrix} \begin{bmatrix} \bar{\boldsymbol{x}}_t^t \\ \bar{\boldsymbol{x}}_t^c \\ \bar{\boldsymbol{x}}_t^s \end{bmatrix} + \begin{bmatrix} \bar{\boldsymbol{K}}^t \\ \bar{\boldsymbol{K}}^c \\ \bar{\boldsymbol{K}}^s \end{bmatrix} \boldsymbol{a}_t \tag{8.41}$$

$$\boldsymbol{z}_t = \begin{bmatrix} \bar{\boldsymbol{H}}^t & \bar{\boldsymbol{H}}^c & \bar{\boldsymbol{H}}^s \end{bmatrix} \begin{bmatrix} \bar{\boldsymbol{x}}_t^t \\ \bar{\boldsymbol{x}}_t^c \\ \bar{\boldsymbol{x}}_t^s \end{bmatrix} + \boldsymbol{a}_t, \tag{8.42}$$

where $\bar{\boldsymbol{x}}_t^t$ is the vector of nonstationary states, $\bar{\boldsymbol{x}}_t^c$ is the vector of stationary (nonseasonal) states, and $\bar{\boldsymbol{x}}_t^s$ is the vector of seasonal states. As such, the structural components are given by the different addends in the observation equation (8.42):

$$\boldsymbol{t}_t = \bar{\boldsymbol{H}}^t \bar{\boldsymbol{x}}_t^t \tag{8.43}$$

$$\boldsymbol{c}_t = \bar{\boldsymbol{H}}^c \bar{\boldsymbol{x}}_t^c \tag{8.44}$$

$$\boldsymbol{s}_t = \bar{\boldsymbol{H}}^s \bar{\boldsymbol{x}}_t^s \tag{8.45}$$

$$\boldsymbol{\varepsilon}_t = \boldsymbol{a}_t. \tag{8.46}$$

8.3.2 Estimation of the structural components

Given the characterization of the structural components detailed in Section 8.3.1, the problem now reduces to obtaining estimates of the state variables and combining them according to (8.43)–(8.46). The method of choice to achieve is the fixed-interval smoother; see Chapter 4.

One must also bear in mind that the stochastic subsystem (8.6)–(8.7) is a SEM, and thus the smoothed estimates of its states converge to exact values with null variances and covariances; see Section 4.6. Accordingly, the estimates for the components will likewise be exact, as they are linear combinations of these smoothed states.

Null variances and covariances are very important for signal extraction due to two reasons. First, null variances ensure that the components do not change when the sample increases[2]. Second, zero covariances ensure that the resulting components are conditionally independent, so that any component can be analyzed and interpreted independently of any other component(s).

8.4 Structure of the method

Building on the ideas discussed in previous sections, the structural decomposition of a vector of time series, z_t, can be organized according to the following steps:

Step 1 Obtain an adequate SS model for z_t and, optionally, the equivalent SEM; see Section 2.2. Obtain the equivalent block-diagonal form; see Subsection 3.1.3.

Step 2 Obtain the implied deterministic and stochastic subsystems as shown in Section 8.1, and reduce them to their minimal dimensions by means of the Staircase algorithm; see Rosenbrock [190].

Step 3 Estimate the deterministic component, z_t^d, following the procedure described in Section 8.2.

Step 4 Determine the stochastic component as $z_t^s = z_t - z_t^d$, and compute smoothed estimates of the states in the stochastic subsystem.

Step 5 Classify the different states following the indications in Table 8.1. Compute estimates of the trend, cycle, and seasonal components by combining the estimates of the states with the corresponding coefficients in the observation equation; see (8.43)–(8.45).

Step 6 Estimate the irregular component by subtracting from z_t the estimates for the trend, cycle, and seasonal components; see (8.42).

8.5 Examples

8.5.1 Comparing different methods with simulated data

This example compares the components extracted from a SEM, according to the method described in previous sections, with those resulting from a MEM, which is

[2]Intuitively, if a magnitude is exactly known, then its estimate will not change, regardless of whether the sample is augmented.

the representation employed by many alternative approaches. To this end, we simulate a stochastic process, defined in Table 8.2, which can be interpreted both, as a Structural Time Series Model (STSM), see Harvey [118], as well as an ARIMA-based decomposition. The components are then extracted using the direct MEM representation and the equivalent SEM[3].

Table 8.2 *Definition of the data generating process in STSM and ARIMA forms.*

Component	STSM	ARIMA representation
Trend	$\mu_{t+1} = \mu_t + \beta_t$ $\beta_{t+1} = \beta_t + \zeta_t$	$(1-B)^2\mu_{t+1} = \zeta_t$
Seasonal	$\gamma_{t+1} = \gamma_t + \gamma_{t-1} + \gamma_{t-2} + \omega_t$	$(1+\mathrm{B}+\mathrm{B}^2+\mathrm{B}^3)\gamma_{t+1} = \omega_t$
Time series	$z_t = \mu_t + \gamma_t + e_t$	$(1-\mathrm{B})(1-\mathrm{B}^4)z_t = (1-0.93\mathrm{B} - 0.09\mathrm{B}^2 - 0.04\mathrm{B}^3 - 0.59\mathrm{B}^4 + 0.55\mathrm{B}^5)a_t$ $\simeq (1-0.93\mathrm{B})(1-0.54\mathrm{B}^4)a_t$

Therefore the trend is an integrated random walk process and the seasonal component follows what Harvey [118] calls a "dummy variable seasonality" model, where the sum of the seasonal components over a year is a random disturbance. The terms ζ_t, ω_t, and e_t are Gaussian white noise errors with the following covariances:

$$\operatorname{cov}\left(\begin{bmatrix}\zeta_t\\ \omega_t\\ e_t\end{bmatrix}\right) = \begin{bmatrix}1/1600 & 0 & 0\\ 0 & 0.1 & 0\\ 0 & 0 & 1\end{bmatrix}. \tag{8.47}$$

Note that the variance ratio, $\sigma_\zeta^2/\sigma_e^2 = 1/1600$, was chosen according to the value assumed in the Hodrick–Prescott [126] filter for quarterly data.

The second column in Table 8.2 reveals how the model can be interpreted as an ARIMA-model-based decomposition, which realizes an ARIMA process for the observable series z_t. Using the results in Chapters 2 and 3, this model can be written in the equivalent block-diagonal SEM form displayed in Table 8.3.

Note that: (a) the states in this representation are clearly separated in (1×1) and (2×2) blocks, and (b) the eigenvalues of the transition matrix are identical to the inverse AR roots of the stochastic processes in Table 8.2. According to their frequencies, the first and second states correspond to the trend, while the third, fourth, and fifth states correspond to the seasonal component.

Taking into account this classification and the coefficients in the last row of Table 8.3, the trend and seasonal components would be given by:

$$t_t = \bar{x}_t^1 \tag{8.48}$$

$$s_t = -0.707\bar{x}_t^4 + 0.577\bar{x}_t^5 \tag{8.49}$$

$$\varepsilon_t = z_t - t_t - s_t. \tag{8.50}$$

[3]Note that this design puts the method described here at a deliberate disadvantage, as the MEM takes into account certain variance proportionality constraints which are then lost in the SEM representation.

Table 8.3 *Structure of the block-diagonal SEM representation of the data generating process and association of the state variables to structural components.*

	Inputs								
	$\bar{x}_t^1$	$\bar{x}_t^2$	$\bar{x}_t^3$	$\bar{x}_t^4$	$\bar{x}_t^5$	a_t	Eig($\bar{\Phi}$)	Freq. f_i	Component
Outputs	$\bar{\Phi}$					$\bar{E}$			
$\bar{x}_{t+1}^1$	1	1	0	0	0	0.188	1	0	Trend
$\bar{x}_{t+1}^2$	0	1	0	0	0	0.019	1	0	
$\bar{x}_{t+1}^3$	0	0	0	−0.577	0	0.135	$\pm i$	0.25	Seasonal
$\bar{x}_{t+1}^4$	0	0	1.732	0	0	0.005			
$\bar{x}_{t+1}^5$	0	0	0	0	−1	−0.203	−1	0.5	
	$\bar{H}$					−			
z_t	1	0	0	−0.707	0.577	1			

Combining the SS model in Table 8.3 with (8.48)–(8.50), we can derive ARIMA representations for these components; see Chapter 9. Table 8.4 compares them with the corresponding models for the data generating process.

Table 8.4 *Comparison between the ARIMA representations of the STSM and the block-diagonal SEM form.*

Component	STSM	Block-diagonal SEM
Trend	$(1-\mathrm{B})^2\mu_{t+1} = \zeta_{t-1}$ $\zeta_{t-1} \sim \mathrm{NID}(0, 1/1600)$	$(1-\mathrm{B})^2 t_t = (1-0.899\mathrm{B})\varepsilon_{t-1}$
Seasonal	$(1+\mathrm{B}+\mathrm{B}^2+\mathrm{B}^3)\gamma_t = \omega_{t-1}$ $\omega_{t-1} \sim \mathrm{NID}(0, 0.1)$	$(1+\mathrm{B}+\mathrm{B}^2+\mathrm{B}^3)s_t =$ $(1+1.402\mathrm{B}+2.347\mathrm{B}^2)$ $(-0.120)\varepsilon_{t-1}$
Irregular	$e_t \sim \mathrm{NID}(0,1)$	$\varepsilon_t \sim \mathrm{NID}(0,1.824)$

Note that the ARIMA representations for the STSM and the equivalent SEM in Table 8.4 have the same autoregressive structures, but different error terms. While the components of the STSM follow pure AR processes, the components implied by the SEM receive shocks from the delayed irregular component, affected by different moving average polynomials and scale parameters.

We now generate 200 random samples for the data generating process and compute the smoothed estimates of the components using both the STSM and the equivalent SEM. Figure 8.1 compares both sets of components.

Note that: (a) the trend and seasonal components obtained in both cases are very similar; (b) the trend estimate resulting from the SEM is more volatile because the smoothness of the STSM trend is a byproduct of the variance ratios in (8.47), which are explicit in the STSM but not in the SEM; and (c) the irregular components have different volatilities, as expected; see Table 8.4.

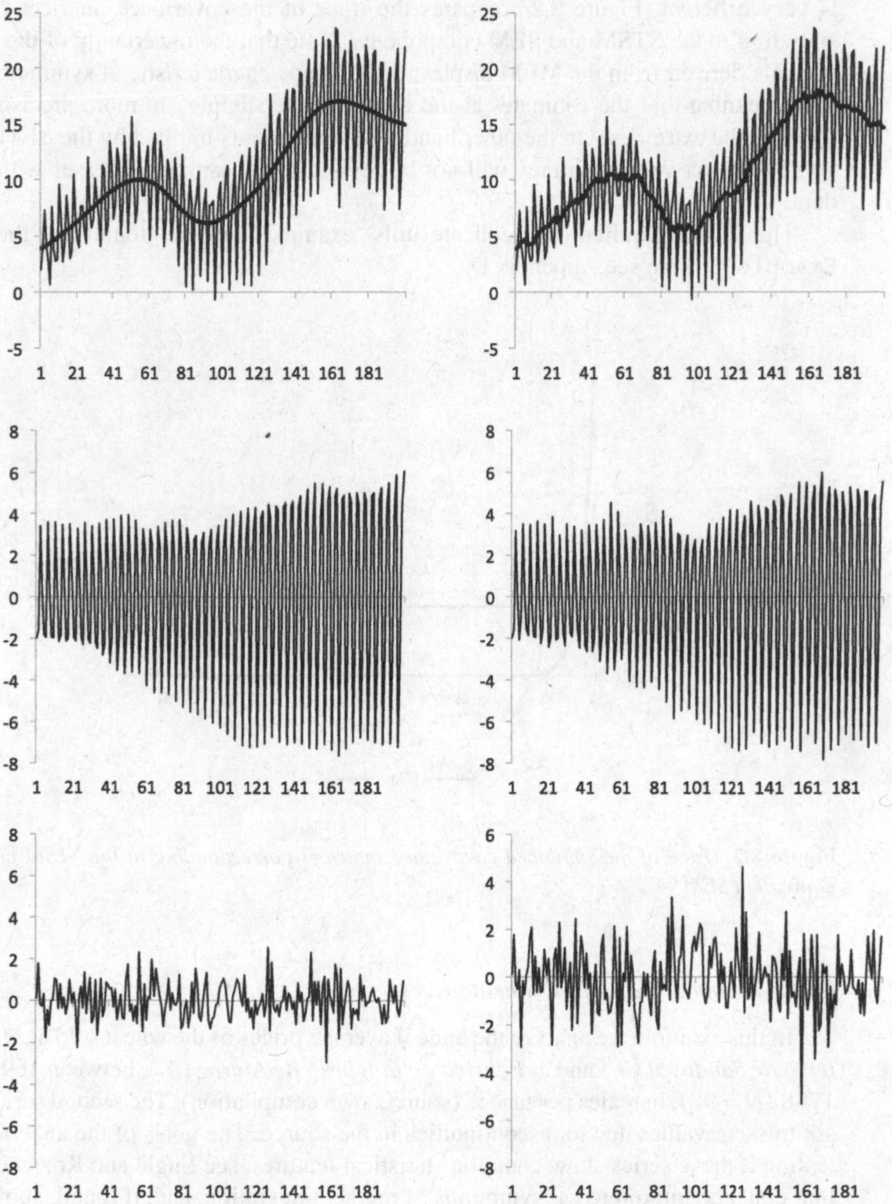

Figure 8.1 *Comparison between the model components obtained from the STSM (left) and the equivalent SEM (right).*

Despite their overall similarity, the uncertainty affecting both sets of components is very different. Figure 8.2 compares the trace of the covariance matrices corresponding to the STSM and SEM components. Note that the uncertainty of the components derived from the MEM display, the U-shape characteristic of symmetric filters, meaning that the estimates at the center of the sample are more precise than those at the extremes. On the other hand, the components implied by the SEM converge to exact values, so they will not be revised if the sample increases with new data.

The code required to replicate this example can be found in the file `Example0851.m`; see Appendix D.

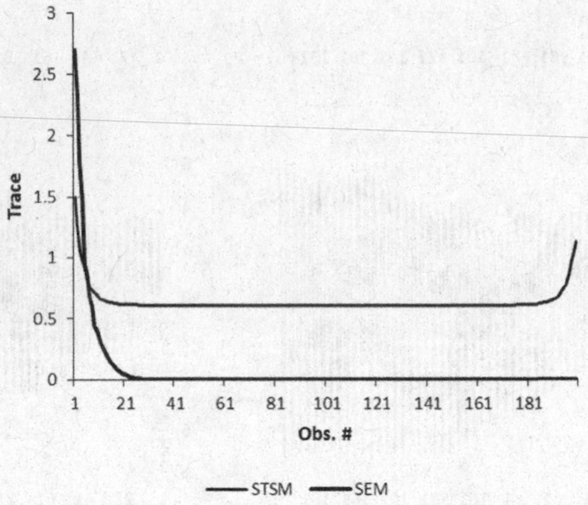

Figure 8.2 *Trace of the smoothed covariance matrices corresponding to the STSM and the equivalent SEM.*

8.5.2 *Common features in wheat prices*

In this example we analyze the annual average prices of the wheat sold at *Monasterio de Sandoval* (S_t) and at *Fábrica de la iglesia de Alaraz* (A_t), between 1691 and 1788 ($N = 98$), in reales per fanega (source: own compilation). The second series has six missing values due to discontinuities in the source. The goals of the analysis are testing if these series show common statistical features, see Engle and Kozicki [87], that could be interpreted as symptoms of market integration, and, if found, apply our method to estimate them.

After a standard univariate analysis, the following models were fitted to the log-transformed series and the differences in log prices:

$$(1 - \underset{(0.097)}{0.042}\text{B} + \underset{(0.097)}{0.296}\text{B}^2)(1-\text{B})\log S_t = \hat{a}_t^S, \qquad \hat{\sigma}_S^2 = 0.077,$$
$$Q(10) = 5.29, \qquad p = 4.10, \qquad df = 0.54, \tag{8.51}$$

$$(1 - \underset{(0.100)}{0.004}\text{B} + \underset{(0.100)}{0.342}\text{B}^2)(1-\text{B})\log A_t = \hat{a}_t^A, \qquad \hat{\sigma}_A^2 = 0.113,$$
$$Q(10) = 10.67, \qquad p = 3.99, \qquad df = 0.59, \tag{8.52}$$

$$(1 - \underset{(0.102)}{0.280}\text{B})(\log A_t - \log S_t) = \underset{(0.031)}{0.107} + \hat{a}_t^S, \qquad \hat{\sigma}^2 = 0.067, \qquad Q(10) = 6.02, \tag{8.53}$$

where the figures in parentheses are standard errors for the corresponding estimates, $Q(10)$ is the Ljung–Box [151] statistic computed with the first 10 lags of the residual sample autocorrelation function, and p and df are, respectively, the period and damping factor of the pseudo-cycle implied by the corresponding AR(2) factors. Because of the missing values in series A_t, the estimates in (8.52) and (8.53) were obtained using the technique described in Subsection 3.3.2.

It therefore appears that individual log prices are adequately represented by ARIMA(2,1,0) processes, implying that both series are nonstationary and have harmonic pseudo-cycles with a period of roughly four years. Also, the model for the difference of log prices is a stationary AR(1) with a constant term, meaning that both series share a common trend and a common harmonic cycle, see Engle and Kozicki [87]. The presence of these cofeatures supports the idea that both markets were substantially integrated in this historical period.

Further evidence can be found by building a multivariate model, constrained to represent these cofeatures. After a standard specification analysis (Jenkins and Alavi [131]) we obtained the model:

$$\begin{bmatrix} 1 - \underset{(0.087)}{0.317}\text{B} & 0 \\ 0 & (1 - \underset{(0.085)}{0.048}\text{B} + \underset{(0.082)}{0.260}\text{B}^2)(1-\text{B}) \end{bmatrix} \begin{bmatrix} 1 & -1 \\ 0 & 1 \end{bmatrix} \begin{bmatrix} \log A_t \\ \log S_t \end{bmatrix}$$
$$= \begin{bmatrix} \underset{(0.026)}{0.098} \\ 0 \end{bmatrix} + \begin{bmatrix} 1 & 0 \\ \underset{(0.091)}{0.602}\text{B} & 0 \end{bmatrix} \begin{bmatrix} \hat{a}_t^1 \\ \hat{a}_t^2 \end{bmatrix}, \tag{8.54}$$

with a noise covariance matrix of

$$\hat{\boldsymbol{\Sigma}}_a = \begin{bmatrix} 0.068 & -0.003 \\ -0.003 & 0.053 \end{bmatrix}, \qquad \boldsymbol{Q}(10) = \begin{bmatrix} 7.54 & 11.65 \\ 11.11 & 9.76 \end{bmatrix},$$

where $\boldsymbol{Q}(10)$ is a matrix of portmanteau statistics for the residual sample autocorrelations and cross-correlations.

Table 8.5 presents the block-diagonal SEM equivalent to (8.54). According to the eigenvalues of the transition matrix, the first state corresponds to the trend, the second state to a cyclical movement, and the third and fourth states to a harmonic pseudo-cycle. The observation equation in Table 8.5 implies the following decomposition of

Table 8.5 *Structure of the block-diagonal SEM equivalent to model (8.54) and association of the state variables to structural components.*

	Inputs								
	$\bar{x}_t^1$	$\bar{x}_t^2$	$\bar{x}_t^3$	$\bar{x}_t^4$	$u_t = 1$	$\hat{a}_t^1$	$\hat{a}_t^2$	eig($\bar{\mathbf{\Phi}}$)	Component
Outputs			$\mathbf{\Phi}$		$\mathbf{\Gamma}$	$\boldsymbol{E}$			
$\bar{x}_{t+1}^1$	1	0	0	0	0	0.497	0.825	1	Trend
$\bar{x}_{t+1}^2$	0	0.317	0	0	0.031	0.317	0	0.317	Cycle
$\bar{x}_{t+1}^3$	0	0	0.024	1.001	0	0.602	0.056	0.024 ±	Cycle
$\bar{x}_{t+1}^4$	0	0	–0.259	0.024	0	0.200	–0.258	0.509 i	(harmonic)
			$\boldsymbol{H}$		$\mathbf{D}$	$\boldsymbol{I}$			
$\log A_t$	1	1	0.202	–0.819	0.098	1	0		
$\log S_t$	1	0	0.202	–0.819	0	0	1		

the output:

$$t_t = \bar{x}_t^1 \quad (8.55)$$
$$c_t = 0.202\bar{x}_t^3 - 0.819\bar{x}_t^4 \quad (8.56)$$
$$c_t^A = \bar{x}_t^2 + 0.098 \quad (8.57)$$
$$\varepsilon_t^A = \hat{a}_t^1 \quad (8.58)$$
$$\varepsilon_t^S = \hat{a}_t^2, \quad (8.59)$$

where t_t denotes the common trend, c_t is the common cycle, and c_t^A is a stationary component capturing all the individual cyclical features of the log price in Alaraz. Note that this component includes a constant term. Finally, the white noise terms $\varepsilon_t^A = a_t^1$ and $\varepsilon_t^S = a_t^2$, are the irregular components affecting the series of Alaraz and Sandoval, respectively.

As the sample has some missing values, the trace of the smoothed covariance matrices drops quickly to zero, but shows transitory peaks of uncertainty in the years following a missing value (Figure 8.3). The estimated cofeatures are shown in Figure 8.4.

The code required to replicate this example can be found in the file `Example0852.m`; see Appendix D.

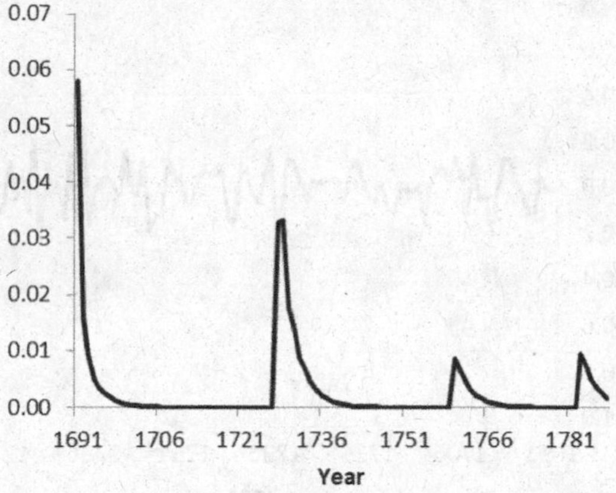

Figure 8.3 *Trace of the smoother covariance of Model (8.54).*

8.5.3 *The effect of advertising on sales*

This example will highlight the manner in which the decompositions proposed in this chapter can be used to gain insight into the relationship between different time series. To this end, we will use the famous monthly series of sales and advertising of the Lydia Pinkham vegetable compound, from January 1954 to June 1960; see Palda

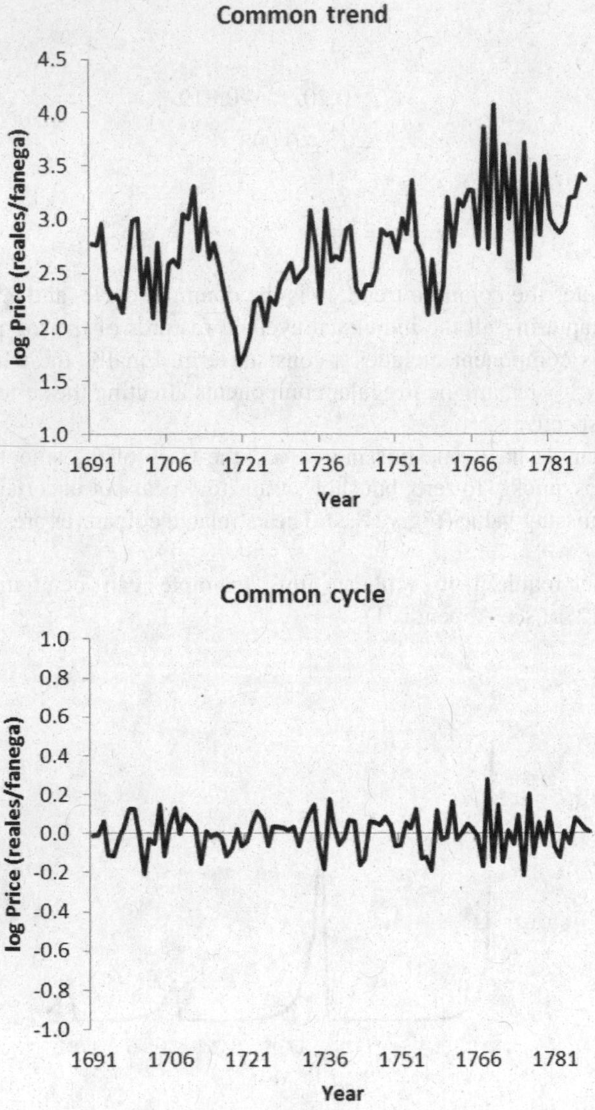

Figure 8.4 *Common features of (log) wheat prices.*

[175]. Simple inspection of the series, displayed in Figure 8.5, shows that both show seasonal fluctuations and a downward drift.

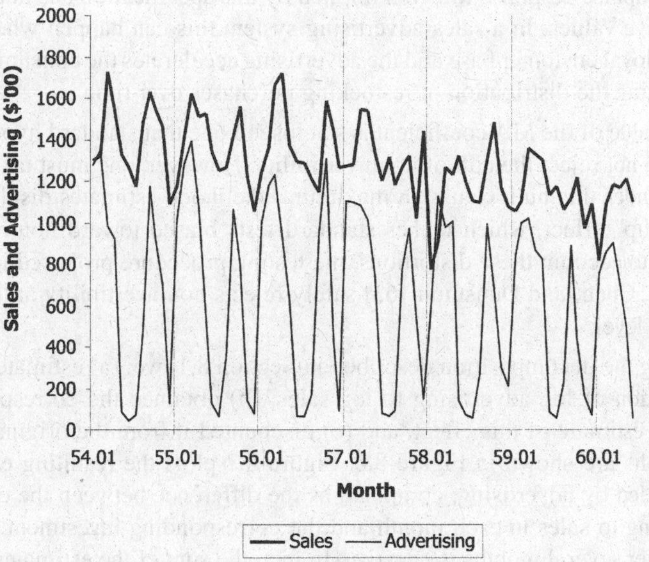

Figure 8.5 *Monthly series of sales and advertising of the Lydia Pinkham vegetable compound from January 1954 to June 1960. Source: Palda [175].*

Building on the methods described by Box, Jenkins, and Reinsel [30], we obtained the following model:

$$\log S_t \times 100 = \frac{\underset{(.022)}{.048} + \underset{(.022)}{.016}\text{B} + \underset{(.026)}{.043}\text{B}^2}{1 - \underset{(.127)}{.713}\text{B} + \underset{(.094)}{.751}\text{B}^2} \log A_t \times 100$$

$$+ \frac{(1 - \underset{(.052)}{.899}\text{B})(1 - \underset{(.094)}{.628}\text{B}^{12})}{(1-\text{B})(1-\text{B}^{12})} \hat{a}_t, \quad (8.60)$$

where S_t and A_t denote, respectively, monthly sales and advertising[4], while the values in parentheses are standard errors for the corresponding estimates. The estimated residual standard deviation is $\hat{\sigma}_a = 7.536$, and the residuals do not show evidence of misspecification, as the portmanteau Q statistics for the residual auto and cross-correlations did not reject the null of no autocorrelation.

Note that:

1. The roots of the polynomials in the numerator and denominator of the transfer function (8.60) are, respectively, $-0.186 \pm 1.040\,i$ and $0.475 \pm 1.052\,i$. Thus, even though this model may be somewhat overparametrized, there are no redundant dynamic factors in this term.

[4]Observations between November 1957 and February 1958 have been treated as missing values because they were outliers.

2. The impulse-response function implied by this specification has both positive and negative values. In a sales-advertising system this can happen when the product has a loyal customer base and the advertising accelerates the consumption, thereby changing the distribution of restocking purchases over time.
3. The value of the MA coefficient is so close to one that standard hypothesis testing would not reject the null of noninvertibility. However, one must take into account that under this null, Gaussian maximum-likelihood estimates display a so-called "pile-up" effect, which makes standard tests biased toward non-rejection. Taking into account these distortions, the testing procedure proposed in Section 3 of Davis, Chen, and Dunsmuir [63] safely rejects noninvertibility at the 5% significance level.

Using the decomposition described in Section 8.1, we: (a) estimated the additive contribution of log advertising to log sales, (b) obtained the corresponding multiplicative estimate of this effect, and (c) discounted it from the original sales series. The results are shown in Figure 8.6. Figure 8.7 plots the resulting estimate of the value added by advertising, computed as the difference between the contribution of advertising to sales in each month and the corresponding investment. Note that the return over several months is negative. In fact, the sum of the estimates in Figure 8.7 is −0.3 million dollars, and thus advertising in this case did not create value for the firm.

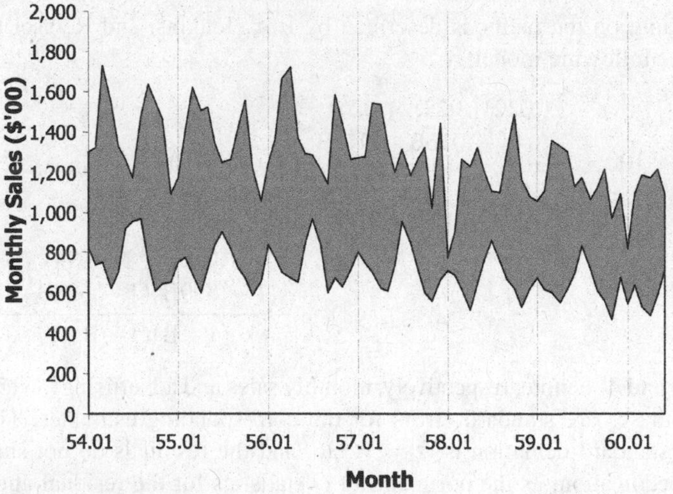

Figure 8.6 *Sales versus the exponential of the stochastic component for the Lydia Pinkham data. The distance between both series (gray area) is an estimate of the effect of advertising over sales.*

Further decomposition of the stochastic components reveals a possible cause of this lack of performance. Figure 8.8 plots the multiplicative components of sales corresponding to seasonality and the estimated effect of advertising. Note that both series have a clear negative correlation, so advertising was systematically increased

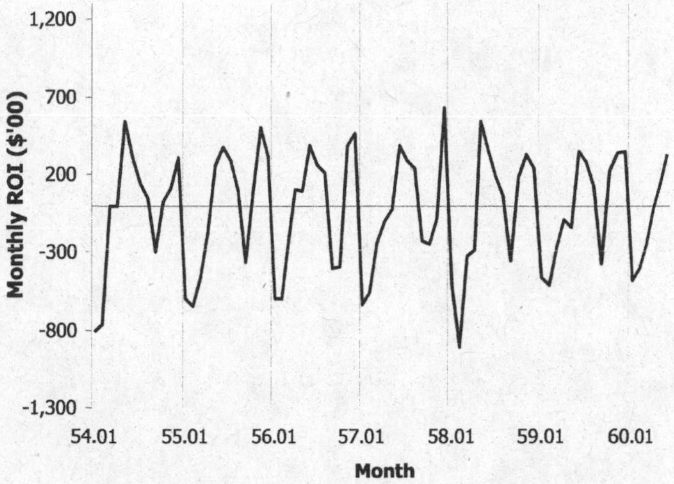

Figure 8.7 *Estimated value added by advertising, computed in each period as the estimate of sales generated by advertising minus the advertising investment (Lydia Pinkham data).*

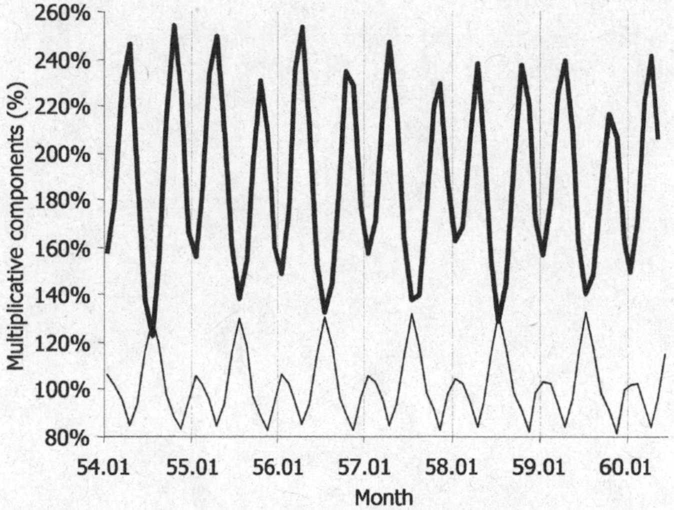

Figure 8.8 *Multiplicative effect of advertising (thick line) versus multiplicative seasonality of sales (thin line) for the Lydia Pinkham data. The values between November 1957 and February 1958 are smoothed interpolations.*

in the seasonal troughs, and decreased in the peaks. Taking into account that both effects are multiplicative, it is immediately apparent that this anti-cyclical management of advertising did not optimize sales.

The code required to replicate this example can be found in the file `Example0853.m`; see Appendix D.

Chapter 9

The VARMAX representation of a State-Space model

9.1 Notation and previous results 130
9.2 Obtaining the VARMAX form of a State-Space model 131
9.2.1 From State-Space to standard VARMAX 132
9.2.2 From State-Space to canonical VARMAX 133
9.3 Practical applications and examples 135
9.3.1 The VARMAX form of some common State-Space models 135
9.3.2 Identifiability and conditioning of the estimates 135
9.3.3 Fitting an errors-in-variables model to Wolf's sunspot series 139
9.3.4 "Bottom-up" modeling of quarterly US GDP trend 140

In Subsection 2.2.4 we presented an analytical procedure to write a generalized VARMAX model in SS form. In this context, "analytical" means that it can be applied to formulations where the model parameters may be generically represented by letters. In this chapter we will discuss a numerical procedure to do the inverse transformation, that is, to obtain the coefficients of the VARMAX representation for any linear SS model. By "numerical" we now mean that this is a computational procedure, which requires numerical values for the parameters.

The fact that there are procedures to go from VARMAX to SS form and vice versa has a clear and interesting consequence: both representations are equally general and choosing between them is just a matter of convenience. Beyond this theoretical implication, these procedures are also useful to solve three practical problems: discussing the identifiability of an SS model, performing diagnostic checks, and calibrating the SS model parameters to ensure that it realizes any given reduced-form VARMAX. Section 9.3 discusses these applications and provides practical examples.

9.1 Notation and previous results

Much applied work in time series analysis is based on variations of the model:

$$\bar{\boldsymbol{F}}(\mathrm{B})\boldsymbol{z}_t = \bar{\boldsymbol{G}}(\mathrm{B})\boldsymbol{u}_t + \bar{\boldsymbol{L}}(\mathrm{B})\boldsymbol{a}_t \tag{9.1}$$

which was already defined in equation (2.33). We will assume that this model is left coprime[1], but the roots of $\bar{\boldsymbol{F}}(\mathrm{B})$ and $\bar{\boldsymbol{L}}(\mathrm{B})$ are allowed to be either inside or outside the unit circle.

Left coprimeness is important, but it is not enough to ensure identifiability of (9.1), as there are still infinitely many parameter sets that realize $\boldsymbol{z}_t$. To see this, note that one can pre-multiply expression (9.1) by any scalar or conformable matrix and obtain an output-equivalent model. Accordingly, identifiability requires additional constraints over $\bar{\boldsymbol{F}}_0$ and $\bar{\boldsymbol{L}}_0$. For instance, $\bar{\boldsymbol{F}}_0 = \bar{\boldsymbol{L}}_0 = \boldsymbol{I}$ yields the standard VARMAX process.

An important feature of model (9.1) is its maximum dynamic order, defined as $p_{\max} = \max\{p, s, q\}$, where p, s, and q are the orders of the polynomials $\bar{\boldsymbol{F}}(\mathrm{B})$, $\bar{\boldsymbol{G}}(\mathrm{B})$, and $\bar{\boldsymbol{L}}(\mathrm{B})$, respectively. We will use this concept often throughout this chapter.

An interesting alternative to standard VARMAX is the widely used VARMAX *echelon* form. The system (9.1) is in echelon form if the triple $\left[\bar{\boldsymbol{F}}(\mathrm{B}) : \bar{\boldsymbol{G}}(\mathrm{B}) : \bar{\boldsymbol{L}}(\mathrm{B})\right]$ is in echelon canonical form, *i.e.*, denoting by $\bar{F}_{kl}(\mathrm{B})$ the kl-th element of $\bar{\boldsymbol{F}}(\mathrm{B})$, and similarly $\bar{G}_{kl}(\mathrm{B})$ for $\bar{\boldsymbol{G}}(\mathrm{B})$ and $\bar{L}_{kl}(\mathrm{B})$ for $\bar{\boldsymbol{L}}(\mathrm{B})$, the polynomial operators are uniquely defined by:

$$\bar{F}_{kk}(\mathrm{B}) = 1 + \sum_{i=1}^{p_k} \bar{F}_{kk}(i)\mathrm{B}^i, \quad \text{for } k = 1, \ldots, m \tag{9.2}$$

$$\bar{F}_{kl}(\mathrm{B}) = \sum_{i=p_k-p_{kl}+1}^{p_k} \bar{F}_{kl}(i)\mathrm{B}^i, \quad \text{for } k \neq l \tag{9.3}$$

$$\bar{G}_{kl}(\mathrm{B}) = \sum_{i=0}^{p_k} \bar{G}_{kl}(i)\mathrm{B}^i, \quad \text{for } k = 1, \ldots, m \tag{9.4}$$

$$\bar{L}_{kl}(\mathrm{B}) = \sum_{i=0}^{p_k} \bar{L}_{kl}(i)\mathrm{B}^i, \quad \text{with } \bar{L}_{kl}(0) = \bar{F}_{kl}(0) \text{ for } k, l = 1, \ldots, m, \tag{9.5}$$

where the integers p_k, $k = 1, \ldots, m$, are known as "Kronecker indices" or "observability indices." They determine the structure of ones and zeros in the echelon form. Equation (9.3) uses the index p_{kl} defined as:

$$p_{kl} = \begin{cases} \min(p_k + 1, p_l), & \text{for } k \geq l, \\ \min(p_k, p_l), & \text{for } k < l, \end{cases} \qquad k, l = 1, \ldots, m. \tag{9.6}$$

where m is the dimension of $\boldsymbol{z}_t$.

[1] For a formal definition of left coprimeness see, e.g. Reinsel [187]. Informally speaking, a process is left coprime if it does not have identical dynamic factors in its autoregressive and moving average polynomials.

Example 9.1.1 Standard VARMAX versus VARMAX echelon

Consider the standard VARMA(2,2) model:

$$(\boldsymbol{I} + \bar{\boldsymbol{F}}_1 \mathrm{B} + \bar{\boldsymbol{F}}_2 \mathrm{B}^2)\boldsymbol{z_t} = (\boldsymbol{I} + \bar{\boldsymbol{L}}_1 \mathrm{B} + \bar{\boldsymbol{L}}_2 \mathrm{B}^2)\boldsymbol{a_t}, \tag{9.7}$$

where,

$$\bar{\boldsymbol{F}}_1 = \begin{bmatrix} -0.70 & 0 & 0 \\ 0.48 & -0.50 & -0.90 \\ -0.02 & 0.30 & -0.20 \end{bmatrix}, \quad \bar{\boldsymbol{F}}_2 = \begin{bmatrix} 0.30 & -0.20 & 0.50 \\ -0.12 & 0.08 & -0.20 \\ 0.18 & -0.12 & 0.30 \end{bmatrix},$$

$$\bar{\boldsymbol{L}}_1 = \begin{bmatrix} -0.20 & 0.40 & 0.70 \\ 0.68 & -0.46 & -0.68 \\ 0.18 & 1.24 & -0.38 \end{bmatrix}, \quad \bar{\boldsymbol{L}}_2 = \begin{bmatrix} 0.30 & 0.50 & -0.80 \\ -0.12 & -0.20 & 0.32 \\ 0.18 & 0.30 & -0.48 \end{bmatrix}.$$

Note that this representation has 34 non-zero parameters, excluding those of the covariance matrix, and $p_{\max} = 2$.

Consider now the VARMA echelon form:

$$(\bar{\boldsymbol{F}}_{\mathbf{0}} + \bar{\boldsymbol{F}}_{\mathbf{1}} \mathrm{B} + \bar{\boldsymbol{F}}_{\mathbf{2}} \mathrm{B}^2)\boldsymbol{z_t} = (\bar{\boldsymbol{L}}_{\mathbf{0}} + \bar{\boldsymbol{L}}_{\mathbf{1}} \mathrm{B} + \bar{\boldsymbol{L}}_{\mathbf{2}} \mathrm{B}^2)\boldsymbol{a_t}, \tag{9.8}$$

with,

$$\bar{\boldsymbol{F}}_0 = \bar{\boldsymbol{L}}_0 = \begin{bmatrix} 1.0 & 0 & 0 \\ 0.4 & 1.0 & 0 \\ -0.6 & 0 & 1.0 \end{bmatrix}, \quad \bar{\boldsymbol{F}}_1 = \begin{bmatrix} -0.7 & 0 & 0 \\ 0.2 & -0.5 & -0.9 \\ 0.4 & 0.3 & -0.2 \end{bmatrix},$$

$$\bar{\boldsymbol{F}}_2 = \begin{bmatrix} 0.3 & -0.2 & 0.5 \\ 0 & 0 & 0 \\ 0 & 0 & 0 \end{bmatrix}, \bar{\boldsymbol{L}}_1 = \begin{bmatrix} -0.2 & 0.4 & 0.7 \\ 0.6 & -0.3 & -0.4 \\ 0.3 & 1.0 & -0.8 \end{bmatrix}, \bar{\boldsymbol{L}}_2 = \begin{bmatrix} 0.3 & 0.5 & -0.8 \\ 0 & 0 & 0 \\ 0 & 0 & 0 \end{bmatrix}.$$

It is easy to see that models (9.7) and (9.8) are equivalent by premultiplying (9.8) by $\bar{\boldsymbol{F}}_{\mathbf{0}}^{-1}$. Kronecker indices in model (9.8) are $p_k = \{2,1,1\}$, corresponding to the maximum dynamic order of each component of $\boldsymbol{z_t}$. Obviously, $p_{\max}$ must be the same in the standard (9.7) and echelon (9.8) representations. Finally, the VARMA echelon: (a) reduces the number of non-zero parameters from 34 to 24 and, (b) is a canonical form, meaning that there are no alternative representations with fewer parameters. These advantages have been pointed out by many authors, Hannan and Deistler [116] and Lutkepohl and Poskitt [159] among others, and more recently by Mélard, Roy, and Saidi [164].

9.2 Obtaining the VARMAX form of a State-Space model

We will now describe two separate procedures for computing the VARMAX coefficients corresponding to a given SS model.

9.2.1 *From State-Space to standard VARMAX*

The first algorithm requires that the dynamics of the SS model comply with two conditions:

Condition C.1: The system order, n, must be an integer multiple of the number of system outputs, m.

Condition C.2: The maximum system order, $p_{\max}$, must be equal to n/m, and the corresponding observability matrix,

$$\boldsymbol{O}_{p_{\max}} = \left[\boldsymbol{H}^T \quad (\boldsymbol{H\Phi})^T \quad (\boldsymbol{H\Phi}^2)^T \quad \ldots \quad (\boldsymbol{H\Phi}^{p_{\max}-1})^T\right]^T,$$

must be of full rank.

Two important special cases that satisfy these conditions are:

1. Every univariate minimal system, as it has $m = 1$ and $\boldsymbol{O}_{p_{\max}} = \boldsymbol{O}_n$, which is of full rank (for any minimal system), and
2. The multivariate minimal systems whose outputs ($z_{k,t}$) have identical Kronecker indices (p_k), since in this case, $n = \sum_{k=1}^{m} p_k$.

Under conditions C.1 and C.2, the algorithm proceeds as follows.

Step 1: If the SS model is not minimal (see Definition 2.4), reduce it to an equivalent minimal realization by the procedure discussed in Subsection 8.1.2.

Step 2: Transform the minimal SS model to the corresponding SEM form (2.19)–(2.20) by the procedure described in Subsection 2.2.4.

Step 3: Obtain the Observable Canonical Form (OCF). To this end, we must take into account that for any minimal SEM fulfilling conditions C.1 and C.2, there exists a nonsingular matrix $\boldsymbol{T}$ such that the similar transformation given by $\boldsymbol{\Phi}^* = \boldsymbol{T}^{-1}\boldsymbol{\Phi}\boldsymbol{T}$ and $\boldsymbol{H}^* = \boldsymbol{H}\boldsymbol{T}$ has the OCF structure, that is:

$$\boldsymbol{\Phi}^* = \begin{bmatrix} -\bar{\boldsymbol{F}}_1 & \boldsymbol{I} & \boldsymbol{0} & \ldots & \boldsymbol{0} \\ -\bar{\boldsymbol{F}}_2 & \boldsymbol{0} & \boldsymbol{I} & \ldots & \boldsymbol{0} \\ \vdots & \vdots & \vdots & \ddots & \vdots \\ -\bar{\boldsymbol{F}}_{p_{\max}-1} & \boldsymbol{0} & \boldsymbol{0} & \ldots & \boldsymbol{I} \\ -\bar{\boldsymbol{F}}_{p_{\max}} & \boldsymbol{0} & \boldsymbol{0} & \ldots & \boldsymbol{0} \end{bmatrix}, \qquad \boldsymbol{H}^* = \left[\boldsymbol{I} \quad \boldsymbol{0} \quad \ldots \quad \boldsymbol{0}\right]. \tag{9.9}$$

where $\bar{\boldsymbol{F}}_i$, $i = 1, 2, \ldots, p_{\max}$, denotes the autoregressive coefficient matrices in Model (9.1). Note that the first component in $\boldsymbol{H}^*$ is the identity matrix because $\bar{\boldsymbol{F}}_0 = \boldsymbol{I}$ in standard VARMAX. A formal proof of this result and a procedure to compute the transformation matrix $\boldsymbol{T}$ can be found in Casals, Garcia–Hiernaux, and Jerez [39].

Step 4: Obtain the polynomial matrices $\bar{\boldsymbol{G}}(\mathrm{B})$ and $\bar{\boldsymbol{L}}(\mathrm{B})$ in Model (9.1). Using the matrices $\bar{\boldsymbol{F}}_i$ in (9.9) the remaining polynomial matrices of the VARMAX representation can be calculated as:

$$\bar{G}_0 = D, \quad \begin{bmatrix} \bar{G}_1 \\ \bar{G}_2 \\ \vdots \\ \bar{G}_{p_{\max}} \end{bmatrix} = T^{-1}\Gamma + \begin{bmatrix} \bar{F}_1\bar{G}_0 \\ \bar{F}_2\bar{G}_0 \\ \vdots \\ \bar{F}_{p_{\max}}\bar{G}_0 \end{bmatrix}, \quad \begin{bmatrix} \bar{L}_1 \\ \bar{L}_2 \\ \vdots \\ \bar{L}_{p_{\max}} \end{bmatrix} = T^{-1}E + \begin{bmatrix} \bar{F}_1 \\ \bar{F}_2 \\ \vdots \\ \bar{F}_{p_{\max}} \end{bmatrix}, \tag{9.10}$$

where D, Γ and E come from the original SEM and T is the transformation matrix.

9.2.2 *From State-Space to canonical VARMAX*

We will now discuss an alternative algorithm which is more general than the previous one, as it does not require any *a priori* conditions. The downside is that it is more complex. In particular, it requires one to determine the Kronecker indices of the system, which in the previous case were directly implied by conditions C.1 and C.2. The algorithm can be divided into two stages: obtaining the Luenberger Canonical Form, and deriving the corresponding VARMAX echelon coefficients.

Stage 1: *Obtaining the LCF.*

Step 1 (Enforcing minimality) and *Step 2* (Obtaining the equivalent SEM) are identical to those in the previous algorithm.

Step 3. Identifying the Kronecker indices. To this end, consider the observability matrix of the SEM:

$$O_n = \left[H^T \quad (H\Phi)^T \quad (H\Phi^2)^T \quad \dots \quad (H\Phi^{n-1})^T \right]^T. \tag{9.11}$$

Minimality assures that this matrix has n linearly independent rows. If these rows are chosen in descending order we can build a base, which after re-ordering can be written in block matrix form as:

$$M = \left[h_1 \quad (h_1^T\Phi)^T \quad \dots \quad (h_1^T\Phi^{p_1-1})^T \right. \\ \left. \dots \quad h_m \quad (h_m^T\Phi)^T \quad \dots \quad (h_m^T\Phi^{p_m-1})^T \right]^T, \tag{9.12}$$

where h_k is the k-th row of H, and p_k $(k = 1, ..., m)$ are the Kronecker indices, so that $\sum_{k=1}^{m} p_k = n$.

Step 4. Computing the LCF matrices. For any minimal SEM with Kronecker indices p_k $(k = 1, ..., m)$, there exists a nonsingular matrix T such that $\Phi^* = T^{-1}\Phi T$ and $H^* = HT$ have the LCF structure, that is:

$$\Phi^* = \begin{bmatrix} F_1 & Q_1 & 0 & \dots & 0 \\ F_2 & 0 & Q_2 & \dots & 0 \\ \vdots & \vdots & \vdots & \ddots & \vdots \\ F_{p_{\max}-1} & 0 & 0 & \dots & Q_{p_{\max}-1} \\ F_{p_{\max}} & 0 & 0 & \dots & 0 \end{bmatrix} \quad \text{and} \quad H^* = \left[F_0 \quad 0 \quad 0 \right]. \tag{9.13}$$

being $\boldsymbol{\Phi}^*$ a companion matrix, where each $\boldsymbol{F}_j$ block $(j = 1, ..., p_{\max})$ has a number of rows equal to the number of Kronecker indices greater or equal to k, has $\bar{m} = \sum_{k=1}^{m} \min\{p_k, 1\}$ columns, and some null elements. In fact, the (k,l)-th element of $\boldsymbol{F}_j$ will be nonzero only if $j \in [p_k - p_{kl} + 1 p_k]$, where p_{kl} is defined as:

$$p_{kl} = \left\{ \begin{array}{l} \min(p_k + 1, p_l) \text{ for } k \geq l \\ \min(p_k, p_l) \text{ for } k < l \end{array} \right\} \quad k, l = 1, 2, ..., m, \tag{9.14}$$

where the integers p_k, $k = 1, ..., m$, are the Kronecker or observability indices. Each $\boldsymbol{Q}_k$ block is a zero/one matrix, with as many columns as the number of observability indices that are greater or equal to k. If the endogenous variables are sorted according to their observability indices, the structure of $\boldsymbol{Q}_k$ will be $\boldsymbol{Q}_k = \begin{bmatrix} \boldsymbol{I}_{k+1} & \mathbf{0} \end{bmatrix}^T$, where $\boldsymbol{I}_{k+1}$ is an identity matrix with the same number of rows as $\boldsymbol{F}_{k+1}$. With respect to $\boldsymbol{H}^*$, $\boldsymbol{F}_0$ is an $m \times \bar{m}$ matrix, such that the rows corresponding to components with nonzero observability indices can be organized in an $\bar{m} \times \bar{m}$ lower triangular matrix with ones in the main diagonal. A formal proof of this result and how to derive the transformation matrix $\boldsymbol{T}$ can be found in Casals, Garcia–Hiernaux, and Jerez [39].

Stage 2: *Identifying the VARMAX echelon coefficients.*
Transforming the SEM (2.19)–(2.20) to the LCF yields:

$$\boldsymbol{x}^*_{t+1} = \boldsymbol{\Phi}^* \boldsymbol{x}^*_t + \boldsymbol{\Gamma}^* \boldsymbol{u}_t + \boldsymbol{E}^* \boldsymbol{a}_t \tag{9.15}$$

$$\boldsymbol{z}_t = \boldsymbol{H}^* \boldsymbol{x}^*_t + \boldsymbol{D} \boldsymbol{u}_t + \boldsymbol{a}_t, \tag{9.16}$$

where the matrices $\boldsymbol{\Phi}^*$ and $\boldsymbol{H}^*$ have the LCF structure characterized in (9.13). In order to write this model in an equivalent polynomial form, it is convenient to augment the system dimension up to $mp_{\max}$ by adding as many non-excited states as needed. Then, the structure of $\boldsymbol{\Phi}^*$ will be as in an LCF, but with: (i) the identity matrix instead of $\boldsymbol{Q}_j$, and (ii) an augmented dimension of matrices $\boldsymbol{F}_j$, now $m \times m$. Note that the constraints on potentially nonzero parameters also affect these augmented matrices. Consequently, the new non-excited states require adding null columns to $\boldsymbol{H}^*$ except for the endogenous variables with a null observability matrix, so that the augmented $\boldsymbol{F}_j$ is an $m \times m$ lower triangular matrix with ones in its main diagonal. This particular structure allows us to write the observation equation as:

$$\boldsymbol{F}_0^{-1} \boldsymbol{z}_t = \boldsymbol{x}^*_{1:m,t} + \boldsymbol{F}_0^{-1} \boldsymbol{D} \boldsymbol{u}_t + \boldsymbol{F}_0^{-1} \boldsymbol{a}_t, \tag{9.17}$$

where $\boldsymbol{x}^*_{1:m,t}$ denotes the first m elements of the state vector $\boldsymbol{x}^*_t$. According to (9.17), one can isolate the $\boldsymbol{x}^*_{1:m,t}$ and substitute them into the state equation.

Finally, taking into account the companion structure of $\mathbf{\Phi}^*$, we obtain the coefficients in the VARMAX form (9.1) as:

$$\bar{F}_0 = F_0^{-1}, \qquad \bar{F}_j = -F_j F_0^{-1}, \tag{9.18}$$
$$\bar{G}_0 = F_0^{-1} D, \qquad \bar{G}_j = \Gamma^* + \bar{F}_j D, \tag{9.19}$$
$$\bar{L}_0 = \bar{F}_0, \qquad \bar{L}_j = E^* + \bar{F}_j, \qquad j = 1, \ldots, p_{\max}, \tag{9.20}$$
$$\bar{F}(\mathrm{B}) = \sum_{j=0}^{p_{\max}} \bar{F}_j \mathrm{B}^j, \qquad \bar{G}(\mathrm{B}) = \sum_{j=0}^{p_{\max}} \bar{G}_j \mathrm{B}^j, \qquad \bar{L}(\mathrm{B}) = \sum_{j=0}^{p_{\max}} \bar{L}_j \mathrm{B}^j. \tag{9.21}$$

Note that this representation has the characteristic structure of a canonical VARMAX echelon model; see Dickinson, Morf, and Kailath [79].

9.3 Practical applications and examples

9.3.1 *The VARMAX form of some common State-Space models*

SS models are adequate for many uses, such as extracting the structural components of a time series or working with nonstandard samples such as *e.g.*, those with missing values, aggregated data, or observation errors. On the other hand, VARMAX representations are more familiar in mainstream time series analysis, and they come equipped with a convenient set of diagnostic checks stemming from the fact that they produce a unique residual vector whose statistical properties are well understood; see *e.g.* Li [148]. To get the best of both worlds, one therefore needs the ability to derive the VARMAX form corresponding to a given SS model, bearing in mind that the literature describes how to write a VARMAX model in SS form; see Subsection 2.2.4.

Table 9.1 illustrates the results of the methods described in Section 9.2 by showing the ARIMAX structure corresponding to some common SS models. All the models in Table 9.1 are univariate, so these results could have been obtained via other approaches such as, *e.g.*, by identifying the autocorrelation function of the endogenous variable, see Chapter 2 of Harvey [118], or using the pseudo-spectrum implied by the unobserved components and reduced form models, see Bujosa, Garcia–Ferrer, and Young [33].

Table 9.2 shows that the methods described in the previous section can also be applied to vector models. Even in the simplest cases, it would be difficult to obtain the corresponding VARMAX form via the above mentioned autocorrelation and spectral approaches.

The code required to replicate this example can be found in the files `Example0931a.m` and `Example0931b.m`; see Appendix D.

9.3.2 *Identifiability and conditioning of the estimates*

The methods described in Section 9.2 are also useful to analyze two important issues: model identifiability and conditioning.

Table 9.1 *Structural and ARIMAX forms for some common univariate specifications. Both representations are observationally equivalent.*

Model	Structural representation	ARIMAX representation
(i) Dynamic regression with errors in variables	$x_{t+1} = .5x_t + .7u_t + w_t, \quad w_t \sim \text{IID}(0, 1.5)$ $z_t = x_t + v_t, \quad v_t \sim \text{IID}(0, 1)$	$(1+.5\text{B})z_t = .7u_{t-1} + (1-.188\text{B})a_t,$ $\sigma_a^2 = 2.660$
(ii) Random walk plus noise model	$\mu_{t+1} = \mu_t + \zeta_t, \quad \zeta_t \sim \text{IID}(0, 1/100)$ $z_t = \mu_t + \varepsilon_t, \quad \varepsilon_t \sim \text{IID}(0, 1)$	$(1-\text{B})z_t = (1-.905\text{B})a_t,$ $\sigma_a^2 = 1.105$
(iii) Integrated random walk plus noise model	$\mu_{t+1} = \mu_t + \beta_t$ $\beta_{t+1} = \beta_t + \zeta_t, \quad \zeta_t \sim \text{IID}(0, 1/100)$ $z_t = \mu_t + \varepsilon_t, \quad \varepsilon_t \sim \text{IID}(0, 1)$	$(1-\text{B})^2 z_t = (1-1.558\text{B}+.638\text{B}^2)a_t,$ $\sigma_a^2 = 1.567$
(iv) Model (iii) with one input	$\mu_{t+1} = \mu_t + \beta_t$ $\beta_{t+1} = \beta_t + \zeta_t, \quad \zeta_t \sim \text{IID}(0, 1/100)$ $z_t = \mu_t + .5u_t + \varepsilon_t, \quad \varepsilon_t \sim \text{IID}(0, 1)$	$(1-\text{B})^2 z_t = .5(1-\text{B})^2 u_t + (1-1.558\text{B}+.638\text{B}^2)a_t,$ $\sigma_a^2 = 1.567$
(v) Model (iii) with quarterly dummy variable seasonality	$\mu_{t+1} = \mu_t + \beta_t$ $\beta_{t+1} = \beta_t + \zeta_t, \quad \zeta_t \sim \text{IID}(0, 1/100)$ $\gamma_{t+1} = -\gamma_t - \gamma_{t-1} - \gamma_{t-2} + w_t, \; w_t \sim \text{IID}(0, 1/10)$ $z_t = \mu_t + \varepsilon_t, \quad \varepsilon_t \sim \text{IID}(0, 1)$	$(1-\text{B})(1-\text{B}^4)z_t =$ $(1-.714\text{B}+.114\text{B}^2-.010\text{B}^3-.563\text{B}^4+.438\text{B}^5)a_t,$ $\sigma_a^2 = 2.283$

Note: In all cases the errors affecting the state and observation equations are assumed to be independent.

Table 9.2 *Structural and VARMAX forms for some common bivariate specifications. Both representations are observationally equivalent.*

Model	Representations: Structural and VARMAX
(vi) Seemingly unrelated time series equations with inputs	$$\begin{bmatrix}\mu_{t+1}^1\\ \mu_{t+1}^2\end{bmatrix} = \begin{bmatrix}1 & 0\\ 0 & 1\end{bmatrix}\begin{bmatrix}\mu_t^1\\ \mu_t^2\end{bmatrix} + \begin{bmatrix}\zeta_t^1\\ \zeta_t^2\end{bmatrix}, \quad \begin{bmatrix}\zeta_t^1\\ \zeta_t^2\end{bmatrix} \sim \text{IID}\left(\begin{bmatrix}0\\0\end{bmatrix}, \begin{bmatrix}1/100 & 5/1000\\ 5/1000 & 2/100\end{bmatrix}\right);$$ $$\begin{bmatrix}z_t^1\\ z_t^2\end{bmatrix} = \begin{bmatrix}1 & 0\\ 0 & 1\end{bmatrix}\begin{bmatrix}\mu_t^1\\ \mu_t^2\end{bmatrix} + \begin{bmatrix}.5\\ .7\end{bmatrix} u_t + \begin{bmatrix}\varepsilon_t^1\\ \varepsilon_t^2\end{bmatrix}, \quad \begin{bmatrix}\varepsilon_t^1\\ \varepsilon_t^2\end{bmatrix} \sim \text{IID}\left(\begin{bmatrix}0\\0\end{bmatrix}, \begin{bmatrix}1 & .2\\ .2 & .5\end{bmatrix}\right).$$ $$\begin{bmatrix}1-\text{B} & 0\\ 0 & 1-\text{B}\end{bmatrix}\begin{bmatrix}z_t^1\\ z_t^2\end{bmatrix} = \begin{bmatrix}.5\\ .7\end{bmatrix}(1-\text{B})u_t + \begin{bmatrix}1-.910\text{B} & -.012\text{B}\\ -.009\text{B} & 1-.816\text{B}\end{bmatrix}\begin{bmatrix}a_t^1\\ a_t^2\end{bmatrix},$$ $$\begin{bmatrix}a_t^1\\ a_t^2\end{bmatrix} \sim \text{IID}\left(\begin{bmatrix}0\\0\end{bmatrix}, \begin{bmatrix}1.104 & .232\\ .232 & .610\end{bmatrix}\right).$$
(vii) Two series with a common random walk trend	$$\mu_{t+1} = \mu_t + \zeta_t, \quad \zeta_t \sim \text{IID}(0, 1/100);$$ $$\begin{bmatrix}z_t^1\\ z_t^2\end{bmatrix} = \begin{bmatrix}1\\ 1\end{bmatrix}\mu_t + \begin{bmatrix}\varepsilon_t^1\\ \varepsilon_t^2\end{bmatrix}, \quad \begin{bmatrix}\varepsilon_t^1\\ \varepsilon_t^2\end{bmatrix} \sim \text{IID}\left(\begin{bmatrix}0\\0\end{bmatrix}, \begin{bmatrix}1 & .2\\ .2 & .5\end{bmatrix}\right).$$ $$\begin{bmatrix}1-\text{B} & 0\\ -1 & 1\end{bmatrix}\begin{bmatrix}z_t^1\\ z_t^2\end{bmatrix} = \begin{bmatrix}1-.961\text{B} & .104\text{B}\\ 0 & 1\end{bmatrix}\begin{bmatrix}a_t^1\\ a_t^2\end{bmatrix}, \quad \begin{bmatrix}a_t^1\\ a_t^2\end{bmatrix} \sim \text{IID}\left(\begin{bmatrix}0\\0\end{bmatrix}, \begin{bmatrix}1.070 & .270\\ .270 & .570\end{bmatrix}\right).$$
(viii) Two series with a common trend and an autonomous random walk component	$$\begin{bmatrix}\mu_{t+1}^1\\ \beta_{t+1}\\ \mu_{t+1}^2\end{bmatrix} = \begin{bmatrix}1 & 1 & 0\\ 0 & 1 & 0\\ 0 & 0 & 1\end{bmatrix}\begin{bmatrix}\mu_t^1\\ \beta_t\\ \mu_t^2\end{bmatrix} + \begin{bmatrix}0 & 0\\ 1 & 0\\ 0 & 1\end{bmatrix}\begin{bmatrix}\zeta_t^1\\ \zeta_t^2\end{bmatrix}, \quad \begin{bmatrix}\zeta_t^1\\ \zeta_t^2\end{bmatrix} \sim \text{IID}\left(\begin{bmatrix}0\\0\end{bmatrix}, \begin{bmatrix}1/100 & 5/1000\\ 5/1000 & 2/100\end{bmatrix}\right);$$ $$\begin{bmatrix}z_t^1\\ z_t^2\end{bmatrix} = \begin{bmatrix}1 & 1 & 0\\ 0 & 1 & 1\end{bmatrix}\begin{bmatrix}\mu_t^1\\ \beta_t\\ \mu_t^2\end{bmatrix} + \begin{bmatrix}\varepsilon_t^1\\ \varepsilon_t^2\end{bmatrix}, \quad \begin{bmatrix}\varepsilon_t^1\\ \varepsilon_t^2\end{bmatrix} \sim \text{IID}\left(\begin{bmatrix}0\\0\end{bmatrix}, \begin{bmatrix}1 & .2\\ .2 & .5\end{bmatrix}\right).$$ $$\begin{bmatrix}(1-\text{B})^2 & 0\\ 0 & 1-\text{B}\end{bmatrix}\begin{bmatrix}z_t^1\\ z_t^2\end{bmatrix} = \begin{bmatrix}1-1.582\text{B}+.650\text{B}^2 & .060\text{B}-.021\text{B}^2\\ .036\text{B} & 1-.777\text{B}\end{bmatrix}\begin{bmatrix}a_t^1\\ a_t^2\end{bmatrix},$$ $$\begin{bmatrix}a_t^1\\ a_t^2\end{bmatrix} \sim \text{IID}\left(\begin{bmatrix}0\\0\end{bmatrix}, \begin{bmatrix}1.549 & .329\\ .329 & .659\end{bmatrix}\right).$$

Note: In all these models the errors affecting the state and observation equations are assumed to be independent.

A parametric model is said to be *identifiable* if no two parameter settings yield the same distribution of observations. By definition, canonical VARMAX models are always identifiable, while, in general, there are infinitely many SS models realizing the same reduced-form VARMAX (see discussion on similar transformations in Subsection 2.1.2). Therefore, SS models are unidentifiable unless one imposes an adequate structure on its parameter matrices.

Glover and Willem [108] provide analytical conditions for local and global identifiability of a linear SS model. These conditions are scientifically valuable, but very difficult to apply in practice. On the other hand, the algorithms described in Section 9.2 can be applied to discuss the local identifiability of a given SS model via the following numerical procedure.

Step 1: Compute the response of VARMAX coefficients to small perturbations in the SS parameters. These values would be finite-difference approximations to the corresponding partial derivatives.

Step 2: Organize these derivatives into a Jacobian matrix, $\boldsymbol{J}$, with as many columns as the number of free parameters in the SS model (N_{SS}), and as many rows as the number of parameters in the VARMAX form (N_V).

Step 3: Compute the rank of $\boldsymbol{J}$, denoted by rank($\boldsymbol{J}$).

Step 4: Characterize the identifiability of the system according to the procedure described in Table 9.3.

Table 9.3 *Characterization of the identifiability of an SS model using the rank of the Jacobian ($\boldsymbol{J}$) measuring how a change in the structural model parameters affects the parameters in the reduced form.*

Condition	Identifiability
$\text{rank}(\boldsymbol{J}) < N_{SS}$	The SS model parameters are underidentified
$\text{rank}(\boldsymbol{J}) = N_{SS} = N_V$	The SS model parameters are exactly identified
$\text{rank}(\boldsymbol{J}) = N_{SS} < N_V$	The SS model parameters are overidentified

Note: N_{SS} and N_V are, respectively, the number of parameters in the SS and VARMAX representation.

An efficient way to perform Step 3 above consists of computing the singular-value decomposition of $\boldsymbol{J}$ in Step 2; see Appendix A.4. Note that a null singular value corresponds to a linear combination of the VARMAX parameters that is not affected by perturbations in the SS model parameters, thereby pointing to a specific source of non-identifiability. Accordingly, the rank of $\boldsymbol{J}$ is the number of non-zero singular values.

Moreover, by defining the transformation as the function $P_V = f(P_{SS})$, being P_V and P_{SS} vectorizations of the set of VARMAX and SS parameters, respectively, one can use the Jacobian $\boldsymbol{J}$ to compute its condition number as $c[f(P_{SS})] = ||\boldsymbol{J}|| \cdot ||P_{SS}||/||P_V||$. This value measures the sensitivity of the VARMAX coefficients with respect to changes in the SS model parameters, so that a small conditioning number means that the SS coefficients are ill-identified because they are insensitive to changes in the VARMAX parameters[2].

Consider for example the following SS model:

$$x_{t+1} = -\phi x_t + w_t, \qquad w_t \sim \text{IID}(0, 0.1) \tag{9.22}$$

$$z_t = x_t + v_t, \qquad v_t \sim \text{IID}(0, 1), \tag{9.23}$$

with $\text{E}(w_t, v_t) = 0$, which has an ARMA(1,1) reduced form:

$$(1 + \phi \text{B}) z_t = (1 + \theta \text{B}) a_t, \quad a_t \sim \text{IID}(0, \sigma_a^2) \tag{9.24}$$

where the AR parameter in (9.24) coincides with the opposite of the transition scalar in (9.22). Figure 9.1 displays the values of the θ and σ_a^2 parameters associated with different values of ϕ, as well as the smallest singular value of the Jacobian defined above in each case. Note that, when $\phi = 0$, the corresponding singular value is null. In this case, the structural model degenerates to the sum of two white noise processes, and is therefore unidentifiable.

The code required to replicate this example can be found in the file `Example0932.m`; see Appendix D.

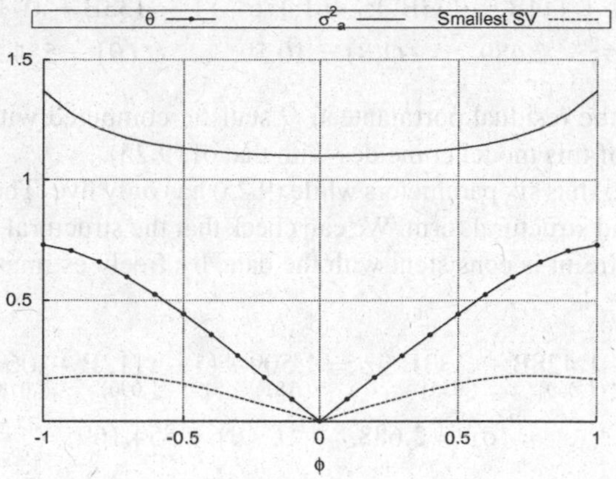

Figure 9.1 *Parameters of the ARMA model (9.24) and smallest singular value (SV) of the Jacobian for several values of ϕ in the AR(1) plus error model.*

9.3.3 Fitting an errors-in-variables model to Wolf's sunspot series

This example illustrates the use of the procedures described in Section 9.2 to perform diagnostic checking on a given estimated SS model.

[2]The term "ill-identification" refers here to a relaxation of the idea of strict non-identification, this being a situation where a statistical model has more than one set of parameters generating the same distribution of observations, implying that multiple parameterizations are observationally equivalent.

Consider the annual series of Wolf's Sunspot Numbers 1700–1988, taken from [210]. Several prior analyses of this series conclude that it has a harmonic cycle with a period of about 11 years. On the other hand, the dataset is compoosed of records compiled by human observers using optical devices of diverse qualities, so it seems natural to infer that the recorded values are affected by observation errors.

Building on these two ideas, we fitted an AR(2) plus white noise errors model to the original data divided by 10, to improve the scaling of the parameters. The resulting estimates are:

$$(1 - \underset{(.048)}{1.444}\mathrm{B} + \underset{(.047)}{.743}\mathrm{B}^2)\hat{z}_t^* = \underset{(.145)}{1.476} + \hat{w}_t, \qquad \hat{\sigma}_w^2 = 2.205,$$
$$z_t = \hat{z}_t^* + \hat{v}_t, \qquad \hat{\sigma}_v^2 = 0.147, \qquad \ell^*(\hat{\theta}) = 554.246, \tag{9.25}$$

where z_t and z_t^* are, respectively, the scaled Wolf number at year t, and the underlying "error free" value. $\ell^*(\hat{\theta})$ denotes the minus log-likelihood value corresponding to the estimates. Note that the latent AR(2) structure has complex roots, which implies that the data follows a damped cycle with a period of 10.87 years.

Using the algorithm described in Subsection 9.2.1, the SS representation of (9.25) can be written as the ARMA(2,2) model:

$$(1 + 1.444\mathrm{B} + .743\mathrm{B}^2)z_t = 1.476 + (1 - .133\mathrm{B} + .041\mathrm{B}^2)\hat{a}_t,$$
$$\sigma_{\hat{a}}^2 = 2.689, \qquad Q(8) = 10.59, \qquad \ell^*(\hat{\theta}) = 554.246, \tag{9.26}$$

where $Q(8)$ is the residual portmanteau Q statistic computed with 8 lags. Note that the likelihood of this model coincides with that of (9.25).

Model (9.26) has six parameters while (9.25) has only five. Therefore, the latter is an overidentified structural form. We can check that the structural model's (overidentification) constraint is consistent with the data, by freely estimating the parameters in (9.26):

$$(1 + \underset{(.069)}{1.428}\mathrm{B} + \underset{(.055)}{.733}\mathrm{B}^2)z_t = \underset{(.181)}{1.509} + (1 - \underset{(.090)}{.112}\mathrm{B} + \underset{(.078)}{.064}\mathrm{B}^2)\hat{a}_t,$$
$$\hat{\sigma}_{\hat{a}}^2 = 2.688, \qquad \ell^*(\hat{\theta}) = 554.169. \tag{9.27}$$

Therefore, models (9.26) and (9.27) are almost identical. Their statistical consistency can be formally tested by computing a likelihood ratio statistic; for which the value of 0.154 confirms that the structural constraint is consistent with the data.

The code required to replicate this example can be found in the file `Example0933.m`; see Appendix D.

9.3.4 *"Bottom-up" modeling of quarterly US GDP trend*

The model-building sequence adopted in the above Wolf's sunspot series example can be described as "top-down," meaning that we first fit a structural ("top") model and then obtain the corresponding ARIMA ("down") reduced form.

We will now demonstrate that the methods presented in Section 9.2 can be used to implement a "bottom-up" modeling strategy, whereby one first fits a reduced-form model to the data, and then computes the structural model parameters which (either exactly or approximately) realize the reduced form. This approach, originally proposed by Nerlove, Grether, and Carvalho [166], is justified if one wants to combine the advantages of a structural SS model with the ability of reduced-form models to capture the data sample properties.

Note also that this idea is closely related to the notion of ARIMA-based time series decomposition, originally suggested by Hillmer and Tiao [125], where a reduced form ARIMA model is decomposed into the sum of several structural components; see Chapter 8.

Now consider the quarterly and seasonally adjusted series of US Gross Domestic Product (GDP_t), from the 1st quarter of 1947 to the 3rd quarter of 2008, in constant 2000 US Dollars. The GDP trend is often extracted with the filter proposed by Hodrick and Prescott [126], which is equivalent to the smoothed trend obtained from the SS model:

$$\begin{aligned}
\mu_{t+1} &= \mu_t + \beta_t \\
\beta_{t+1} &= \beta_t + \zeta_t \\
z_t &= \mu_t + \varepsilon_t \\
\begin{bmatrix} \zeta_t \\ \varepsilon_t \end{bmatrix} &\sim \text{IID}\left(\begin{bmatrix} \zeta_t \\ \varepsilon_t \end{bmatrix}, \begin{bmatrix} \sigma_\zeta^2 & 0 \\ 0 & \sigma_\varepsilon^2 \end{bmatrix} \right),
\end{aligned} \tag{9.28}$$

with a signal-to-noise variance ratio such that $\sigma_\zeta^2/\sigma_\varepsilon^2 = 1/1600$; see Harvey and Trimbur [120].

While the Hodrick–Prescott filter is a simple and effective tool to extract a smooth long-term trend component, it does not capture the dynamics of the data well. In this case, if we fit model (9.28) to the series $z_t = \log(GDP_t) \times 100$, we obtain the ML variance estimates $\sigma_\varepsilon^2 = 1.359$ and $\sigma_\zeta^2 = 1.359/1600$. Applying our method to this model yields the reduced form ARIMA model:

$$\begin{aligned}
&(1-\text{B})^2 \log(GDP_t) \times 100 = (1 - 1.777\text{B} + 0.799\text{B}^2)\hat{a}_t, \\
&\sigma_{\hat{a}}^2 = 1.699, \qquad Q(15) = 239.82, \qquad \ell^*(\hat{\theta}) = 174.8,
\end{aligned} \tag{9.29}$$

where the large value of the residual Q-statistic suggests that a strict Hodrick-Prescott specification does not capture all the autocorrelation in this series. Therefore, we may want to adjust a trend model with the dynamic structure of (9.28) so that it realizes a previously fitted ARIMA model. This modeling strategy can be implemented with the following procedure:

Step (1) Fit an ARIMA (or VARMAX, in a multiple time series framework) model to the dataset.

Step (2) Compute the SS model parameters which realize more closely the model previously fitted. This requires a non-linear iterative procedure to minimize a given loss function. In this example we specify this function as the square root of the difference between the parameters of:

Step (2.a) ...the model fitted in Step (1), and

Step (2.b) ...those of the reduced-form corresponding to the SS model[3].

Table 9.4 summarizes the results of the bottom-up sequence applied to the GDP data. In Step (1) we fitted an ARIMA model to $z_t = \log(GDP_t) \times 100$. Note that its parameters are very different from those of model (9.29).

In Step (2.a) we estimated the two variances of an integrated random-walk model by minimizing the loss function defined above and the corresponding reduced-form model. Note that the latter is similar but not identical to the model in Step (1), so both models are not strictly exactly equivalent.

On the other hand, comparing the models in Steps (1) and (2.a) we note immediately that the latter is overidentified, as it only has two free parameters. In Step (2.b) we freed the null constraint imposed on the model covariance, in order to improve the fit between it and the reduced-form. The results show clearly that both are equivalent so we can conclude that, without the zero-correlation overidentifying constraint, the dynamic structure underlying the HP filter model could be flexible enough to capture most of the data autocorrelation.

The code required to replicate this example can be found in the file `Example0934.m`; see Appendix D.

[3]Note that there are many valid specifications for the loss function employed in Step (2). For example, one could minimize the squared sum of the difference between the log-likelihood of both models, or the residual series generated by both models. These alternative functions would be particularly useful if the SS model cannot realize exactly the reduced form model.

Table 9.4 *Step-by-step results from a bottom-up modeling of the US GDP series.*

Step	Structural model	ARIMAX model
(1) Fit a reduced-form model to the series	NONE	$(1-\mathrm{B})^2\log(GDP_t)\times 100 = (1-\underset{(.075)}{.661}\mathrm{B}+.000^*\mathrm{B}^2)\hat{a}_t$; $\sigma^2_{\hat{a}} = .250,\quad Q(15) = 16.54,\quad \ell^*(\hat{\theta}) = 75.371$
(2.a) Estimate the integrated random walk constraining the covariance to zero	$\hat{\mathrm{cov}}\begin{bmatrix}\zeta_t\\ \varepsilon_t\end{bmatrix} = \begin{bmatrix}.083 & 0^*\\ 0^* & .044\end{bmatrix}$	$(1-\mathrm{B})^2\log(GDP_t)\times 100 = (1-.600\mathrm{B}+.176\mathrm{B}^2)\hat{a}_t$; $\sigma^2_{\hat{a}} = .250,\quad Q(15) = 22.64,\quad \ell^*(\hat{\theta}) = 77.173$
(2.b) Estimate the integrated random walk freeing the covariance constraint	$\hat{\mathrm{cov}}\begin{bmatrix}\zeta_t\\ \varepsilon_t\end{bmatrix} = \begin{bmatrix}.029 & -.083\\ -.083 & .083\end{bmatrix}$	$(1-\mathrm{B})^2\log(GDP_t)\times 100 = (1-.661\mathrm{B}+.000\mathrm{B}^2)\hat{a}_t$; $\sigma^2_{\hat{a}} = .250,\quad Q(15) = 16.54,\quad \ell^*(\hat{\theta}) = 75.371$

Notes: The figure in parenthesis is the standard error of the estimate; some ARIMAX parameters do not have standard errors because they have been derived from the structural SS model estimates. The values $Q(15)$ and $\ell^*(\hat{\theta})$ represent, respectively, the Ljung–Box [151] Q statistic computed with 15 lags of the residual autocorrelation function, and the minus log (Gaussian) likelihood upon convergence. The figures with an asterisk correspond to constrained parameters.

Chapter 10

Aggregation and disaggregation of time series

10.1 The effect of aggregation on an SS model 146

10.1.1 The high-frequency model in stacked form 146

10.1.2 Aggregation relationships 148

10.1.3 Relationships between the models for high, low, and mixed-frequency data 149

10.1.4 The effect of aggregation on predictive accuracy 150

10.2 Observability in the aggregated model 150

10.2.1 Unobservable modes 150

10.2.2 Observability and fixed-interval smoothing 152

10.2.3 An algorithm to aggregate a linear model: theory and examples 153

10.3 Specification of the high-frequency model 154

10.3.1 Enforcing approximate consistency 154

10.3.2 "Bottom-up" determination of the quarterly model 158

10.4 Empirical example 158

10.4.1 Annual model 159

10.4.2 Decomposition of the quarterly indicator 159

10.4.3 Specification and estimation of the quarterly model 160

10.4.4 Diagnostics 160

10.4.5 Forecast accuracy and non-conformable samples 162

10.4.6 Comparison with alternative methods 163

This chapter discusses how to model, interpolate, and forecast a vector of time series sampled at different frequencies. To avoid cumbersome wording, we will refer to the low-frequency variables as "annual" and to the high-frequency variables as "quarterly." The results, however, are valid for any combination of sampling frequencies.

Our starting point consists of breaking the global problem into four basic questions: How could one specify a quarterly model from a sample combining quarterly and annual data? How could such a model be estimated? How can we compute the within-sample estimates of the unobserved quarterly values? How can we forecast the annual variables, exploiting the quarterly information available? All these issues, except the first, have been addressed in previous chapters. In particular, Chapter 5 discussed the computation of the Gaussian likelihood for any fixed parameter SS model, thus solving the problem of estimation, and Chapter 4 presented the fixed-interval smoothing algorithm, which can be used after estimation to estimate and forecast the quarterly values.

Therefore, in this chapter we concentrate on the specification problem. To this end we study how aggregation over time affects the dynamic structure of a vector of time series and its observability. We find that the basic dynamic components remain unchanged but some of them, in particular those related to seasonality, become unobservable. Building on these results, we present a structured specification method based on the idea that the models relating the variables in high and low sampling frequencies should be mutually consistent. Once a consistent and observable high-frequency model has been specified, estimation, diagnostic checking, interpolation, and forecasting can be addressed by standard SS methods. An example with national accounting data illustrates the practical application of these ideas.

10.1 The effect of aggregation on an SS model

Let z_t be an $m \times 1$ random vector of quarterly values, realized by the minimal SEM:

$$\boldsymbol{x}_{t+1} = \boldsymbol{\Phi}\boldsymbol{x}_t + \boldsymbol{\Gamma}\boldsymbol{u}_t + \boldsymbol{E}\boldsymbol{a}_t \tag{10.1}$$

$$\boldsymbol{z}_t = \boldsymbol{H}\boldsymbol{x}_t + \boldsymbol{D}\boldsymbol{u}_t + \boldsymbol{a}_t. \tag{10.2}$$

10.1.1 The high-frequency model in stacked form

Let S be the seasonal frequency, defined as the number of high-frequency sampling periods (quarters) corresponding to a single low-frequency (annual) observation. It is difficult to discuss aggregation over time using the quarterly representation (10.1)–(10.2). To this end, it is more convenient to use a representation where the values of $\boldsymbol{z}_t$, $\boldsymbol{u}_t$, and $\boldsymbol{a}_t$ corresponding to a whole seasonal cycle are "stacked" into the following vectors:

$$\boldsymbol{z}_{t:t+S-1} = \begin{bmatrix} \boldsymbol{z}_t \\ \boldsymbol{z}_{t+1} \\ \vdots \\ \boldsymbol{z}_{t+S-1} \end{bmatrix}, \quad \boldsymbol{u}_{t:t+S-1} = \begin{bmatrix} \boldsymbol{u}_t \\ \boldsymbol{u}_{t+1} \\ \vdots \\ \boldsymbol{u}_{t+S-1} \end{bmatrix}, \quad \boldsymbol{a}_{t:t+S-1} = \begin{bmatrix} \boldsymbol{a}_t \\ \boldsymbol{a}_{t+1} \\ \vdots \\ \boldsymbol{a}_{t+S-1} \end{bmatrix}. \tag{10.3}$$

Without loss of generality we assume that the aggregation period coincides with the seasonal frequency, so the stacked vectors in (10.3) include all the values subject to aggregation. Under these conditions, the following proposition holds.

Proposition 10.1 *Model (10.1)–(10.2) can be written equivalently as:*

$$\boldsymbol{x}_{T+1} = \overline{\boldsymbol{\Phi}}\boldsymbol{x}_T + \overline{\boldsymbol{\Gamma}}\boldsymbol{u}_{t:t+S-1} + \overline{\boldsymbol{E}}\boldsymbol{a}_{t:t+S-1} \tag{10.4}$$

$$\boldsymbol{z}_{t:t+S-1} = \overline{\boldsymbol{H}}\boldsymbol{x}_T + \overline{\boldsymbol{D}}\boldsymbol{u}_{t:t+S-1} + \overline{\boldsymbol{C}}\boldsymbol{a}_{t:t+S-1} \tag{10.5}$$

where the time indexes T and t refer respectively to the annual and quarterly time units so that, if $\boldsymbol{x}_T = \boldsymbol{x}_t$, then $\boldsymbol{x}_{T+1} = \boldsymbol{x}_{t+S}$, and the matrices in (10.4)–(10.5) are related to those in (10.1)–(10.2) by the expressions:

$$\overline{\boldsymbol{\Phi}} = \boldsymbol{\Phi}^S \tag{10.6}$$

$$\overline{\boldsymbol{\Gamma}} = \left[\boldsymbol{\Phi}^{S-1}\boldsymbol{\Gamma}, \boldsymbol{\Phi}^{S-2}\boldsymbol{\Gamma}, \ldots, \boldsymbol{\Gamma}\right] \tag{10.7}$$

$$\overline{\boldsymbol{E}} = \left[\boldsymbol{\Phi}^{S-1}\boldsymbol{E}, \boldsymbol{\Phi}^{S-2}\boldsymbol{E}, \ldots, \boldsymbol{E}\right] \tag{10.8}$$

$$\overline{\boldsymbol{H}} = \begin{bmatrix} \boldsymbol{H} \\ \boldsymbol{H}\boldsymbol{\Phi} \\ \vdots \\ \boldsymbol{H}\boldsymbol{\Phi}^{S-1} \end{bmatrix} \tag{10.9}$$

$$\overline{\boldsymbol{D}} = \begin{bmatrix} \boldsymbol{D} & \boldsymbol{0} & \ldots & \boldsymbol{0} \\ \boldsymbol{H}\boldsymbol{\Gamma} & \boldsymbol{D} & \ldots & \boldsymbol{0} \\ \vdots & \vdots & \ddots & \vdots \\ \boldsymbol{H}\boldsymbol{\Phi}^{S-2}\boldsymbol{\Gamma} & \boldsymbol{H}\boldsymbol{\Phi}^{S-3}\boldsymbol{\Gamma} & \ldots & \boldsymbol{D} \end{bmatrix} \tag{10.10}$$

$$\overline{\boldsymbol{C}} = \begin{bmatrix} \boldsymbol{I} & \boldsymbol{0} & \ldots & \boldsymbol{0} \\ \boldsymbol{H}\boldsymbol{E} & \boldsymbol{I} & \ldots & \boldsymbol{0} \\ \vdots & \vdots & \ddots & \vdots \\ \boldsymbol{H}\boldsymbol{\Phi}^{S-2}\boldsymbol{E} & \boldsymbol{H}\boldsymbol{\Phi}^{S-3}\boldsymbol{E} & \ldots & \boldsymbol{I} \end{bmatrix} \tag{10.11}$$

$$\overline{\boldsymbol{Q}} = \boldsymbol{I} \otimes \boldsymbol{Q}. \tag{10.12}$$

Proof. Equation (10.5) is obtained by successively substituting (10.1) in (10.2) and writing the resulting system in matrix form. On the other hand (10.4) is immediately obtained by propagating (10.1) from t to $t+S$. ■

Note that, according to expression (10.6), the size of the state vector in the stacked model (10.4)–(10.5) is the same as that in the SEM (10.1)–(10.2). This is a critical feature of the stacked model, as any increase in the size of the state vector would imply that the representation (10.4)–(10.5) has either redundant states or spurious dynamic structures, not present in the original high-frequency SEM.

10.1.2 Aggregation relationships

Now consider the aggregation relationships:

$$\boldsymbol{z}_{\boldsymbol{T}}^{\boldsymbol{A}} = \boldsymbol{J}^{\boldsymbol{A}} \boldsymbol{z}_{t:t+S-1}, \tag{10.13}$$

$$\boldsymbol{z}_{\boldsymbol{T}}^{\boldsymbol{P}} = \boldsymbol{J}^{\boldsymbol{P}} \boldsymbol{z}_{t:t+S-1}, \tag{10.14}$$

where $\boldsymbol{z}_{\boldsymbol{T}}^{\boldsymbol{A}}$ and $\boldsymbol{z}_{\boldsymbol{T}}^{\boldsymbol{P}}$ denote, respectively, the vectors of annual and partially aggregated data observed in year T, including the quarterly values at time $t, t+1, \ldots, t+S-1$. By "partially aggregated data" we mean a mixed-frequency sample combining all of the observed annual and quarterly values; see Examples 10.1.2.1 and 10.1.2.2 below.

Note also that there always exists a relationship between the annual and partially aggregated series:

$$\boldsymbol{z}_{\boldsymbol{T}}^{\boldsymbol{A}} = \boldsymbol{J}^* \boldsymbol{z}_{\boldsymbol{T}}^{\boldsymbol{P}}, \tag{10.15}$$

where (10.13)–(10.14) imply $\boldsymbol{J}^* \boldsymbol{J}^{\boldsymbol{P}} = \boldsymbol{J}^{\boldsymbol{A}}$; see Example 10.1.2.3 below.

The aggregation matrices in (10.13), (10.14) and (10.15) depend on the aggregation constraints and the nature of the variables. Following Di Fonzo [76], there are four basic types of quarterly variables: flows, indices, beginning-of-period stocks, or end-of-period stocks. Accordingly, the annual samples will be sums of quarterly values, averages, or discrete beginning-of-period/end-of-period values. The following examples illustrate some common aggregation structures:

Example 10.1.2.1

If all the variables considered are flows then $\boldsymbol{J}^{\boldsymbol{A}} = [\boldsymbol{I}, \boldsymbol{I}, \ldots, \boldsymbol{I}]$, meaning that each annual figure is the sum of the corresponding quarterly values. On the other hand, $\boldsymbol{J}^{\boldsymbol{A}} = [\boldsymbol{0}, \boldsymbol{0}, \ldots, \boldsymbol{I}]$ would imply that the variables are end-of-period stocks, such that the value $\boldsymbol{z}_{t+S-1}$ is observed while the values at $t, t+1, \ldots, t+S-2$ are not.

Example 10.1.2.2

Assume that the first m_1 variables in the vector $\boldsymbol{z}_{t:t+S-1}$ of (10.3) are annual flows, while the remaining m_2 variables ($m_1 + m_2 = m$) are quarterly values. In this case the partial aggregation matrix would be:

$$\boldsymbol{J}^{\boldsymbol{P}} = \begin{bmatrix} \boldsymbol{I}_{m_1} & \boldsymbol{0} & \boldsymbol{I}_{m_1} & \boldsymbol{0} & \cdots & \boldsymbol{0} \\ \boldsymbol{0} & \boldsymbol{I}_{m_2} & \boldsymbol{0} & \boldsymbol{0} & \cdots & \boldsymbol{0} \\ \boldsymbol{0} & \boldsymbol{0} & \boldsymbol{0} & \boldsymbol{I}_{m_2} & \cdots & \boldsymbol{0} \\ \vdots & \vdots & \vdots & \vdots & \ddots & \vdots \\ \boldsymbol{0} & \boldsymbol{0} & \boldsymbol{0} & \boldsymbol{0} & \cdots & \boldsymbol{I}_{m_2} \end{bmatrix}_{(m_1+S\cdot m_2)\times(S\cdot m_1+S\cdot m_2)}, \tag{10.16}$$

where $\boldsymbol{I}_m$ is the $m \times m$ identity and the vectors of quarterly and partially observed values have the following structures:

$$\boldsymbol{z}_{t:t+S-1} = \begin{bmatrix} \boldsymbol{z}_t^{m_1} \\ \boldsymbol{z}_t^{m_2} \\ \boldsymbol{z}_{t+1}^{m_1} \\ \boldsymbol{z}_{t+1}^{m_2} \\ \vdots \\ \boldsymbol{z}_{t+S-1}^{m_1} \\ \boldsymbol{z}_{t+S-1}^{m_2} \end{bmatrix}_{(S \cdot m_1 + S \cdot m_2) \times 1} \tag{10.17}$$

$$\boldsymbol{z}_T^P = \begin{bmatrix} \boldsymbol{z}_t^{m_1} + \boldsymbol{z}_{t+1}^{m_1} + \ldots + \boldsymbol{z}_{t+S-1}^{m_1} \\ \boldsymbol{z}_t^{m_2} \\ \boldsymbol{z}_{t+1}^{m_2} \\ \vdots \\ \boldsymbol{z}_{t+S-1}^{m_2} \end{bmatrix}_{(m_1 + S \cdot m_2) \times 1}. \tag{10.18}$$

Example 10.1.2.3

Assuming again that all the variables are flows, the matrix $\boldsymbol{J}^*$ in (10.15), transforming partially aggregated data into annual values, would be:

$$\boldsymbol{J}^* = \begin{bmatrix} \boldsymbol{I}_{m_1} & \boldsymbol{0} & \boldsymbol{0} & \cdots & \boldsymbol{0} \\ \boldsymbol{0} & \boldsymbol{I}_{m_2} & \boldsymbol{I}_{m_2} & \cdots & \boldsymbol{I}_{m_2} \end{bmatrix}_{(m_1+m_2) \times (m_1 + S \cdot m_2)}. \tag{10.19}$$

10.1.3 *Relationships between the models for high, low, and mixed-frequency data*

Under the previous conditions, the following result formalizes the relationship between the quarterly data model in the form (10.4)–(10.5) and the corresponding models for the annual and partially observed data:

Proposition 10.2 *The models for $\boldsymbol{z}_T^A$ and $\boldsymbol{z}_T^P$ have the state equation (10.4) with the observation equations:*

$$\boldsymbol{z}_T^A = \boldsymbol{H}^A \boldsymbol{x}_T + \boldsymbol{D}^A \boldsymbol{u}_{t:t+S-1} + \boldsymbol{C}^A \boldsymbol{a}_{t:t+S-1} \tag{10.20}$$

$$\boldsymbol{z}_T^P = \boldsymbol{H}^P \boldsymbol{x}_T + \boldsymbol{D}^P \boldsymbol{u}_{t:t+S-1} + \boldsymbol{C}^P \boldsymbol{a}_{t:t+S-1}, \tag{10.21}$$

where:

$$\boldsymbol{H}^A = \boldsymbol{J}^A \bar{\boldsymbol{H}}, \quad \boldsymbol{D}^A = \boldsymbol{J}^A \bar{\boldsymbol{D}}, \quad \boldsymbol{C}^A = \boldsymbol{J}^A \bar{\boldsymbol{C}}, \tag{10.22}$$

$$\boldsymbol{H}^P = \boldsymbol{J}^P \bar{\boldsymbol{H}}, \quad \boldsymbol{D}^P = \boldsymbol{J}^P \bar{\boldsymbol{D}}, \quad \boldsymbol{C}^P = \boldsymbol{J}^P \bar{\boldsymbol{C}}. \tag{10.23}$$

Proof. Pre-multiplying (10.4) by $\boldsymbol{J}^A$ and $\boldsymbol{J}^P$ immediately yields (10.20) and (10.21) respectively. ■

As a corollary to proposition 10.2, note that the observer of $\boldsymbol{z}_T^A$ can be alternatively obtained by aggregation of $\boldsymbol{z}_T^P$. Pre-multiplying (10.21) by $\boldsymbol{J}^*$ yields:

$$\boldsymbol{z}_T^A = \boldsymbol{J}^* \boldsymbol{z}_T^P = \boldsymbol{J}^* \boldsymbol{H}^P \boldsymbol{x}_T + \boldsymbol{J}^* \boldsymbol{D}^P \boldsymbol{u}_{t:t+S-1} + \boldsymbol{J}^* \boldsymbol{C}^P \boldsymbol{a}_{t:t+S-1}. \tag{10.24}$$

10.1.4 The effect of aggregation on predictive accuracy

The following proposition states that a model combining annual and quarterly data may predict the annual values better than the corresponding annual model.

Proposition 10.3 *The forecast error variance of the annual model, given by the state equation (10.4) and the observation equation (10.20), is greater than or equal to the error variance of the annual forecasts computed by:*

- *Case (1): aggregation of the forecasts for $\boldsymbol{z}_t$ derived from the quarterly model (10.4)–(10.5), and*
- *Case (2): aggregation of the forecasts for $\boldsymbol{z}_T^P$ derived from the partially aggregated model given by the state equation (10.4) and the observation equation (10.21).*

Proof: See Appendix A in Casals, Sotoca, and Jerez [46][1].

Proposition 10.3 compares the ability of three models to predict the observable annual values: (a) the model for the annual sample, (b) the "true" quarterly data generating process, and (c) the model for the partially aggregated sample. If these models are mutually consistent, given the aggregation constraint, model (b) would provide optimal forecasts, but cannot be empirically specified. Model (a) can be specified by standard methods, but cannot predict better than (b) or (c). Last, model (c) can be inferred from data using, e.g. the method defined in the next Section, and may provide better forecasts than model (a). It is also more flexible, because it has the ability to update the annual forecast as new quarterly information becomes available.

10.2 Observability in the aggregated model

10.2.1 Unobservable modes

As we saw previously, the SS models for quarterly, annual and partially aggregated data have the same state equation (10.4), so aggregation does not affect the states governing a dynamic system. However, it may reduce its observability.

[1]This proposition is an extension of previous results. In particular, Wei [220] proved Case (1) for univariate processes, also showing that the loss in forecasting efficiency due to aggregation can be substantial if the nonseasonal component of the model is nonstationary. Conversely, there is no loss in efficiency if the quarterly model is a pure seasonal process. Lutkepohl [155] also discussed Case (1) in a multivariate stationary framework.

Comparing the observation equations (10.5) and (10.20)–(10.21) it is immediate to see that observability loss occurs in two ways: (a) the quarterly observer (10.5) realizes more signals than the aggregated observers (10.20) or (10.21), and (b) the matrices in (10.20)–(10.21) are non-linear functions of the matrices in (10.5), potentially decreasing the observability of some states.

To further discuss this issue, the following definitions particularize the general concept of observability (see Definition 2.4 in Chapter 2) to the notation previously defined.

Definition 10.1 (observability in the quarterly model). All the components in the quarterly model (10.1)–(10.2) are said to be observable if and only if there is no real vector $\boldsymbol{w} \neq \mathbf{0}$, with $\boldsymbol{\Phi}\,\boldsymbol{w} = \lambda\,\boldsymbol{w}$ such that $\boldsymbol{H}\,\boldsymbol{w} = \mathbf{0}$, where λ is the eigenvalue associated to the eigenvector $\boldsymbol{w}$.

Definition 10.2 (unobservable modes in the annual model). Assuming that the variables are flows, so $\boldsymbol{J}^A = [\boldsymbol{I}, \boldsymbol{I}, \ldots, \boldsymbol{I}]$, the annual model given by (10.4) and (10.20) has unobservable modes if and only if there exists a vector $\boldsymbol{w} \neq \mathbf{0}$, with $\boldsymbol{\Phi}^S \boldsymbol{w} = \lambda \boldsymbol{w}$ such that $\boldsymbol{H}^A \boldsymbol{w} = \mathbf{0}$ or, equivalently, $\boldsymbol{H} \sum_{i=0}^{S-1} \boldsymbol{\Phi}^i \boldsymbol{w} = \mathbf{0}$; see (10.9) and (10.22).

Under these conditions, the following Theorem characterizes which components of the quarterly model become unobservable after annual aggregation.

Theorem 10.1 ***(loss of observability in the annual model).*** *Assuming that the quarterly model (10.4)–(10.5) is observable, the annual model given by the state equation (10.4) and the observation equation (10.20) includes unobservable modes in any of the following cases.*

- *Case (1): when the annual variables are flows, such that* $\boldsymbol{J}^A = [\boldsymbol{I}, \boldsymbol{I}, \ldots, \boldsymbol{I}]$, *if there exists an eigenvalue* λ *of* $\boldsymbol{\Phi}$ *such that* $1 + \lambda + \lambda^2 + \cdots + \lambda^{S-1} = 0$.
- *Case (2): if there is a subset of k eigenvalues of* $\boldsymbol{\Phi}$, *denoted by* $\lambda_k(\boldsymbol{\Phi})$, *such that for any pair* $\lambda_i, \lambda_j \in \lambda_k(\boldsymbol{\Phi})$, *with* $\lambda_i \neq \lambda_j$, *it holds that* $(\lambda_i)^S = (\lambda_j)^S$.
- *Case (3): when* $\boldsymbol{\Phi}$ *has a* $k \times k$ *Jordan block with null eigenvalues, such that the geometric multiplicity of this block grows with its S-th order power.*

Proof: See Appendix B in Casals, Sotoca, and Jerez [46][2].

Theorem 10.1 characterizes the situation where some modes of the quarterly model, associated to different transition eigenvalues, collapse to a smaller number of modes in the annual model. Consider, e.g., a quarterly univariate model where all the dynamics of a flow variable are given by a seasonal difference $\nabla_S =$

[2]Two important works connected with this Theorem are those by Stram and Wei [204] and Wei and Stram [222]. In particular, the loss of observability is related with the "hidden periodicity" notion defined by Stram and Wei [204] for univariate models. In a univariate framework, Case (2) of Theorem 10.1 is equivalent to hidden periodicity because all the indistinguishable modes created by aggregation affect a single time series. However, in the multivariate case hidden periodicity does not implies loss of observability in all cases. For example, observability is not lost if two modes with hidden periodicity affect two different time series. Note also that the mode elimination rules discussed for cases (2) and (3) are coherent in the univariate case with Theorems 4.1 and 3.1, respectively, in Stram and Wei [204].

$(1-\mathrm{B})(1+\mathrm{B}+\mathrm{B}^2+\cdots+\mathrm{B}^{S-1})$. Case (1) implies that, after aggregation, this structure is indistinguishable from a nonseasonal difference $\nabla=(1-\mathrm{B})$ in the annual model. Then, observability of S-1 dynamic components of the quarterly model is lost, and the dynamics of the corresponding annual model simplify accordingly.

Under Case (2), aggregation reduces the number of system modes, while maintaining the dimension of the state vector. The number of unobservable modes is $k-\mathrm{rank}(\boldsymbol{H}\cdot\boldsymbol{W})$, where k is the number of elements in $\lambda_k(\boldsymbol{\Phi})$ and $\boldsymbol{W}$ is a matrix whose columns are the eigenvectors $\boldsymbol{w_i}$, $i=1,\ldots,k$, that generate the k-order subspace $\boldsymbol{S_k}$. In the univariate case aggregation keeps only one mode, which corresponds with the larger Jordan block. In the multivariate case the reduction in the number of modes depends on how they affect the different observable time series so, in this case, there is no way to compute *a priori* how many modes will become unobservable, but the maximum number of unobservable modes is $k-1$.

Case (3) occurs when the seasonal structure is a pure moving average. In this situation a systematic rule for mode elimination only exists in the univariate case, where there are $1+[(k-1)/S]$ distinguishable modes, with $[\cdot]$ denoting the integer part operator and k the dimension of the corresponding Jordan block.

After discussing the effect of aggregation on the quarterly model dynamics and its observability, it is important to characterize the correspondence between the models for disaggregated and aggregated data. This is done in the following theorem.

Theorem 10.2 ***(correspondence between quarterly and annual models).*** *If the quarterly model (10.4)–(10.5) is minimal and all its components are observable from annual data, then the annual model given by (10.4) and (10.20) is unique, allowing for similar transformations (minimal and observable).*

Proof: See Appendix 10.3 in Casals, Sotoca, and Jerez [46][3].

The converse of this result is not true in general, as there are minimal and observable annual models which necessarily correspond to quarterly models with unobservable components. Assume e.g. that the annual model is an AR(1) with a negative parameter and that the seasonal frequency S is even. In this case, there is no high-frequency ARMA(1,1) process that adds up to the annual AR(1) model because the S-th power of the transition matrix will always be positive; see (10.6).

10.2.2 *Observability and fixed-interval smoothing*

In an SS framework, the optimal method to estimate the system states is the fixed-interval smoother; see Chapter 4. The uncertainty of smoothed estimates depends on the system detectability, see Chapter 2 and, building on this idea, the following result characterizes the effect of non-detectable modes on smoothed estimates.

Proposition 10.4 *The variance of the smoothed estimates of the states in the models given by the state equation (10.4) and the observation equation (10.20) or (10.21) is finite if and only if all the states are detectable.*

[3] In the case of ARIMA models this result collapses to Lemma 2 in Wei and Stram [222].

Proof: See Appendix 10.3 in Casals, Sotoca, and Jerez [46].

In time series disaggregation, smoothing is typically applied to estimate the unobserved quarterly values. Therefore detectability is a necessary and sufficient condition for these estimates to have a bounded uncertainty, while observability is a sufficient (not necessary) condition.

Infinite smoothed variances arise, for example, when the target variables are flows and their quarterly model includes seasonal roots in the unit circle. In this case the aggregated model has non-detectable components which would have infinite variances. In practice this means that if quarterly indicators have seasonal components, it is advisable to remove them before disaggregation because they can induce spurious dynamics in the disaggregated series.

10.2.3 An algorithm to aggregate a linear model: theory and examples

The VARMAX annual model corresponding to any linear quarterly model can be obtained by combining Propositions 10.1 and 10.2 with Theorem 10.1 and the algorithms described in Chapter 9. The step-by-step procedure to do this is as follows.

Step (1) Quarterly VARMAX Consider any linear and fixed-coefficients quarterly model. Write the model in the SEM form (10.1)–(10.2). If required, first write the MEM form and then compute the parameter matrices of the equivalent SEM; see Subsection 2.2.3. Finally, obtain the VARMAX model equivalent to the SEM using the procedures described in Chapter 9.

Step (2) Annual model Obtain the equivalent quarterly representation (10.4)–(10.5) and the corresponding annual representation, given by (10.4) and (10.20).

Step (3): Annual VARMAX Apply either the algorithm described in Subsection 9.2.1 to obtain the standard VARMAX representation, or the one described in Subsection 9.2.2, to obtain the corresponding canonical VARMAX echelon model.

Tables 10.1 and 10.2 illustrate the application of this algorithm and some previous results by showing the aggregation of several univariate and bivariate models. In particular:

1. Models 1–3 exemplify the observability loss described in Theorem 10.1 cases (1), (2) and (3), respectively. In particular, aggregation of Model 1 shows how a seasonal difference collapses to a nonseasonal unit root. This result is consistent with Granger and Siklos [113] and Stram and Wei [204].
2. Models 1, 4, 5, and 7–11 show that aggregation does not affect the number of unit roots in a time series, so I(0), I(1) or I(2) quarterly flows yield I(0), I(1) or I(2) annual aggregates. A straightforward implication of this is that if the quarterly variables are cointegrated, the corresponding annual aggregates will also be cointegrated, this result being consistent with Pierse and Snell [183] and Marcellino [161].

3. Models 4–5 have regular and seasonal unit roots in the quarterly frequency. In this very frequent case, the corresponding annual model has two unit roots and a MA structure induced by aggregation[4].
4. Aggregation induces additional MA structure and preserves the order of the stationary AR structure, typically reducing its persistence. Models 2 and 8 are clear examples of this. This result is consistent in the univariate case with Amemiya and Wu [5], Wei [220] and Stram and Wei [204]. In the multivariate stationary case, it is consistent with the findings in Chapters 4 and 6 of Lutkepohl [155] and in Marcellino [161].
5. Models 8–11 show that a feedback in the quarterly frequency is maintained in the annual frequency.
6. Model 6 is the annual model corresponding to a quarterly regression with AR(1) errors; see Chow–Lin [52]. Therefore, this method is empirically justified only if the model relating the target variable with the indicator in the annual frequency is a static regression with ARMA(1,1) errors.
7. Models 7 and 11 show that the aggregation algorithm is not restricted to VARMAX or transfer functions, as it can be applied to structural time series models, see Harvey [118], and VARMAX echelon models; see Hannan and Deistler [116]. In general, it supports any model with an equivalent SS representation.
8. Finally, model 10 shows how a monthly model aggregates to a quarterly VARMA, thus exemplifying that the aggregation algorithm can be applied to a general combination of sampling frequencies.

The code required to replicate these examples can be found in the file `Example1033.m`, see Appendix D.

10.3 Specification of the high-frequency model

Assume that a linear model has been fitted to all the available variables in the annual frequency. We will now discuss a systematic method to enforce consistency between this annual model and the unknown quarterly model, given the aggregation constraint and the partially aggregated sample. Without loss of generality, we will refer to the process of specifying a VARMA model, consisting of the successive determination of unit roots, AR and MA dynamics. The basic ideas can be easily applied to other model-building methods such as, e.g., structural time series modeling; see Harvey [118].

10.3.1 Enforcing approximate consistency

It is relatively easy to specify the structure of the quarterly model which realizes approximately a given annual model and, after estimating this high-frequency struc-

[4]This suggests that many annual variables should be I(2) while, in practice, annual models are often specified with a single unit root. This apparent contradiction is easy to explain because aggregation induces MA structures which could compensate an AR unit root if some MA root is close to unity.

Table 10.1 *Aggregation of several univariate models. The columns labelled "states" show the number of dynamic components in the minimal SS representation of the corresponding model. Therefore, the difference between the number of states in the quarterly and annual models is the number of dynamic components that become unobservable after aggregation. The quarterly variables z_t, z_{1t} and z_{2t} are assumed to be flows, so their annual aggregate is the sum of the corresponding quarterly values.*

#	Quarterly model	States	Annual model	States
1	$(1-B^4)z_t = a_t; \sigma_a^2 = 1$	4	$(1-B)z_t^A = a_t^A; \sigma_A^2 = 4.00$	1
2	$(1-0.8B)(1+0.8B)z_t = a_t; \sigma_a^2 = 1$	2	$(1-0.410B)z_t^A = (1+0.160B)a_t^A; \sigma_A^2 = 7.99$	1
3	$z_t = (1-0.6B^4)a_t; \sigma_a^2 = 1$	4	$z_t^A = (1-0.600B)a_t^A; \sigma_A^2 = 4.00$	1
4	$(1-B)(1-B^4)z_t = a_t; \sigma_a^2 = 1$	5	$(1-B)^2 z_t^A = (1+0.240B)a_t^A; \sigma_A^2 = 41.60$	2
5	$(1-B)(1-B^4)z_t = (1-0.8B)(1-0.6B^4)a_t; \sigma_a^2 = 1$	5	$(1-B)^2 z_t^A = (1-0.997B+0.238B^2)a_t^A; \sigma_A^2 = 7.05$	2
6	$z_{1t} = 0.5z_{2t} + \frac{1}{1-0.8B}a_t; \sigma_a^2 = 1$	1	$z_{1t}^A = 0.500z_{2t}^A + \frac{1+0.228B}{1-0.410B}a_t^A; \sigma_A^2 = 23.103$	1
7	$(1-B)T_{t+1} = u_t; \sigma_u^2 = 0.01$	4	$(1-B)z_t^A = (1-0.669B)a_t^A; \sigma_A^2 = 5.844$	1
	$(1+B+B^2+B^3)S_{t+1} = v_t; \sigma_v^2 = 0.01$			
	$z_t = T_t + S_t + \varepsilon_t; \sigma_\varepsilon^2 = 1$			

Table 10.2 *Aggregation of several bivariate models. All the quarterly variables* z_{1t} *and* z_{2t} *are assumed to be flows, so their annual aggregate is the sum of the corresponding quarterly values. If a variable were to be aggregated as an average of the quarterly values, the model would require an appropriate re-scaling.*

#	High-frequency model	States	Low-frequency model	States
8	$\begin{bmatrix} 1-0.7\mathrm{B} & -0.8\mathrm{B} \\ 0 & 1 \end{bmatrix} \begin{bmatrix} (1-\mathrm{B}^4)z_{1t} \\ (1-\mathrm{B}^4)z_{2t} \end{bmatrix} = \begin{bmatrix} a_{1t} \\ a_{2t} \end{bmatrix};$	10	$\begin{bmatrix} 1-0.240\mathrm{B} & 0 \\ 0 & 1 \end{bmatrix} \begin{bmatrix} (1-\mathrm{B})z^A_{1t} \\ (1-\mathrm{B})z^A_{2t} \end{bmatrix} = \begin{bmatrix} 1-0.039\mathrm{B} & 1.461 \\ 0 & 1 \end{bmatrix} \begin{bmatrix} a^A_{1t} \\ a^A_{2t} \end{bmatrix};$	3
	$\boldsymbol{\Sigma}_a = \begin{bmatrix} 1 & 0.5 \\ 0.5 & 1 \end{bmatrix}$		$\boldsymbol{\Sigma}_A = \begin{bmatrix} 35.36 & 7.62 \\ 7.62 & 4.00 \end{bmatrix}$	
9	$\begin{bmatrix} (1-\mathrm{B}^4)z_{1t} \\ (1-\mathrm{B}^4)z_{2t} \end{bmatrix} = \begin{bmatrix} 1 & -0.8\mathrm{B} \\ 1.2\mathrm{B} & 1 \end{bmatrix} \begin{bmatrix} a_{1t} \\ a_{2t} \end{bmatrix};$	8	$\begin{bmatrix} (1-\mathrm{B})z^A_{1t} \\ (1-\mathrm{B})z^A_{2t} \end{bmatrix} = \begin{bmatrix} 1-0.080\mathrm{B} & -0.053\mathrm{B} \\ 0.289\mathrm{B} & 1+0.015\mathrm{B} \end{bmatrix} \begin{bmatrix} a^A_{1t} \\ a^A_{2t} \end{bmatrix};$	2
	$\boldsymbol{\Sigma}_a = \begin{bmatrix} 1 & 0.5 \\ 0.5 & 1 \end{bmatrix}$		$\boldsymbol{\Sigma}_A = \begin{bmatrix} 4.09 & 1.41 \\ 1.41 & 13.00 \end{bmatrix}$	
10	$\begin{bmatrix} (1-\mathrm{B}^{12})z_{1t} \\ (1-\mathrm{B}^{12})z_{2t} \end{bmatrix} = \begin{bmatrix} 1 & -0.8\mathrm{B} \\ 1.2\mathrm{B} & 1 \end{bmatrix} \begin{bmatrix} a_{1t} \\ a_{2t} \end{bmatrix};$	24	$\begin{bmatrix} (1-\mathrm{B}^4)z^A_{1t} \\ (1-\mathrm{B}^4)z^A_{2t} \end{bmatrix} = \begin{bmatrix} 1-0.100\mathrm{B} & -0.075\mathrm{B} \\ 0.365\mathrm{B} & 1+0.024\mathrm{B} \end{bmatrix} \begin{bmatrix} a^A_{1t} \\ a^A_{2t} \end{bmatrix};$	8
	$\boldsymbol{\Sigma}_a = \begin{bmatrix} 1 & 0.5 \\ 0.5 & 1 \end{bmatrix}$		$\boldsymbol{\Sigma}_A = \begin{bmatrix} 3.22 & 1.03 \\ 1.03 & 9.27 \end{bmatrix}$	
11	$\begin{bmatrix} 1-\mathrm{B} & 0 \\ -0.5 & 1 \end{bmatrix} \begin{bmatrix} z_{1t} \\ z_{2t} \end{bmatrix} = \begin{bmatrix} 1-0.4\mathrm{B} & 0.8\mathrm{B} \\ 0 & 1 \end{bmatrix} \begin{bmatrix} a_{1t} \\ a_{2t} \end{bmatrix};$	1	$\begin{bmatrix} 1-\mathrm{B} & 0 \\ -0.5 & 1 \end{bmatrix} \begin{bmatrix} z^A_{1t} \\ z^A_{2t} \end{bmatrix} = \begin{bmatrix} 1-0.745\mathrm{B} & 1.808\mathrm{B} \\ 0 & 1 \end{bmatrix} \begin{bmatrix} a^A_{1t} \\ a^A_{2t} \end{bmatrix};$	1
	$\boldsymbol{\Sigma}_a = \begin{bmatrix} 1 & 0.5 \\ 0.5 & 1 \end{bmatrix}$		$\boldsymbol{\Sigma}_A = \begin{bmatrix} 51.91 & 29.55 \\ 29.55 & 19.58 \end{bmatrix}$	

ture, to assess whether the quarterly model obtained is statistically adequate or not. To these ends, one can use the following heuristic procedure.

Step (1) Annual modeling Specify and estimate a model relating the target annual variable(s) with the annualized values of the quarterly indicator(s). Any model having an equivalent linear SS representation, such as e.g., a transfer function or VARMAX, is adequate for this purpose.

Step (2) Decomposition of the quarterly indicator Adjust the quarterly indicator(s) to suppress undesired features, such as seasonality and calendar effects.

Step (3) Model specification Specify the VAR and VMA orders by the following steps:

Step (3.1) Set the VAR factor order of the quarterly model to be equal to that of the annual model and, particularly, constrain the number of unit roots to be the same. The foundation of this step results from comparing the quarterly model (10.4)–(10.5) and the annual model given by (10.4) and (10.20). As both models share the same state equation, they have the same (stationary and nonstationary) autoregressive components.

Step (3.2) Add a VMA(q) structure, with $q \leq n$, being n the size of the state vector in the annual model. This upper bound on the MA dynamics results from the fact that, in a minimal SS representation, the size of the state vector is $n = \max\{p,q\}m$, being p and q the orders of the VAR and VMA factors, respectively.

Step (4) Estimation Estimate the model specified in Steps (2) and (3) by maximum likelihood and prune insignificant parameters to obtain a parsimonious parameterization.

Step (5) Diagnostic checking Check the final quarterly model by obtaining the corresponding annual representation, using the algorithm described in the previous Section, and then:

Step (5.1) ...compare this model with the one specified in Step (1), and

Step (5.2) ...check whether it filters the annual data to white noise residuals.

Step (6) Check forecasting accuracy If the sample is long enough, compare out-of-the-sample forecasts for the annual values produced by both the tentative quarterly model and the annual model specified in Step (1).

Our experience in applying this method provides some clues about how to implement it:

- First, a good characterization of unit roots in Step (1) is critical, as misspecification of these components severely impacts the quality of the final results; see Tiao [207]. When in doubt, we think that over-differencing is safer than under-differencing.
- Second, the number of MA parameters specified in Step (3.2) may be excessive, depending on the sample size and number of time series. In this case, it is a good idea to constrain the MA matrices to be diagonal and, later, add off-diagonal parameters in a sequence of overfitting experiments.

- Third, as Nunes [169] points out, there are many situations where missing observations arise in a time series disaggregation framework. Some of these are: different release dates for some variables, changes in the sampling frequency or nonconformable samples, where the series considered have different starting dates. In these cases the ability of SS methods to take care of missing values is an important asset, as it allows for using all the information available.
- Last, the forecasting accuracy check proposed in Step (6) can be implemented by setting some within-sample values to missing, and estimating them afterwards. We have found this alternative useful when the sample is too short to reserve some values for out-of-sample forecasting.

10.3.2 *"Bottom-up" determination of the quarterly model*

On the other hand, one can adopt a "bottom-up" approach, similar to the one discussed in Subsection 9.3.4, to maximize the consistency between the high and low-frequency models. To this end, one must solve the equations relating the known annual model and the unknown quarterly model, using the algorithm described in Subsection 10.2.3. This can be done by using the following procedure.

Perform steps (1)–(4) of Subsection 10.3.1. The resulting outcomes will be: (a) an annual data model, whose parameters will be denoted by $\hat{\theta}_a$, and (b) a preliminary quarterly model, whose parameter vector will be denoted by $\hat{\Theta}_q^{\mathbf{0}}$. Obtain the corresponding annual model parameters as $\hat{\theta}_a^{\mathbf{0}} = \boldsymbol{A}(\hat{\Theta}_q^{\mathbf{0}})$, where the function $\boldsymbol{A}(\cdot)$ denotes the aggregation procedure discussed in Subsection 10.2.3. Then execute:

Step (5) Iteration Estimate the model specified in Steps (2) and (3) above as

$$\hat{\Theta}^* = \arg\min \left[\boldsymbol{A}(\hat{\Theta}_q) - \hat{\theta}_a\right]^T \left[\boldsymbol{A}(\hat{\Theta}_q) - \hat{\theta}_a\right],$$

where the optimization must be performed iteratively because the problem is nonlinear. The estimates obtained in previous steps will be useful to initialize this iteration.

After convergence, a null value for the loss function implies that the quarterly and annual models are perfectly consistent, so the annual aggregation of the former realizes the latter exactly. On the other hand, a nonzero optimal loss means that it is not possible to achieve an exact match between the quarterly and annual models[5]. In this case it would be convenient to perform the diagnostic testing checks described in the previous subsection to formally validate the quarterly model.

10.4 Empirical example

We will now illustrate the application of the methods proposed in previous sections by disaggregating and forecasting the annual series of Value Added by the Industry in Spain (VAI), from 1980 to 2001, using as an indicator the quarterly values

[5] Bear in mind that, as stated in the discussion of Theorem 10.2, a statistically adequate model for the annual data may not have a perfectly consistent quarterly representation.

of a re-balanced Industrial Production Index (IND), from the 1st quarter of 1980 to the 4th quarter of 2001. The latter series is the indicator actually employed by the agency in charge of the Spanish national accounts.

To clarify when a series is expressed in annual or quarterly frequency we will use an uppercase/lowercase notation, so VAI_T^A and IND_T^A are the values of VAI and the annual average of the indicator in year T ($T = 1980, 1981, \ldots, 2001$), while vai_t and ind_t denote the values of both variables in quarter t ($t = 1980.1, 1980.2, \ldots, 2001.4$).

10.4.1 Annual model

The first step in the analysis consists of modeling the annual values of the target variable and the indicator. A standard analysis, see Jenkins and Alavi [131], resulted in the following model for $T = 1980, \ldots, 2001$:

$$\begin{bmatrix} (1-\mathrm{B})^2 & 0 \\ 0 & (1-\mathrm{B})^2 \end{bmatrix} \begin{bmatrix} VAI_T^A \\ IND_T^A \end{bmatrix} = \begin{bmatrix} 1-\underset{(0.06)}{1.85}\mathrm{B} & \underset{(0.02)}{0.74}\mathrm{B} \\ -\underset{(0.05)}{1.48}\mathrm{B} & 1 \end{bmatrix} \begin{bmatrix} \hat{a}_T^{VAI} \\ \hat{a}_T^{IND} \end{bmatrix}, \tag{10.25}$$

$$\hat{\boldsymbol{\Sigma}}_{\boldsymbol{a}} = \begin{bmatrix} 2.64 & 5.53 \\ 5.53 & 13.08 \end{bmatrix}, \qquad \boldsymbol{Q}(5) = \begin{bmatrix} 4.84 & 4.31 \\ 6.23 & 4.17 \end{bmatrix},$$

where the values in parentheses are the standard errors of the estimates, $\hat{\boldsymbol{\Sigma}}_{\boldsymbol{a}}$ represents the estimate of the error covariance matrix, and $\boldsymbol{Q}(5)$ is the matrix of portmanteau statistics.

Model (10.25) provides precise clues about the dynamics of VAI and IND in the quarterly frequency. Following the guidelines in Subsection 10.3.1 a consistent quarterly model must have a nonstationary AR structure with two unit roots for each series and an MA term with a maximum order of 4, because the minimal SS representation of (10.25) requires four state variables.

The code required to replicate these results can be found in the file `Example104Step1.m`, see Appendix D.

10.4.2 Decomposition of the quarterly indicator

The second step requires modeling the quarterly indicator to extract the components useful for disaggregation. To this end, we will use the following model for $t = 1980.1, 1980.2, \ldots, 2001.4$:

$$\begin{aligned} ind_t &= \underset{(0.03)}{1.02}L_t - \underset{(0.35)}{1.85}E_t - \underset{(1.72)}{5.94}S_t^{92.4} + \hat{N}_t, \\ (1-\mathrm{B})(1-\mathrm{B}^4)\hat{N}_t &= (1-\underset{(0.06)}{0.83}\mathrm{B}^4)\hat{a}_t, \\ \hat{\sigma}_a^2 = 3.21, &\qquad Q(15) = 10.90, \end{aligned} \tag{10.26}$$

where L_t is the number of non-holidays in quarter t, E_t is a dummy variable to account for Easter effects, and $S_t^{92.4}$ is an intervention variable capturing a persistent change from the 4th quarter of 1992 onwards.

According to the signal extraction ideas explained in Chapter 8, model (10.26) implies that the indicator can be decomposed into: (a) trend, (b) seasonal component, (c) irregular component, (d) calendar effects associated to the number of non-holidays and Easter; and (e) the 1992 4th quarter effect. Figure 10.1 shows these components. In the light of previous results, components (b) and (d) are useless for disaggregation. By adding the remaining components we obtain a seasonally and calendar-adjusted quarterly indicator series.

The code required to replicate these results can be found in the file `Example104Step2.m`, see Appendix D.

10.4.3 Specification and estimation of the quarterly model

The third and fourth steps consist of specifying and estimating a model for the quarterly data. Building on the results from the previous steps, we estimate a doubly integrated VMA(4) process for the value added and the adjusted indicator in the quarterly frequency. After pruning insignificant parameters we obtained the following model for $t = 1980.1, 1980.2, \ldots, 2001.4$:

$$
\begin{bmatrix} (1-\mathrm{B})^2 & 0 \\ 0 & (1-\mathrm{B})^2 \end{bmatrix} \begin{bmatrix} vai_t \\ \widetilde{ind}_t \end{bmatrix} = \begin{bmatrix} 1 - \underset{(0.04)}{1.48}\mathrm{B} + \underset{(0.02)}{0.36}\mathrm{B}^2 & \underset{(0.003)}{0.044}\mathrm{B} \\ \underset{(0.11)}{-0.84}\mathrm{B} & 1 - \underset{(0.02)}{0.78}\mathrm{B} \end{bmatrix} \begin{bmatrix} \hat{a}_t^{vai} \\ \hat{a}_t^{\widetilde{ind}} \end{bmatrix},
$$

$$
\hat{\mathbf{\Sigma}}_{\boldsymbol{a}} = \begin{bmatrix} 0.06 & 0.40 \\ 0.40 & 3.64 \end{bmatrix}, \tag{10.27}
$$

where vai_t denotes the unobserved value of VAI in quarter t and $\widetilde{ind}_t$ is the corresponding adjusted indicator. Finally, applying to the sample the fixed-interval smoothing described in Chapter 4, we obtain the disaggregates and forecasts shown in Figure 10.2.

The code required to replicate these results can be found in the file `Example104Step3and4.m`, see Appendix D.

10.4.4 Diagnostics

The fifth step of the analysis consists of assessing whether the quarterly model is empirically consistent with the annual data. To this end, we obtain the annual representation corresponding to model (10.27) by means of the aggregation algorithm described in Section 10.3.1. It is:

$$
\begin{bmatrix} (1-\mathrm{B})^2 & 0 \\ 0 & (1-\mathrm{B})^2 \end{bmatrix} \begin{bmatrix} VAI_T^A \\ \widetilde{IND}_T^A \end{bmatrix} = \begin{bmatrix} 1 - 1.52\mathrm{B} - .23\mathrm{B}^2 & .70\mathrm{B} + .10\mathrm{B}^2 \\ -1.18\mathrm{B} - 0.19\mathrm{B}^2 & 1 + .14\mathrm{B} - .07\mathrm{B}^2 \end{bmatrix} \begin{bmatrix} \hat{a}_T^{VAI} \\ \hat{a}_T^{\widetilde{IND}} \end{bmatrix},
$$

$$
\hat{\mathbf{\Sigma}}_{\boldsymbol{a}} = \begin{bmatrix} 2.54 & 5.53 \\ 5.53 & 13.79 \end{bmatrix}, \quad Q(5) = \begin{bmatrix} 7.38 & 7.81 \\ 9.49 & 8.48 \end{bmatrix}, \tag{10.28}
$$

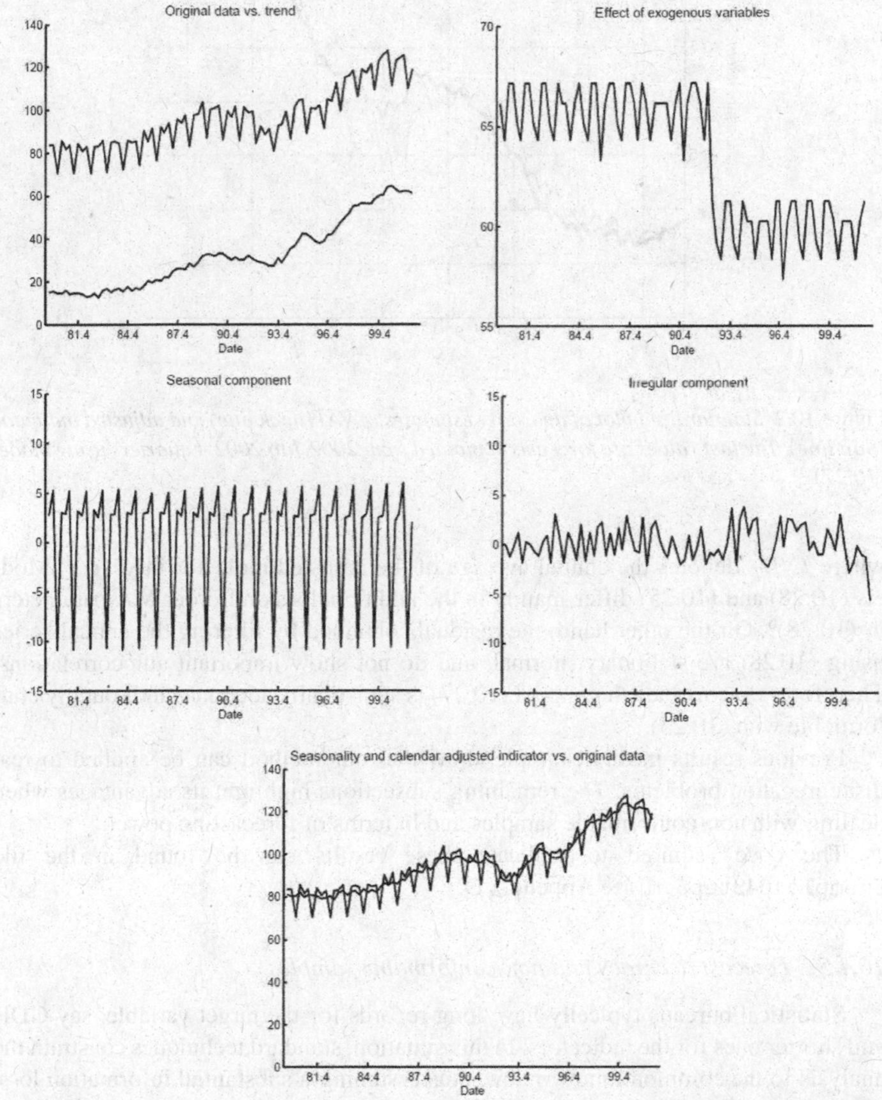

Figure 10.1 *Decomposition of the quarterly indicator. The adjusted indicator includes the trend, irregular, and level change components, and excludes seasonality and calendar affects.*

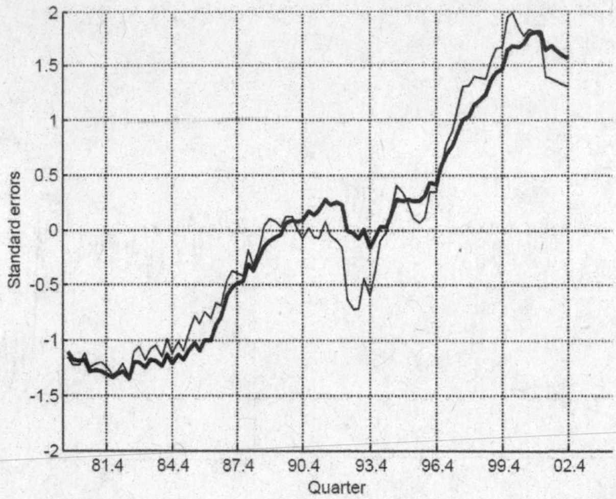

Figure 10.2 *Standardized plot of quarterly estimates of VAI (thick line) and adjusted indicator (thin line). The last values are forecasts computed from 2002.1 to 2002.4 quarters using model (10.27).*

where $\widetilde{IND}_T^A$ denotes the annual average of the adjusted indicator in year T. Models (10.28) and (10.25) differ mainly in the additional second-order MA parameters in (10.28)[6]. On the other hand, the residuals obtained by filtering the annual series using (10.28) are stationary, normal, and do not show important autocorrelations. Therefore, we conclude that model (10.27) is statistically adequate and roughly conformable with (10.25).

Previous results in this example show that our method can be applied to real disaggregation problems. The remaining subsections highlight its advantages when dealing with non-conformable samples and in terms of forecasting power.

The code required to replicate these results can be found in the file `Example104Step5.m`, see Appendix D.

10.4.5 *Forecast accuracy and non-conformable samples*

Statistical bureaus typically have long records for the target variable, say GDP, and shorter ones for the indicators. In this situation, standard techniques constrain the analysis to the common time window, thus assuming a substantial information loss. However the SS methods employed here allow for missing values, so all the available information can be used. To illustrate this idea, we delete the first four observations of the quarterly indicator and the last annual value of VAI. Re-estimating model (10.27)

[6]We tried to fit an annual model including these parameters and the corresponding estimates resulted insignificant. Perhaps this additional MA structure is due to the disaggregated indicator information included in (10.27) and excluded in (10.25)

with this non-conformable sample yields the following results:

$$
\begin{bmatrix} (1-\mathrm{B})^2 & 0 \\ 0 & (1-\mathrm{B})^2 \end{bmatrix} \begin{bmatrix} \widetilde{vai_t} \\ \widetilde{ind_t} \end{bmatrix} = \begin{bmatrix} 1 - \underset{(0.06)}{1.48}\mathrm{B} + \underset{(0.05)}{0.34}\mathrm{B}^2 & \underset{(0.006)}{0.045}\mathrm{B} \\ -\underset{(0.04)}{0.89}\mathrm{B} & 1-\underset{(0.03)}{0.78}\mathrm{B} \end{bmatrix} \begin{bmatrix} \hat{a}_t^{vai} \\ \widetilde{\hat{a}_t^{ind}} \end{bmatrix},
$$

$$
\hat{\boldsymbol{\Sigma}}_{\boldsymbol{a}} = \begin{bmatrix} 0.06 & 0.41 \\ 0.41 & 3.74 \end{bmatrix}. \tag{10.29}
$$

Estimates in (10.29) are remarkably close to those in (10.27). Table 10.3 shows the original sample information and the numerical results obtained with models (10.27) and (10.29). Note that: (a) the forecast provided by (10.29) for the 2001 annual value is very accurate, (b) quarterly interpolations and out-of-sample forecasts obtained with both models are very similar, and (c) the indicator retropolations computed with (10.29) and the non-conformable sample have acceptable errors.

10.4.6 Comparison with alternative methods

Table 10.4 shows the root mean squared errors (RMSEs) obtained by forecasting the last five end-of-year values of VAI with Model (10.27) and some alternative methods. All the forecasts are conditional on the true indicator values[7]. Note that Model (10.27) produced the best forecasts, with an 8–9% advantage over the second-best, and much larger gains in comparison with the remaining methods. Obviously, this comparison is not perfectly fair, as Model (10.27) was carefully fitted to the data while the alternative procedures are mechanical. Therefore, these results seem to reinforce the idea that a model fitted by a trained human is usually able to forecast better than an automatic system.

The code required to replicate these results can be found in the files `Example104Step6.m` and `Example104Step6-Chow-Lin.m`, see Appendix D.

[7]For example, to compute the 1997 values each model was estimated using annual VAI and quarterly indicator values up to 1996, and was then used to predict the end-of-year value of VAI using: (a) the past of VAI up to 1996, and (b) past values of the indicator up to 1997. The same procedure was applied for 1998, 1999, 2000, and 2001 extending the sample in each case with one, two, three, and four additional years of data.

Table 10.3 *Original data and interpolations obtained with full and non-conformable samples. Underlined values correspond to interpolations, retropolations or forecasts. The column "Indicator" refers to the calendar and seasonally adjusted values computed with model (10.26).*

Obs.	Original sample		Full sample			Non-conformable sample		
	Annual VAI	Indicator	Interpolation	Annual sum	Indicator	Interpolation	Annual sum	Indicator
1980.1		81.421	14.993		81.421	15.009		82.869
1980.2		80.451	14.810		80.451	14.896		82.237
1980.3		80.524	14.744		80.524	14.778		81.584
1980.4	59.341	81.877	14.794	59.341	81.877	14.658	59.341	80.968
1981.1		80.196	14.474		80.196	14.502		80.196
1981.2		80.710	14.531		80.710	14.519		80.710
1981.3		80.823	14.483		80.823	14.474		80.823
1981.4	57.895	80.299	14.407	57.895	80.299	14.400	57.895	80.299
...								
2000.1		123.520	23.326		123.520	23.342		123.520
2000.2		124.005	23.417		124.005	23.431		124.005
2000.3		122.033	23.374		122.033	23.371		122.033
2000.4	93.620	121.147	23.502	93.620	121.147	23.476	93.620	121.147
2001.1		121.823	23.753		121.823	23.690		121.823
2001.2		121.550	23.801		121.550	23.718		121.550
2001.3		121.006	23.830		121.006	23.734		121.006
2001.4	94.711	115.986	23.327	94.711	115.986	23.217	94.359	115.986
2002.1			23.414		115.664	23.311		115.723
2002.2			23.295		115.343	23.199		115.461
2002.3			23.177		115.021	23.088		115.198
2002.4			23.059	92.944	114.699	22.976	92.574	114.936

Table 10.4 *Ranking of RMSEs for end-of year forecasts of annual VAI from 1997 to 2001. RMSEs are computed for forecasts of both, VAI levels and growth rates. The columns RMSE% show the corresponding RMSEs normalized so that the RMSE of Model (10.29) is 100%.*

Rig	Model	Levels		Annual growth rates	
		RMSE	RMSE %	RMSE	RMSE %
1	Model (10.29)	0.65	100.0	0.71	100.0
2	Litterman [149] with constant	0.72	109.6	0.77	108.4
3	Fernandez [95] with constant	0.96	147.5	1.03	145.2
4	Chow–Lin [52] AR(1) with constant	1.21	185.2	1.3	182.5
5	Fernandez [95]) without constant	1.44	219.8	1.54	216.0
6	Litterman [149] without constant	1.47	224.9	1.57	220.9
7	ADL(1,0)[a] without constant	1.53	234.0	1.73	242.4
8	ADL(1,0) with constant	1.61	247.0	1.75	245.3
9	Chow–Lin [52] AR(1) without constant	1.82	278.9	1.98	277.2
10	Boot, Feibes and Lisman [26]	3.04	466.2	3.39	475.2
11	Denton [75]	18.01	2758.4	20.26	2842.3

[a]The ADL (Autoregressive Distributed Lag model) of orders p and q, denoted by ADL(p,q), is defined by the equation: $y_t = \beta_0 + \beta_1 y_{t-1} + \beta_2 y_{t-2} + ... + \beta_p y_{t-p} + \delta_0 x_t + \delta_1 x_{t-1} + \delta_2 x_{t-2} + ... + \delta_q x_{t-q} + \varepsilon_t$

Chapter 11

Cross-sectional extension: longitudinal and panel data

11.1 Model formulation 168
11.2 The Kalman Filter 169
11.2.1 Case of uncorrelated state and observational errors 170
11.2.2 Case of correlated state and observational errors 171
11.3 The linear mixed model in SS form 171
11.4 Maximum likelihood estimation 174
11.5 Missing data modifications 176
11.5.1 Missingness in responses only 176
11.5.2 Missingness in both responses and covariates: method 1 178
11.5.3 Missingness in both responses and covariates: method 2 180
11.6 Real data examples 182
11.6.1 A LMM for the mare ovarian follicles data 182
11.6.2 Smoothing and prediction of missing values for the beluga whales data 183

Multiple time series on independent *units* occur routinely in biostatistics and econometrics, among other disciplines. In the classical longitudinal study, multivariate quantitative outcomes may be collected over time on multiple *subjects* (living organisms) and their contemporaneously observed explanatory variables or *covariates*. Cause and effect relationships among the variables are typically of interest here. Similarly, econometrics panel data is comprised of multivariate quantitative and/or qualitative responses on several *entities* (countries, firms, individuals, etc.), frequently collected with the intent of using any available explanatory variables (now called *exogenous variables*) to aid in making future predictions.

This chapter discusses the cross-sectional extension that is required in the classical SS formulation of Chapter 2, in order to accommodate these types of data. The common linear mixed model for longitudinal analyses will be seen to be expressible in SS form. Appropriate Kalman smoothing, filtering, and prediction recursions in this context will be discussed, as well as ML estimation[1]. A particularly attractive and powerful feature of this methodology is its ability to account for missing or irregularly-spaced time series data, together with the ease with which smoothing and forecasting is accomplished. Since this "missingness" is often present in longitudinal analyses, a substantial portion of this chapter is devoted to techniques for accommodating such features. The methods will be illustrated on some real datasets.

Finally, we note that, since the cross-sectional extension is essentially a "replication" of the SS model across independent units, all the methods discussed here apply equally well to the situation of a single unit that has been the subject of all the previous chapters. For this reason (but also in order to follow standard longitudinal presentation of the material), we have strived to provide a somewhat different slant on things. However, some of the sections, and in particular the ML estimation and missing data adjustments, discuss approaches that have not been presented in earlier chapters.

11.1 Model formulation

The modeling of longitudinal or panel type data can be accommodated by a *cross-sectional* extension to the standard SS model. It necessitates the addition of an extra subscript so that $\boldsymbol{z}_{i,t}$ and $\boldsymbol{x}_{i,t}$ now represent the observation and state variables for the i-th unit, $i = 1,\ldots,k$, at time t, $t = 1,\ldots,n_i$. Using similar notation as in Chapter 2, the state equation for a simplified MEM becomes

$$\boldsymbol{x}_{i,t} = \boldsymbol{\Phi}\boldsymbol{x}_{i,t-1} + \boldsymbol{\Gamma}\boldsymbol{u}_{i,t-1} + \boldsymbol{w}_{i,t}, \tag{11.1}$$

where $\boldsymbol{w}_{i,t} \sim NID(\boldsymbol{0},\boldsymbol{Q})$, with initial condition $\boldsymbol{x}_{i,0} \sim \mathrm{N}(\boldsymbol{\mu},\boldsymbol{P})$. An alternative formulation that admits a contemporaneous (instead of lagged) relationship between the state and exogenous variables can also be used,

$$\boldsymbol{x}_{i,t} = \boldsymbol{\Phi}\boldsymbol{x}_{i,t-1} + \boldsymbol{\Upsilon}\boldsymbol{u}_{i,t} + \boldsymbol{w}_{i,t}. \tag{11.2}$$

The corresponding observation equation for either of these is

$$\boldsymbol{z}_{i,t} = \boldsymbol{H}_{i,t}\boldsymbol{x}_{i,t} + \boldsymbol{D}\boldsymbol{u}_{i,t} + \boldsymbol{v}_{i,t}, \tag{11.3}$$

where $\boldsymbol{v}_{i,t} \sim NID(\boldsymbol{0},\boldsymbol{R})$ represents the observational noise vector. For simplicity in likelihood computations, $\{\boldsymbol{w}_{i,t}\}$ and $\{\boldsymbol{v}_{i,t}\}$ are assumed to be uncorrelated, but a more general setup is to allow for

$$\mathrm{cov}(\boldsymbol{w}_{i,s},\boldsymbol{v}_{i,t}) = \begin{cases} \boldsymbol{S}, & s = t+1, \\ \boldsymbol{0}, & s \neq t+1. \end{cases} \tag{11.4}$$

The notation $\boldsymbol{H}_{i,t}$ in (11.3) reflects the fact that the coefficient matrix $\boldsymbol{H}$, from the formulation in Chapter 6, is now allowed to vary over both unit i and time t. This

[1] Quantification of the variability in ML estimates via asymptotic results is discussed in Appendix B.

is an important extension that allows for the accommodation of missing data; to be discussed shortly.

A similar class of models popular for handling longitudinal data in the social and behavioral sciences, are the so-called *structural equation models*, see Bollen [24]. In these fields, the essential problem is the estimation of structural relationships between quantitative observed variables, and such models have become a prominent form of data analysis only in the last twenty years or so (thanks in part to the availability of the software LISREL). Although SS models were originally developed from a time series perspective (long series on one unit), and structural equation models from a panel data perspective (short series on multiple units), there has been a gradual convergence of the two, see Oud and Jansen [172], with Chow *et al.* [53] pointing out equivalences and complementary aspects. The recent monograph by Little [150] is a state-of-the-art authoritative guide to applied longitudinal structural equation modeling.

An early version of the cross-sectional extension to the basic SS model seems to have been first proposed by Goodrich and Caines [112], who also investigated maximum (Gaussian) likelihood estimation and derived the corresponding asymptotic properties in a series of three papers, the results of which appear in Caines [37, ch. 9]. Jones [133], Fong [98], Shumway and Stoffer [195, ch. 6], and Hsiao [127], comprise the short list of subsequent attempts at popularizing the use of SS models for longitudinal data analysis. Fahrmeir and Tutz [90] report that the large number of parameters involved in multivariate modeling can make this a difficult endeavor.

In the terminology of Diggle *et al.* [80], the SS approach is categorized as a *transition* (or Markov) model, wherein the response vector $\boldsymbol{z}_{i,t}$ is affected by earlier values of itself as well as the exogenous $\boldsymbol{u}_{i,t}$, in the fashion of a classical VARMAX model. Two other (currently more popular) generalized linear model based approaches include *marginal* models and *random effects* models. The former models the marginal low-order moments and cross moments of $\boldsymbol{z}_{i,t}$ directly in a parametrized functional form, whereas correlation in the responses over time is introduced less directly in the latter by adding random effects to a linear model structure for the systematic component that is then related to the mean via the link function.

11.2 The Kalman Filter

The KF enables one to *optimally* (mean-squared error sense) smooth, filter, and forecast the series. Before presenting a version that applies to the situation of multiple subjects, define the following notation. Let $Z_{i,s} = (\boldsymbol{z}_{i,1}, \ldots, \boldsymbol{z}_{i,s})$ and $U_{i,s} = (\boldsymbol{u}_{i,1}, \ldots, \boldsymbol{u}_{i,s})$ denote the responses and covariates observed up to time s for subject i. Throughout we will assume the covariates to be *exogenous*, implying conditional independence of all preceding values and of all responses up to the present time. However, the results established here will depend only on the weaker implied condition,

$$\mathrm{E}(U_{i,t} \,|\, Z_{i,t}, U_{i,t-1}) = \mathrm{E}(U_{i,t} \,|\, U_{i,t-1}).$$

Similar to Shumway and Stoffer [195, ch. 6], we adopt the following notation for the mean of the state vector at time t given the responses and covariates up to time s, and its error covariance matrix at times t_1 and t_2,

$$\boldsymbol{x}_{i,t}^{s} = \mathrm{E}\left(\boldsymbol{x}_{i,t} \,|\, Z_{i,s}, U_{i,s}\right), \tag{11.5}$$

$$\boldsymbol{P}_{i,t_1,t_2}^{s} = \mathrm{E}\left[\left(\boldsymbol{x}_{i,t_1} - \boldsymbol{x}_{i,t_1}^{s}\right)\left(\boldsymbol{x}_{i,t_2} - \boldsymbol{x}_{i,t_2}^{s}\right)^{T} \,\middle|\, Z_{i,s}, U_{i,s}\right]. \tag{11.6}$$

When $t_1 = t_2 = t$, we will write simply $\boldsymbol{P}_{i,t}^{s}$ for the conditional variance in (11.6). With this notation defining the information set upon which the conditioning is to be performed, the appropriate Kalman filtering relations for each subject i are then identical to those already presented in Chapter 4. In what follows, we cast these in a slightly different form, and adapted specifically for the SS model defined by (11.2)–(11.3)

Remark 11.1 *The optimality conferred by Kalman relations is in the minimum mean-squared error sense, within the class of all linear predicted/filtered/smoothed values. In this case, the conditional expectation operator in (11.5) can more generally be interpreted as the* projection *operator of $\boldsymbol{x}_{i,t}$ on the linear span of $Z_{i,s}$ and $U_{i,s}$, with $\boldsymbol{P}_{i,t}^{s}$ as the corresponding mean-squared error.*

11.2.1 Case of uncorrelated state and observational errors

In the special case of $\boldsymbol{S} = \boldsymbol{0}$ in (11.4), the Kalman *smoothing* ($t > s$) and Kalman *filtering* ($t = s$) relations are given respectively by the recursions

$$\boldsymbol{x}_{i,t}^{t-1} = \boldsymbol{\Phi}\boldsymbol{x}_{i,t-1}^{t-1} + \boldsymbol{\Upsilon}\boldsymbol{u}_{i,t}, \tag{11.7}$$

$$\boldsymbol{P}_{i,t}^{t-1} = \boldsymbol{\Phi}\boldsymbol{P}_{i,t-1}^{t-1}\boldsymbol{\Phi}^{T} + \boldsymbol{Q}, \tag{11.8}$$

and

$$\boldsymbol{x}_{i,t}^{t} = \boldsymbol{x}_{i,t}^{t-1} + \boldsymbol{K}_{i,t}\left(\boldsymbol{z}_{i,t} - \boldsymbol{H}_{i,t}\boldsymbol{x}_{i,t}^{t-1} - \boldsymbol{D}\boldsymbol{u}_{i,t}\right), \tag{11.9}$$

$$\boldsymbol{P}_{i,t}^{t} = \left(I - \boldsymbol{K}_{i,t}\boldsymbol{H}_{i,t}\right)\boldsymbol{P}_{i,t}^{t-1}, \tag{11.10}$$

$$\boldsymbol{K}_{i,t} = \boldsymbol{P}_{i,t}^{t-1}\boldsymbol{H}_{i,t}^{T}\left(\boldsymbol{H}_{i,t}\boldsymbol{P}_{i,t}^{t-1}\boldsymbol{H}_{i,t}^{T} + \boldsymbol{R}\right)^{-1}, \tag{11.11}$$

for $i = 1, \ldots, k$ and $t = 1, \ldots, n_i$, with initial conditions $\boldsymbol{x}_{i,0}^{0} = \boldsymbol{\mu}$ and $\boldsymbol{P}_{i,0}^{0} = \boldsymbol{P}$.

If desired, h-step ahead prediction, $h \geq 1$, of a particular subject beyond the range of observed values $t = 1, \ldots, n_i := n$, can then be accomplished with initial conditions $\boldsymbol{x}_{i,n}^{n}$ and $\boldsymbol{P}_{i,n}^{n}$, assuming known values for the future exogenous $\{\boldsymbol{u}_{i,n+h}\}$. Additional equations can also be introduced if prediction of the observed process $\{\boldsymbol{z}_{i,t}\}$ is of interest. In this case, letting

$$\boldsymbol{z}_{i,n+h}^{n} = \mathrm{E}\left(\boldsymbol{z}_{i,n+h} \,|\, Z_{i,n}, U_{i,n}\right),$$

and

$$\boldsymbol{B}_{i,n+h}^{n} = \mathrm{E}\left[\left(\boldsymbol{z}_{i,n+h} - \boldsymbol{z}_{i,n+h}^{n}\right)\left(\boldsymbol{z}_{i,n+h} - \boldsymbol{z}_{i,n+h}^{n}\right)^{T} \,\middle|\, Z_{i,n}, U_{i,n}\right],$$

denote respectively the prediction and prediction mean squared error of $z_{i,n+h}$, the appropriate Kalman recursions for h-step ahead prediction are then

$$\begin{aligned}
\boldsymbol{x}_{i,n+h}^{n} &= \boldsymbol{\Phi}\boldsymbol{x}_{i,n+h-1}^{n} + \boldsymbol{\Upsilon}\boldsymbol{u}_{i,n+h}, & (11.12)\\
z_{i,n+h}^{n} &= \boldsymbol{H}_{i,n+h}\boldsymbol{x}_{i,n+h}^{n} + \boldsymbol{D}\boldsymbol{u}_{i,n+h}, & (11.13)\\
\boldsymbol{P}_{i,n+h}^{n} &= \boldsymbol{\Phi}\boldsymbol{P}_{i,n+h-1}^{n}\boldsymbol{\Phi}^{T} + \boldsymbol{Q}, & (11.14)\\
\boldsymbol{B}_{i,n+h}^{n} &= \boldsymbol{H}_{i,n+h}\boldsymbol{P}_{i,n+h}^{n}\boldsymbol{H}_{i,n+h}^{T} + \boldsymbol{R}. & (11.15)
\end{aligned}$$

11.2.2 *Case of correlated state and observational errors*

In the most general case of $\boldsymbol{S} \neq \boldsymbol{0}$ in (11.4), we express the Kalman smoothing recursions in prediction error form very similar to those of Section 4.2 as

$$\begin{aligned}
\boldsymbol{x}_{i,t+1}^{t} &= \boldsymbol{\Phi}\boldsymbol{x}_{i,t}^{t-1} + \boldsymbol{\Upsilon}\boldsymbol{u}_{i,t+1} + \boldsymbol{K}_{i,t}\tilde{z}_{i,t}, & (11.16)\\
\tilde{z}_{i,t} &= z_{i,t} - \boldsymbol{H}_{i,t}\boldsymbol{x}_{i,t}^{t-1} - \boldsymbol{D}\boldsymbol{u}_{i,t}, & (11.17)\\
\boldsymbol{P}_{i,t+1}^{t} &= \boldsymbol{\Phi}\boldsymbol{P}_{i,t}^{t-1}\boldsymbol{\Phi}^{T} + \boldsymbol{Q} - \boldsymbol{K}_{i,t}\boldsymbol{B}_{i,t}\boldsymbol{K}_{i,t}^{T}, & (11.18)\\
\boldsymbol{K}_{i,t} &= \left(\boldsymbol{\Phi}\boldsymbol{P}_{i,t}^{t-1}\boldsymbol{H}_{i,t}^{T} + \boldsymbol{S}\right)\boldsymbol{B}_{i,t}^{-1}, & (11.19)\\
\boldsymbol{B}_{i,t} &= \boldsymbol{H}_{i,t}\boldsymbol{P}_{i,t}^{t-1}\boldsymbol{H}_{i,t}^{T} + \boldsymbol{R}, & (11.20)
\end{aligned}$$

and the Kalman filtering recursions

$$\begin{aligned}
\boldsymbol{x}_{i,t}^{t} &= \boldsymbol{x}_{i,t}^{t-1} + \boldsymbol{P}_{i,t}^{t-1}\boldsymbol{H}_{i,t}^{T}\boldsymbol{B}_{i,t}^{-1}\tilde{z}_{i,t}, & (11.21)\\
\boldsymbol{P}_{i,t}^{t} &= \boldsymbol{P}_{i,t}^{t-1} - \boldsymbol{P}_{i,t}^{t-1}\boldsymbol{H}_{i,t}^{T}\boldsymbol{B}_{i,t}^{-1}\boldsymbol{H}_{i,t}\boldsymbol{P}_{i,t}^{t-1}, & (11.22)\\
& & (11.23)
\end{aligned}$$

for $i = 1,\ldots,k$ and $t = 1,\ldots,n_i$, with initial conditions $\boldsymbol{x}_{i,1}^{0} = \boldsymbol{\Phi}\mu + \boldsymbol{\Upsilon}\boldsymbol{u}_{i,1}$ and $\boldsymbol{P}_{i,1}^{0} = \boldsymbol{\Phi}\boldsymbol{P}\boldsymbol{\Phi}^{T} + \boldsymbol{Q}$. Prediction h-steps ahead beyond the range of observed values is identical to (11.12)–(11.15).

Result 11.1 *This discussion assumes a time-invariant state coefficient matrix, i.e.* $\boldsymbol{\Phi}_t \equiv \boldsymbol{\Phi}$ *for all* t*, but all the results in this Section 11.2 hold with the obvious modification of* $\boldsymbol{\Phi} = \boldsymbol{\Phi}_t$ *for step* t *in the case of time-varying state coefficient matrices.*

11.3 The linear mixed model in SS form

The *linear mixed model* (LMM) is perhaps the most frequently employed type of random effects model in longitudinal data. This is sometimes also called the Laird-Ware model after Laird and Ware [146] who first formulated it, and it has a SS representation. The classical LMM in its most general compact matrix form is given by

$$\boldsymbol{y} = \boldsymbol{X\beta} + \boldsymbol{Zb} + \boldsymbol{\varepsilon}, \qquad \mathrm{E}\left(\begin{bmatrix} \boldsymbol{b} \\ \boldsymbol{\varepsilon} \end{bmatrix}\right) = \begin{bmatrix} \boldsymbol{0} \\ \boldsymbol{0} \end{bmatrix}, \qquad \mathrm{cov}\left(\begin{bmatrix} \boldsymbol{b} \\ \boldsymbol{\varepsilon} \end{bmatrix}\right) = \begin{bmatrix} \boldsymbol{Q} & \boldsymbol{0} \\ \boldsymbol{0} & \boldsymbol{R} \end{bmatrix}, \tag{11.24}$$

which implies the following mean and covariance structure for the response vector, $\boldsymbol{y} \sim (\boldsymbol{X\beta}, \boldsymbol{ZQZ}^T + \boldsymbol{R})$. The $\boldsymbol{X}$ and $\boldsymbol{Z}$ are known design matrices corresponding to the *fixed* and *random* effects, respectively, being entertained. The former are viewed as exogenous deterministic inputs into the system, which therefore makes their accommodation by the corresponding exogenous part of (11.3) the obvious choice in a SS framework. Time does not necessarily enter into (11.24); this basic model can accommodate either a strictly cross-sectional or strictly longitudinal setup, as well as both.

In a longitudinal version of the above LMM, the vector of observations for subject i, $\boldsymbol{y}_i = [y_{i,1}, \ldots, y_{i,n_i}]^T, i = 1, \ldots, k$, aggregated across times $t = 1, \ldots, n_i$, is given as

$$\boldsymbol{y}_i = \boldsymbol{X}_i\boldsymbol{\beta} + \boldsymbol{Z}_i\boldsymbol{b}_i + \boldsymbol{\varepsilon}_i, \qquad \boldsymbol{b}_i \sim \mathrm{NID}(\boldsymbol{0}, \boldsymbol{G}), \quad \boldsymbol{\varepsilon}_i = [\varepsilon_{i,1}, \ldots, \varepsilon_{i,n_i}]^T \sim \mathrm{NID}(\boldsymbol{0}, \boldsymbol{\Sigma}_i), \tag{11.25}$$

so that $\boldsymbol{y}_i \sim \mathrm{N}(\boldsymbol{X}_i\boldsymbol{\beta}, \boldsymbol{Z}_i\boldsymbol{G}\boldsymbol{Z}_i^T + \boldsymbol{\Sigma}_i)$. The assumptions of normality on the random effects $\boldsymbol{b}_i$ (between-subjects) and $\boldsymbol{\varepsilon}_i$ (within-subjects), are for convenience in likelihood based inference. Decomposed further into subject i at time t, (11.25) becomes

$$y_{i,t} = \boldsymbol{x}_{i,t}^T\boldsymbol{\beta} + \boldsymbol{z}_{i,t}^T\boldsymbol{b}_i + \varepsilon_{i,t}, \tag{11.26}$$

where $\boldsymbol{x}_{i,t}$ and $\boldsymbol{z}_{i,t}$ are n_i vectors containing the t-th rows of $\boldsymbol{X}_i$ and $\boldsymbol{Z}_i$, respectively.

One way to cast (11.26) in SS form is to place all random components into a large state vector. We can do this in a two-step process. Consider first the case of (11.26) with uncorrelated within-subject errors. Now turn $\boldsymbol{b}_i \equiv \boldsymbol{b}_{i,t}$ into a state vector by introducing (vacuous) time dependence with initial condition, $\boldsymbol{b}_{i,0} \sim \mathrm{N}(\boldsymbol{0}, \boldsymbol{G})$. The observation and state equations then become

$$y_{i,t} = \boldsymbol{z}_{i,t}^T\boldsymbol{b}_{i,t} + \boldsymbol{x}_{i,t}^T\boldsymbol{\beta} + \varepsilon_{i,t}, \qquad \varepsilon_{i,t} \sim \mathrm{NID}(0, \sigma^2) \tag{11.27}$$

$$\boldsymbol{b}_{i,t} = \boldsymbol{b}_{i,t-1}, \tag{11.28}$$

where we note that the $\boldsymbol{x}_{i,t}^T\boldsymbol{\beta}$ term naturally plays the part of the exogenous $\boldsymbol{Du}_{i,t}$ term in (11.3). Note also that the $w_{i,t}$ noise term is absent from (11.28) since it has null variance, this situation being similar to the Recursive Least Squares example in Subsection 4.7.1.

In longitudinal settings however, it will rarely be the case that the $\{\varepsilon_{i,t}\}$ will be serially uncorrelated, and thus appropriate specification of the within-subject error covariance matrix $\boldsymbol{\Sigma}_i$ is key. Letting $\sigma_i(t, t+h)$ denote the row t and column $t+h$ entry of $\boldsymbol{\Sigma}_i$, note that $\mathrm{E}(y_{i,t}) = \boldsymbol{x}_{i,t}^T\boldsymbol{\beta}$, with covariance

$$\mathrm{cov}(y_{i,t}, y_{j,t+h}) = \begin{cases} \boldsymbol{z}_{i,t}^T\boldsymbol{G}\boldsymbol{z}_{i,t+h} + \sigma_i(t, t+h), & \text{if } i = j, \\ 0, & \text{if } i \neq j. \end{cases}$$

With the hope that any nonstationary ehavior has been accounted for by the exogenous term and the random effects $\{z_{i,t}^T b_{i,t}\}$, a stationary within-subject error covariance structure, such as that modeled by an ARMA, is frequently assumed for $\sigma_i(t,t+h) \equiv \sigma_i(h)$. The second step of the two-step process to cast (11.26) in SS form then renders the ARMA model for $\{\varepsilon_{i,t}\}$ in SS form, in the absence of random effects. Finally, these two steps are combined by assembling the separate state vectors into a larger state vector.

The following example illustrates some of the ideas discussed above in converting an LMM into SS form. It focuses on demonstrating the various ways that can potentially be used to effect this conversion, and the trade-offs involved with each. A second example is given in Subsection 11.6.1, and it demonstrates an application of the standard two-step procedure to a real dataset.

Example 11.3.1 A mean with AR(1) errors

Suppose the vector of observations for subject i, $i = 1,\ldots,k$, is of length n and follows a constant mean β with AR(1) errors,

$$\boldsymbol{y}_i = \mathbf{1}\beta + \boldsymbol{\varepsilon}_i, \qquad \mathbf{1} = [1,\ldots,1]^T, \qquad \boldsymbol{\varepsilon}_i \sim \text{NID}(\mathbf{0}, \boldsymbol{\Sigma}_i),$$

where $\boldsymbol{\Sigma}_i$ is an $n \times n$ Toeplitz matrix whose (j,l) entry is, $\boldsymbol{\Sigma}_i(j,l) = \sigma^2\phi^{|j-l|}/(1-\phi^2)$, and $|\phi| < 1$. The most natural SS formulation is to treat the mean as an exogenous input, leading to the following model for each subject i:

$$y_t = \beta + \varepsilon_t, \qquad \varepsilon_t = \phi\varepsilon_{t-1} + w_t, \qquad w_t \sim \text{NID}(0,\sigma^2), \tag{11.29}$$

with initial condition $\varepsilon_0 \sim \text{N}(0,\sigma^2/(1-\phi^2))$. A potential problem with this approach is that β must be estimated, along with ϕ and σ^2, via nonlinear optimization. To circumvent this difficulty, an alternative route proceeds by including the mean in the (now two-dimensional) state vector by specifying $\beta \equiv \beta_t$ for all t, and defining the system

$$y_t = \varepsilon_t, \qquad \begin{bmatrix}\varepsilon_t\\ \beta_t\end{bmatrix} = \begin{bmatrix}\phi & 1\\ 0 & 1\end{bmatrix}\begin{bmatrix}\varepsilon_{t-1}\\ \beta_{t-1}\end{bmatrix} + \begin{bmatrix}w_t\\ 0\end{bmatrix}, \qquad \begin{bmatrix}w_t\\ 0\end{bmatrix} \sim \text{NID}(\mathbf{0},\boldsymbol{W}), \tag{11.30}$$

with initial condition $[\varepsilon_0,\beta_0]^T \sim \text{N}(0,\boldsymbol{G})$ on the state vector. The structure of the two covariance matrices, $\boldsymbol{G}$ and $\boldsymbol{W}$, is similar,

$$\boldsymbol{G} = \begin{bmatrix}\frac{\sigma^2}{1-\phi^2} & 0\\ 0 & 0\end{bmatrix}, \qquad \text{and} \qquad \boldsymbol{W} = \begin{bmatrix}\sigma^2 & 0\\ 0 & 0\end{bmatrix}.$$

In comparison with the first, the problem with this second formulation is that a more complex initial condition has to be entertained, along with a singular structure for the two covariance matrices above.

11.4 Maximum likelihood estimation

This section discusses ML estimation for the set of parameters $\Theta = (\boldsymbol{\mu}, \boldsymbol{P}, \boldsymbol{\Phi}, \boldsymbol{\Upsilon}, \boldsymbol{Q}, \boldsymbol{D}, \boldsymbol{R})$ in SS model (11.2)–(11.3). The asymptotic properties of these estimates are discussed in Appendix B. We extend the notation already introduced in Section 11.2 for the concatenation of observation vectors $Z_{i,s}$ and exogenous vectors $U_{i,s}$ up to time s, to include also the state vectors $X_{i,s} = (\boldsymbol{x}_{i,0}, \boldsymbol{x}_{i,1}, \ldots, \boldsymbol{x}_{i,s})$. In addition, let $\boldsymbol{X} = (X_{1,n_1}, \ldots, X_{k,n_k})$, $\boldsymbol{Z} = (Z_{1,n_1}, \ldots, Z_{k,n_k})$, and $\boldsymbol{U} = (U_{1,n_1}, \ldots, U_{k,n_k})$, with $\boldsymbol{Y} = (\boldsymbol{X}, \boldsymbol{Z}, \boldsymbol{U})$ denoting the *complete* data. Note that (Z,U) constitutes the *observed* data. If all the states X_{i,n_i} were observable ($i = 1, \ldots, k$), then the complete data joint density could be decomposed into a product of conditional densities for the $(\boldsymbol{x}_{i,t}|\boldsymbol{x}_{i,t-1}, \boldsymbol{u}_{i,t})$, the $(\boldsymbol{z}_{i,t}|\boldsymbol{x}_{i,t}, \boldsymbol{u}_{i,t})$, and the marginals of $\boldsymbol{x}_{i,0}$, yielding the likelihood function

$$\begin{aligned} f_\Theta(Y) &= \prod_{i=1}^{k}\left\{ f_{\boldsymbol{\mu},\boldsymbol{P}}(\boldsymbol{x}_{i,0}) \prod_{t=1}^{n_i} f_{\boldsymbol{\Phi},\boldsymbol{\Upsilon},\boldsymbol{Q}}(\boldsymbol{x}_{i,t} \,|\, \boldsymbol{x}_{i,t-1}, \boldsymbol{u}_{i,t}) \prod_{t=1}^{n_i} f_{\boldsymbol{D},\boldsymbol{R}}(\boldsymbol{z}_{i,t} \,|\, \boldsymbol{x}_{i,t}, \boldsymbol{u}_{i,t}) \right\} \\ &:= L(\Theta \,|\, \boldsymbol{Y}). \end{aligned}$$

Two times the natural logarithm of this *complete* data likelihood, $2\log L(\Theta|\boldsymbol{Y})$, ignoring a constant, assuming joint multivariate normality for the error processes and initial state, and letting $\sum_{i=1}^{k} n_i = N$, can be written as

$$\begin{aligned} \ell(\Theta \,|\, Y) = {} & k\log|\boldsymbol{P}^{-1}| \\ & - \sum_{i=1}^{k} (\boldsymbol{x}_{i,0} - \boldsymbol{\mu})^T \boldsymbol{P}^{-1} (\boldsymbol{x}_{i,0} - \boldsymbol{\mu}) + N\left(\log|\boldsymbol{Q}^{-1}| + \log|\boldsymbol{R}^{-1}|\right) \\ & - \sum_{i=1}^{k}\sum_{t=1}^{n_i} (\boldsymbol{x}_{i,t} - \boldsymbol{\Phi}\boldsymbol{x}_{i,t-1} - \boldsymbol{\Upsilon}\boldsymbol{u}_{i,t})^T \boldsymbol{Q}^{-1} (\boldsymbol{x}_{i,t} - \boldsymbol{\Phi}\boldsymbol{x}_{i,t-1} - \boldsymbol{\Upsilon}\boldsymbol{u}_{i,t}) \\ & - \sum_{i=1}^{k}\sum_{t=1}^{n_i} (\boldsymbol{z}_{i,t} - \boldsymbol{H}_{i,t}\boldsymbol{x}_{i,t} - \boldsymbol{D}\boldsymbol{u}_{i,t})^T \boldsymbol{R}^{-1} (\boldsymbol{z}_{i,t} - \boldsymbol{H}_{i,t}\boldsymbol{x}_{i,t} - \boldsymbol{D}\boldsymbol{u}_{i,t}). \end{aligned} \tag{11.31}$$

The *observed* data counterpart to $\ell(\Theta|\boldsymbol{Y})$ however, is the function that must be maximized in order to produce the ML estimate of Θ. Ignoring a constant, twice the observed data log-likelihood can conveniently be expressed in innovations form as

$$\ell(\Theta \,|\, \boldsymbol{Z}, \boldsymbol{U}) := 2\log L(\Theta \,|\, \boldsymbol{Z}, \boldsymbol{U}) = \sum_{i=1}^{k}\sum_{t=1}^{n_i}\left(\log|\boldsymbol{B}_{i,t}^{-1}| - \tilde{\boldsymbol{z}}_{i,t}^T \boldsymbol{B}_{i,t}^{-1} \tilde{\boldsymbol{z}}_{i,t}\right), \tag{11.32}$$

where the *innovations* (or prediction errors) in (11.32) are defined as

$$\tilde{\boldsymbol{z}}_{i,t} = \boldsymbol{z}_{i,t} - \mathrm{E}(\boldsymbol{z}_{i,t}|Z_{i,t-1}, U_{i,t-1}) = \boldsymbol{z}_{i,t} - \boldsymbol{H}_{i,t}\boldsymbol{x}_{i,t}^{t-1} - \boldsymbol{D}\boldsymbol{u}_{i,t}. \tag{11.33}$$

Clearly, $\mathrm{E}(\tilde{\boldsymbol{z}}_{i,t}) = \boldsymbol{0}$, so that the innovations sequence $\{\tilde{\boldsymbol{z}}_{i,t}\}$, $t = 1, \ldots, n_i$, comprise an independent Gaussian process with covariance matrix

$$\boldsymbol{B}_{i,t} := \mathrm{cov}(\tilde{\boldsymbol{z}}_{i,t}) = \mathrm{cov}\left[\boldsymbol{H}_{i,t}\left(\boldsymbol{x}_{i,t} - \boldsymbol{x}_{i,t}^{t-1}\right) + \boldsymbol{v}_{i,t}\right] = \boldsymbol{H}_{i,t}\boldsymbol{P}_{i,t}^{t-1}\boldsymbol{H}_{i,t}^T + \boldsymbol{R}.$$

Assuming independence among subjects i, $i = 1,\ldots,k$, leads to (11.32).

As discussed in Chapter 5, the usual method of maximizing (11.32) is to use an iterative procedure that combines propagation of the Kalman recursions (11.7)–(11.11) for a given parameter value, $\Theta^{(j-1)}$, with a Newton-Raphson algorithm in order to produce an updated parameter value, $\Theta^{(j)}$. This is the usual procedure advocated by say, Jones [132]. However, since the advent of the Expectation-Maximization algorithm, see Dempster, Laird and Rubin [73], an EM based optimization scheme originally devised by Shumway and Stoffer [194] for the standard SS model (without cross-sectional extension) has gained popularity, partly due to the fact that it does not require computation of the gradient vector and Hessian matrix of the log-likelihood (see, also, Shumway and Stoffer [195, ch. 6]). For the cross-sectionally extended model (11.2)–(11.3), Naranjo, Trindade and Casella [165] provide complete closed-form expressions for the E and M-steps as we will now detail.

The EM algorithm consecutively maximizes the expectation of the complete data likelihood given the observations to find the ML estimate of Θ based on the observed data $(\boldsymbol{Z},\boldsymbol{U})$, since the complete data $Y = (\boldsymbol{X},\boldsymbol{Z},\boldsymbol{U})$ are unknown. It begins with implementation of the E-step. At iteration j, let

$$\varphi\left(\Theta\middle|\Theta^{(j-1)}\right) = \mathrm{E}\left[-\ell(\Theta|\boldsymbol{Y})\,\middle|\,\boldsymbol{Z},\boldsymbol{U},\Theta^{(j-1)}\right], \qquad \text{and} \qquad \boldsymbol{\Psi} = [\boldsymbol{\Phi},\boldsymbol{\Upsilon}],$$

define the relevant E-step function, and the matrix formed by concatenating $\boldsymbol{\Phi}$ and $\boldsymbol{\Upsilon}$, in order to simplify notation. A straightforward computation then yields

$$\begin{aligned}\varphi\left(\Theta\middle|\Theta^{(j-1)}\right) &= k\log|\boldsymbol{P}| + N(\log|\boldsymbol{Q}| + \log|\boldsymbol{R}|)\\ &\quad + \sum_{i=1}^{k}\mathrm{tr}\left\{\boldsymbol{P}^{-1}\left[\boldsymbol{P}_{i,0}^{n_i} + \left(\boldsymbol{x}_{i,0}^{n_i}-\boldsymbol{\mu}\right)\left(\boldsymbol{x}_{i,0}^{n_i}-\boldsymbol{\mu}\right)^T\right]\right\}\\ &\quad + \mathrm{tr}\left\{\boldsymbol{Q}^{-1}\left(\boldsymbol{S}_{11} - \boldsymbol{S}_{10}\boldsymbol{\Psi}^T - \boldsymbol{\Psi}\boldsymbol{S}_{10}^T + \boldsymbol{\Psi}\boldsymbol{S}_{00}\boldsymbol{\Psi}^T\right)\right\}\\ &\quad + \mathrm{tr}\left\{\boldsymbol{R}^{-1}\left(\boldsymbol{T}_{11} - \boldsymbol{T}_{10}\boldsymbol{D}^T - \boldsymbol{D}\boldsymbol{T}_{10}^T + \boldsymbol{D}\boldsymbol{T}_{00}\boldsymbol{D}^T\right)\right\},\end{aligned} \tag{11.34}$$

where we define the following "sums of squares" type terms

$$\begin{aligned}\boldsymbol{S}_{11} &= \sum_{i=1}^{k}\sum_{t=1}^{n_i}\left(\boldsymbol{P}_{i,t}^{n_i} + \boldsymbol{x}_{i,t}^{n_i}\boldsymbol{x}_{i,t}^{n_i\mathsf{T}}\right),\\ \boldsymbol{S}_{10} &= \sum_{i=1}^{k}\sum_{t=1}^{n_i}\left[\boldsymbol{P}_{i,t,t-1}^{n_i} + \boldsymbol{x}_{i,t}^{n_i}\boldsymbol{x}_{i,t-1}^{n_i\mathsf{T}} \quad \boldsymbol{x}_{i,t}^{n_i}\boldsymbol{u}_{i,t}^T\right],\\ \boldsymbol{S}_{00} &= \sum_{i=1}^{k}\sum_{t=1}^{n_i}\begin{bmatrix}\boldsymbol{P}_{i,t-1}^{n_i} + \boldsymbol{x}_{i,t-1}^{n_i}\boldsymbol{x}_{i,t-1}^{n_i\mathsf{T}} & \boldsymbol{x}_{i,t-1}^{n_i}\boldsymbol{u}_{i,t}^T\\ \boldsymbol{u}_{i,t}\boldsymbol{x}_{i,t-1}^{n_i\mathsf{T}} & \boldsymbol{u}_{i,t}\boldsymbol{u}_{i,t}^T\end{bmatrix},\end{aligned}$$

and

$$
\begin{aligned}
T_{00} &= \sum_{i=1}^{k}\sum_{t=1}^{n_i} u_{i,t}u_{i,t}^T, \\
T_{10} &= \sum_{i=1}^{k}\sum_{t=1}^{n_i} \left(z_{i,t} - H_{i,t}x_{i,t}^{n_i}\right) u_{i,t}^T, \\
T_{11} &= \sum_{i=1}^{k}\sum_{t=1}^{n_i} \left[\left(z_{i,t} - H_{i,t}x_{i,t}^{n_i}\right)\left(z_{i,t} - H_{i,t}x_{i,t}^{n_i}\right)^T + H_{i,t}P_{i,t}^{n_i}H_{i,t}^T\right].
\end{aligned}
$$

Maximization of (11.34) with respect to Θ at iteration j constitutes the M-step. Taking derivatives with respect to each component of Θ, setting to zero, and solving for each parameter, produces the updated estimators

$$
\begin{aligned}
\mu^{(j)} &= \frac{1}{k}\sum_{i=1}^{k} x_{i,0}^{n_i}, \\
P^{(j)} &= \frac{1}{k}\sum_{i=1}^{k}\left[P_{i,0}^{n_i} + \left(x_{i,0}^{n_i} - \mu^{(j)}\right)\left(x_{i,0}^{n_i} - \mu^{(j)}\right)^T\right], \\
\Psi^{(j)} &= S_{10}S_{00}^{-1}, \\
Q^{(j)} &= \frac{1}{N}\left(S_{11} - S_{10}S_{00}^{-1}S_{10}^T\right), \\
D^{(j)} &= T_{10}T_{00}^{-1}, \\
R^{(j)} &= \frac{1}{N}\left(T_{11} - T_{10}T_{00}^{-1}T_{10}^T\right).
\end{aligned}
$$

11.5 Missing data modifications

Missing data is an almost *de facto* consequence of longitudinal studies of any complexity, especially when biological variables are involved. The SS model can, with certain modifications, accommodate missing values in both responses ($z_{i,t}$) and covariates ($u_{i,t}$). We first discuss methods for handling missingness in responses only, followed by two methods for handling missingness in both responses and covariates. Strictly speaking, these last two methods apply to different SS model formulations, although they do overlap in a particular instance.

11.5.1 Missingness in responses only

The classical approach to this problem proceeds by imputing the missing values (viewed as irregularly-spaced data) via the Kalman recursions, whence an expression for the complete data likelihood is available; see e.g., Brockwell and Davis [32, ch. 12]. On the other hand, an EM algorithm-based solution offers the additional attraction of providing a procedure for not only imputing, but also maximizing; see Shumway and Stoffer [195, ch. 6]. We focus our attention on the latter.

The necessary modifications to the EM procedure presented in Section 11.4 is most easily discussed by simplifying model (11.2)–(11.3) to exclude covariates,

$$\begin{aligned} \boldsymbol{x}_{i,t} &= \boldsymbol{\Phi}\boldsymbol{x}_{i,t-1} + \boldsymbol{w}_{i,t}, \quad (11.35) \\ \begin{bmatrix} \boldsymbol{z}_{i,t}^{(o)} \\ \boldsymbol{z}_{i,t}^{(u)} \end{bmatrix} &= \begin{bmatrix} \boldsymbol{H}_{i,t}^{(o)} \\ \boldsymbol{H}_{i,t}^{(u)} \end{bmatrix} \boldsymbol{x}_{i,t} + \begin{bmatrix} \boldsymbol{v}_{i,t}^{(o)} \\ \boldsymbol{v}_{i,t}^{(u)} \end{bmatrix}, \quad (11.36) \end{aligned}$$

with

$$\operatorname{cov}\left(\begin{bmatrix} \boldsymbol{v}_{i,t}^{(o)} \\ \boldsymbol{v}_{i,t}^{(u)} \end{bmatrix}\right) = \begin{bmatrix} \boldsymbol{R}_{i,t}^{(o)} & \boldsymbol{R}_{i,t}^{(ou)} \\ \boldsymbol{R}_{i,t}^{(ou)T} & \boldsymbol{R}_{i,t}^{(u)} \end{bmatrix},$$

where we have rearranged the observation vector for unit i at time t in partitioned form, with $\boldsymbol{z}_{i,t}^{(o)}$ denoting the *observed* part and $\boldsymbol{z}_{i,t}^{(u)}$ the *unobserved* part. The KF from Section 11.2 can now be applied by "zeroing out" the appropriate portions of the partitioned vectors and matrices in the observation equation, i.e. $\boldsymbol{z}_{i,t}^{(u)} = \mathbf{0}$, $\boldsymbol{H}_{i,t}^{(u)} = \mathbf{0}$, and $\boldsymbol{R}_{i,t}^{(ou)} = \mathbf{0}$; see Stoffer [200] for a justification of the validity of this approach.

In similar notation to Section 11.4 (with $\boldsymbol{U}$ omitted), if $\boldsymbol{Z}^{(o)} = (\boldsymbol{z}_{1,1}^{(o)}, \ldots, \boldsymbol{z}_{k,n_k}^{(o)})$ denotes the concatenation of all observed responses, then $\boldsymbol{Y} = (\boldsymbol{X}, \boldsymbol{Z})$ is still the complete data, but the observed data is now obviously only $\boldsymbol{Z}^{(o)}$. The expression for twice the complete data log-likelihood (11.31) remains unchanged, but in implementing the E-step we now condition on the portion of $\boldsymbol{Z}$ that is actually observed, so that we must compute

$$\varphi\left(\Theta \middle| \Theta^{(j-1)}\right) = \mathrm{E}\left[-\ell(\Theta|\boldsymbol{Y}) \,\middle|\, \boldsymbol{Z}^{(o)}, \Theta^{(j-1)}\right],$$

which results in the expression

$$\begin{aligned} \varphi\left(\Theta \middle| \Theta^{(j-1)}\right) &= k\log|\boldsymbol{P}| + N(\log|\boldsymbol{Q}| + \log|\boldsymbol{R}|) + \operatorname{tr}\{\boldsymbol{R}^{-1}\boldsymbol{T}_{11}\} \\ &\quad + \sum_{i=1}^{k} \operatorname{tr}\left\{\boldsymbol{P}^{-1}\left[\boldsymbol{P}_{i,0}^{(n_i)} + \left(\boldsymbol{x}_{i,0}^{(n_i)} - \boldsymbol{\mu}\right)\left(\boldsymbol{x}_{i,0}^{(n_i)} - \boldsymbol{\mu}\right)^T\right]\right\} \\ &\quad + \operatorname{tr}\{\boldsymbol{Q}^{-1}(\boldsymbol{S}_{11} - \boldsymbol{S}_{10}\boldsymbol{\Phi}^T - \boldsymbol{\Phi}\boldsymbol{S}_{10}^T + \boldsymbol{\Phi}\boldsymbol{S}_{00}\boldsymbol{\Phi}^T)\}, \end{aligned}$$

where the sums of squares terms are now

$$\begin{aligned} \boldsymbol{S}_{11} &= \sum_{i=1}^{k}\sum_{t=1}^{n_i}\left[\boldsymbol{P}_{i,t}^{(n_i)} + \boldsymbol{x}_{i,t}^{(n_i)}\boldsymbol{x}_{i,t}^{(n_i)T}\right], \\ \boldsymbol{S}_{10} &= \sum_{i=1}^{k}\sum_{t=1}^{n_i}\left[\boldsymbol{P}_{i,t,t-1}^{(n_i)} + \boldsymbol{x}_{i,t}^{(n_i)}\boldsymbol{x}_{i,t-1}^{(n_i)T}\right], \\ \boldsymbol{S}_{00} &= \sum_{i=1}^{k}\sum_{t=1}^{n_i}\left[\boldsymbol{P}_{i,t-1}^{(n_i)} + \boldsymbol{x}_{i,t-1}^{(n_i)}\boldsymbol{x}_{i,t-1}^{(n_i)T}\right], \end{aligned}$$

and

$$T_{11} = \sum_{i=1}^{k}\sum_{t=1}^{n_i}\left\{\left(z_{i,t} - H_{i,t}x_{i,t}^{(n_i)}\right)\left(z_{i,t} - H_{i,t}x_{i,t}^{(n_i)}\right)^T + H_{i,t}P_{i,t}^{(n_i)}H_{i,t}^T + \begin{bmatrix} 0 & 0 \\ 0 & R_{i,t}^{(u)} \end{bmatrix}\right\}. \quad (11.37)$$

The T_{11} matrix now involves the covariance matrix for the noise vector $v_{i,t}$, due to computation of the terms $\mathrm{E}(z_{i,t}^{(u)}|Z^{(o)})$ and $\mathrm{E}(z_{i,t}^{(u)}z_{i,t}^{(u)T}|Z^{(o)})$. Note also that an adjustment to the filtered state values $x_{i,t}^{s}$ and their error covariance matrices $P_{i,t}^{s}$ was necessary, so that these have been replaced by

$$x_{i,t}^{(s)} = \mathrm{E}\left(x_{i,t}\,\middle|\,z_{i,1}^{(o)},\ldots,z_{i,s}^{(o)}\right),$$

and

$$P_{i,t_1,t_2}^{(s)} = \mathrm{E}\left[\left(x_{i,t_1} - x_{i,t_1}^{(s)}\right)\left(x_{i,t_2} - x_{i,t_2}^{(s)}\right)^T \,\middle|\, z_{i,1}^{(o)},\ldots,z_{i,s}^{(o)}\right],$$

respectively. Implementation of the M-step (maximization with respect to Θ at iteration j) produces the updated estimators

$$\begin{aligned}
\mu^{(j)} &= \frac{1}{k}\sum_{i=1}^{k} x_{i,0}^{(n_i)}, \\
P^{(j)} &= \frac{1}{k}\sum_{i=1}^{k}\left[P_{i,0}^{(n_i)} + \left(x_{i,0}^{(n_i)} - \mu^{(j)}\right)\left(x_{i,0}^{(n_i)} - \mu^{(j)}\right)^T\right], \\
\Phi^{(j)} &= S_{10}S_{00}^{-1}, \\
Q^{(j)} &= \frac{1}{N}\left(S_{11} - S_{10}S_{00}^{-1}S_{10}^T\right), \\
R^{(j)} &= \frac{1}{N}M_{i,t}T_{11}M_{i,t}^T,
\end{aligned}$$

where $M_{i,t}$ is a permutation matrix (consisting of zeros and ones) that reorders the response vector for subject i at time t in its original order, i.e., if $z_{i,t}^{*} = [z_{i,t}^{(o)T}, z_{i,t}^{(u)T}]^T$ denotes the missing data rearranged $z_{i,t}$, then $M_{i,t}z_{i,t}^{*} = z_{i,t}$.

11.5.2 Missingness in both responses and covariates: method 1

Accommodation of missing values in both responses ($z_{i,t}$) and covariates ($u_{i,t}$), can most easily be achieved for the SS model formulation of (11.1) and (11.3)

$$x_{i,t}^{a} = \Phi^{a}x_{i,t-1}^{a} + \Gamma u_{i,t-1} + w_{i,t}^{a}, \qquad \{w_{i,t}^{a}\} \sim \mathrm{NID}(0, Q^{a}), \quad (11.38)$$

$$z_{i,t} = H_{i,t}^{a}x_{i,t}^{a} + Du_{i,t} + v_{i,t}^{a}, \qquad \{v_{i,t}^{a}\} \sim \mathrm{NID}(0, R^{a}). \quad (11.39)$$

In order to provide a mechanism for imputing the potentially missing (in whole or in part) exogenous $\boldsymbol{u}_{i,t}$, we now introduce an additional SS model for it

$$\boldsymbol{x}^b_{i,t} = \boldsymbol{\Phi}^b \boldsymbol{x}^b_{i,t-1} + \boldsymbol{w}^b_{i,t}, \qquad \{\boldsymbol{w}^b_{i,t}\} \sim \text{NID}(\boldsymbol{0}, Q^b), \tag{11.40}$$

$$\boldsymbol{u}_{i,t} = \boldsymbol{H}^b_{i,t} \boldsymbol{x}^b_{i,t} + \boldsymbol{v}^b_{i,t}, \qquad \{\boldsymbol{v}^b_{i,t}\} \sim \text{NID}(\boldsymbol{0}, R^b). \tag{11.41}$$

Substituting (11.41) into (11.38) and (11.39), leads to the system

$$\boldsymbol{x}^a_{i,t} = \boldsymbol{\Phi}^a \boldsymbol{x}^a_{i,t-1} + \boldsymbol{\Gamma}\left(\boldsymbol{H}^b_{i,t-1}\boldsymbol{x}^b_{i,t-1} + \boldsymbol{v}^b_{i,t-1}\right) + \boldsymbol{w}^a_{i,t}, \tag{11.42}$$

$$\boldsymbol{z}_{i,t} = \boldsymbol{H}^a_{i,t}\boldsymbol{x}^a_{i,t} + \boldsymbol{D}\left(\boldsymbol{H}^b_{i,t}\boldsymbol{x}^b_{i,t} + \boldsymbol{v}^b_{i,t}\right) + \boldsymbol{v}^a_{i,t}. \tag{11.43}$$

We can now combine (11.42) and (11.40) into a single SS equation,

$$\begin{bmatrix} \boldsymbol{x}^a_{i,t} \\ \boldsymbol{x}^b_{i,t} \end{bmatrix} = \begin{bmatrix} \boldsymbol{\Phi}^a & \boldsymbol{\Gamma}\boldsymbol{H}^a_{i,t-1} \\ \boldsymbol{0} & \boldsymbol{\Phi}^b \end{bmatrix} \begin{bmatrix} \boldsymbol{x}^a_{i,t-1} \\ \boldsymbol{x}^b_{i,t-1} \end{bmatrix} + \begin{bmatrix} \boldsymbol{\Gamma}\boldsymbol{v}^b_{i,t-1} + \boldsymbol{w}^a_{i,t} \\ \boldsymbol{w}^b_{i,t} \end{bmatrix}, \tag{11.44}$$

and (11.41) and (11.43) into the single observation equation,

$$\begin{bmatrix} \boldsymbol{z}_{i,t} \\ \boldsymbol{u}_{i,t} \end{bmatrix} = \begin{bmatrix} \boldsymbol{H}^a_{i,t} & \boldsymbol{D}\boldsymbol{H}^b_{i,t} \\ \boldsymbol{0} & \boldsymbol{H}^b_{i,t} \end{bmatrix} \begin{bmatrix} \boldsymbol{x}^a_{i,t} \\ \boldsymbol{x}^b_{i,t} \end{bmatrix} + \begin{bmatrix} \boldsymbol{D}\boldsymbol{v}^b_{i,t} + \boldsymbol{v}^a_{i,t} \\ \boldsymbol{v}^b_{i,t} \end{bmatrix}. \tag{11.45}$$

Note that this reformulation essentially aggregates (11.38)–(11.41), into an equivalent system that is a SS model without exogenous variables, as in (11.35)–(11.36). In so doing, the exogenous variables become endogenous, i.e., the left-hand side of (11.45) is the new response variable. This "trick" of *endogeneization* (already introduced in Section 3.2) has been known since at least Mehra [163], but was more visibly presented by Terceiro [206] and others.

In order to deal with missing values, we now reorder the (new) response vector in (11.45),

$$\begin{bmatrix} \boldsymbol{z}_{i,t} \\ \boldsymbol{u}_{i,t} \end{bmatrix} \longrightarrow \begin{bmatrix} \boldsymbol{z}^{(o)}_{i,t} \\ \boldsymbol{u}^{(o)}_{i,t} \\ \boldsymbol{z}^{(u)}_{i,t} \\ \boldsymbol{u}^{(u)}_{i,t} \end{bmatrix},$$

where $\boldsymbol{z}^{(o)}_{i,t}$ and $\boldsymbol{u}^{(o)}_{i,t}$ denote the observed parts of the response and covariate, and $\boldsymbol{z}^{(u)}_{i,t}$ and $\boldsymbol{u}^{(u)}_{i,t}$ the unobserved parts. After corresponding reordering of the state vector, the newly reordered system in now seen to be in the form of (11.35)–(11.36), whence the missing portion of the "response" vector, $[\boldsymbol{z}^{(u)}_{i,t}, \boldsymbol{u}^{(u)}_{i,t}]^T$, can be handled through the imputation method outlined in Subsection 11.5.1.

Remark 11.2 *A byproduct of this approach for handling missing data is that the resulting state equation (11.44) now has a time-varying state coefficient matrix. As discussed in Result 11.1, this does not pose a problem from the standpoint of Kalman filtering. However, from an empirical perspective, it does raise the issue of how to "estimate" a time-varying state coefficient matrix; see Chapter 6.*

11.5.3 Missingness in both responses and covariates: method 2

If instead of (11.1), the alternative formulation (11.2) admitting a contemporaneous relationship between the state and exogenous variables is used,

$$\boldsymbol{x}_{i,t}^a = \boldsymbol{\Phi}^a \boldsymbol{x}_{i,t-1}^a + \boldsymbol{\Upsilon} \boldsymbol{u}_{i,t} + \boldsymbol{w}_{i,t}^a, \qquad \{\boldsymbol{w}_{i,t}^a\} \sim \text{NID}(\boldsymbol{0}, Q^a), \tag{11.46}$$

$$\boldsymbol{z}_{i,t} = \boldsymbol{H}_{i,t}^a \boldsymbol{x}_{i,t}^a + \boldsymbol{D} \boldsymbol{u}_{i,t} + \boldsymbol{v}_{i,t}^a, \qquad \{\boldsymbol{v}_{i,t}^a\} \sim \text{NID}(\boldsymbol{0}, R^a), \tag{11.47}$$

the trick of endogeneization will not be able to render the resulting system in the form of a new SS model as in (11.44)–(11.45). This is due to the time index match between the state variable $\boldsymbol{x}_{i,t}$ and covariate $\boldsymbol{u}_{i,t}$ in (11.46), which prevents a reformulation of the system into an aggregated state equation as in (11.44). Nevertheless, one can follow the analogous idea of introducing an extra SS model in order to provide a mechanism for imputing the missing covariates.

The details of this argument are as follows. We again partition the observation vector into observed and unobserved components, respectively, as $\boldsymbol{z}_{i,t} = (\boldsymbol{z}_{i,t}^{(o)T}, \boldsymbol{z}_{i,t}^{(u)T})^T$, and similarly for the exogenous variables, $\boldsymbol{u}_{i,t} = (\boldsymbol{u}_{i,t}^{(o)T}, \boldsymbol{u}_{i,t}^{(u)T})^T$. The state and observation equations can then be re-expressed as

$$\boldsymbol{x}_{i,t}^a = \boldsymbol{\Phi} \boldsymbol{x}_{i,t-1}^a + \begin{bmatrix} \boldsymbol{\Upsilon}^{(o)} & \boldsymbol{\Upsilon}^{(u)} \end{bmatrix} \begin{bmatrix} \boldsymbol{u}_{i,t}^{(o)} \\ \boldsymbol{u}_{i,t}^{(u)} \end{bmatrix} + \boldsymbol{w}_{i,t}^a, \tag{11.48}$$

and

$$\begin{bmatrix} \boldsymbol{z}_{i,t}^{(o)} \\ \boldsymbol{z}_{i,t}^{(u)} \end{bmatrix} = \begin{bmatrix} \boldsymbol{H}_{i,t}^{a(o)} \\ \boldsymbol{H}_{i,t}^{a(u)} \end{bmatrix} \boldsymbol{x}_{i,t}^a + \begin{bmatrix} \boldsymbol{D}_{11}^{(o)} & \boldsymbol{D}_{12}^{(u)} \\ \boldsymbol{D}_{21}^{(o)} & \boldsymbol{D}_{22}^{(u)} \end{bmatrix} \begin{bmatrix} \boldsymbol{u}_{i,t}^{(o)} \\ \boldsymbol{u}_{i,t}^{(u)} \end{bmatrix} + \begin{bmatrix} \boldsymbol{v}_{i,t}^{a(o)} \\ \boldsymbol{v}_{i,t}^{a(u)} \end{bmatrix}. \tag{11.49}$$

The covariance between the observed and unobserved portions of the observational error vector is given by

$$\text{cov}\left(\begin{bmatrix} \boldsymbol{v}_{i,t}^{a(o)} \\ \boldsymbol{v}_{i,t}^{a(u)} \end{bmatrix} \right) = \begin{bmatrix} \boldsymbol{R}_{11}^a & \boldsymbol{R}_{12}^a \\ \boldsymbol{R}_{21}^a & \boldsymbol{R}_{22}^a \end{bmatrix}.$$

In (11.48)–(11.49), $\boldsymbol{x}_{i,t}^a$ and $\boldsymbol{z}_{i,t}$ depend on the unobserved portion of the exogenous variable vector, $\boldsymbol{u}_{i,t}^{(u)}$. To model $\boldsymbol{u}_{i,t}$ with a second SS model, we treat $\boldsymbol{u}_{i,t}$ as the observation vector for an underlying state process, $\boldsymbol{x}_{i,t}^b$, $i = 1, \ldots, k$ and $t = 1, \ldots, n_i$. This second (imputation) model is given by

$$\boldsymbol{x}_{i,t}^b = \boldsymbol{\Phi}^b \boldsymbol{x}_{i,t-1}^b + \boldsymbol{w}_{i,t}^b, \tag{11.50}$$

$$\boldsymbol{u}_{i,t} = \boldsymbol{H}_{i,t}^b \boldsymbol{x}_{i,t}^b + \boldsymbol{v}_{i,t}^b, \tag{11.51}$$

where $\boldsymbol{\Phi}^b$ denotes the state transition matrix, with $\boldsymbol{w}_{i,t}^b \sim NID(\boldsymbol{0}, \boldsymbol{Q}^b)$ the corresponding noise vector. $\boldsymbol{H}_{i,t}^b$ is now the observation matrix, and $\boldsymbol{v}_{i,t}^b \sim NID(\boldsymbol{0}, \boldsymbol{R}^b)$ the observational noise vector, with initial condition $\boldsymbol{x}_{i,0}^b \sim NID\left(\boldsymbol{\mu}^b, \boldsymbol{P}^b\right)$. For simplicity, $\{\boldsymbol{w}_{i,t}^a\}$, $\{\boldsymbol{v}_{i,t}^a\}$, $\{\boldsymbol{w}_{i,t}^b\}$, and $\{\boldsymbol{v}_{i,t}^b\}$ are assumed to be independent.

Now, the observation equation in (11.51) can be partitioned into observed and missing parts, yielding

$$\begin{bmatrix} \boldsymbol{u}_{i,t}^{(o)} \\ \boldsymbol{u}_{i,t}^{(u)} \end{bmatrix} = \begin{bmatrix} \boldsymbol{H}_{i,t}^{b(o)} \\ \boldsymbol{H}_{i,t}^{b(u)} \end{bmatrix} \boldsymbol{x}_{i,t}^{b} + \begin{bmatrix} \boldsymbol{v}_{i,t}^{b(o)} \\ \boldsymbol{v}_{i,t}^{b(u)} \end{bmatrix}, \tag{11.52}$$

with the covariance between the observed and unobserved portions of the observational error vector $\boldsymbol{v}_{i,t}^{b}$ given by

$$\operatorname{cov}\left(\begin{bmatrix} \boldsymbol{v}_{i,t}^{b(o)} \\ \boldsymbol{v}_{i,t}^{b(u)} \end{bmatrix}\right) = \begin{bmatrix} \boldsymbol{R}_{11}^{b} & \boldsymbol{R}_{12}^{b} \\ \boldsymbol{R}_{21}^{b} & \boldsymbol{R}_{22}^{b} \end{bmatrix}.$$

Finally, the likelihood must be modified to account for the second SS model. Define

$$\begin{aligned} 2\log L\left(\Theta^{b}\middle| X^{b},U\right) &= k\log|\boldsymbol{P}^{b}|^{-1} + N\left(\log|(\boldsymbol{Q}^{b})^{-1}| + \log|(\boldsymbol{R}^{b})^{-1}|\right) \\ &\quad - \sum_{i=1}^{k}\left(\boldsymbol{x}_{i,0}^{b} - \boldsymbol{\mu}^{b}\right)^{T}\left(\boldsymbol{P}^{b}\right)^{-1}\left(\boldsymbol{x}_{i,0}^{b} - \boldsymbol{\mu}^{b}\right) \\ &\quad - \sum_{i=1}^{k}\sum_{t=1}^{n_i}\left(\boldsymbol{x}_{i,t}^{b} - \boldsymbol{\Phi}^{b}\boldsymbol{x}_{i,t-1}^{b}\right)^{T}(\boldsymbol{Q}^{b})^{-1}\left(\boldsymbol{x}_{i,t}^{b} - \boldsymbol{\Phi}^{b}\boldsymbol{x}_{i,t-1}^{b}\right) \\ &\quad - \sum_{i=1}^{k}\sum_{t=1}^{n_i}\left(\boldsymbol{u}_{i,t} - \boldsymbol{H}_{i,t}^{b}\boldsymbol{x}_{i,t}^{b}\right)^{T}(\boldsymbol{R}^{b})^{-1}\left(\boldsymbol{u}_{i,t} - \boldsymbol{H}_{i,t}^{b}\boldsymbol{x}_{i,t}^{b}\right), \end{aligned} \tag{11.53}$$

where $\Theta^{b} = (\boldsymbol{\mu}^{b}, \boldsymbol{P}^{b}, \boldsymbol{\Phi}^{b}, \boldsymbol{Q}^{b}, \boldsymbol{R}^{b})$ represents the parameters from the second SS formulation used to model the exogenous variables, and $\boldsymbol{X}^{b} = (X_{1,n_1}^{b}, \ldots, X_{k,n_k}^{b})$ denotes the state information for the second SS model, with $X_{i,n_i}^{b} = (\boldsymbol{x}_{i,0}^{b}, \ldots, \boldsymbol{x}_{i,n_i}^{b})$. Ignoring constants in (11.31) and (11.53), the natural logarithm of the complete data likelihood is now

$$2\log L\left(\Theta,\Theta^{b}\middle|\boldsymbol{X},\boldsymbol{Z},\boldsymbol{X}^{b},\boldsymbol{U}\right) = 2\log L(\Theta\,|\,\boldsymbol{X},\boldsymbol{Z},\boldsymbol{U}) + 2\log L\left(\Theta^{b}\middle|\boldsymbol{X}^{b},\boldsymbol{U}\right), \tag{11.54}$$

where Θ are the model parameters to be estimated in (11.46)–(11.47), with $\boldsymbol{Y} = (\boldsymbol{X},\boldsymbol{Z},\boldsymbol{U})$ the corresponding concatenation of the complete data.

The resulting system is non-standard, requiring the derivation of relevant KF equations. This approach, together with the development of an Expectation/Conditional Maximization (ECM) algorithm for parameter estimation, was derived by Naranjo, Trindade and Casella [165]. The non-standard setup therefore makes this a less attractive method of dealing with missingness than the first method outlined in Subsection 11.5.2. To this end, note from Remark 11.3 below that if we only include the exogenous variables in the observation equation of the system (11.46)–(11.47), then the endogeneization trick will be able to accommodate the missingness in both responses and covariates.

Remark 11.3 *Including the exogenous only in the observation equation by setting* $\Upsilon = 0$ *in the system (11.46)–(11.47), one obtains a simplified version of (11.44)–(11.45) where* $\Gamma = 0$*. This is tantamount to fitting a so-called* regression with ARMA errors *(if correlation is allowed between the* $\{w_{i,t}\}$ *and* $\{v_{i,t}\}$ *noise processes).*

11.6 Real data examples

11.6.1 A LMM for the mare ovarian follicles data

The dataset `Ovary` in the `nlme` library of R, see [186], contains the number of ovarian follicles ($> 10mm$ in diameter) detected in 11 different mares at several times in their estrus cycle. The follicles present in each mare were counted over one complete cycle, starting 3 days before ovulation and ending 3 days after the next ovulation. The times of measurement vary by animal, and are scaled so that ovulation occurs at times 0 and 1.

Pinheiro and Bates [184, ch. 5] consider a LMM with sinusoidal components to explain the oscillatory nature of the data. If $y_{i,t}$ denotes the number of follicles in mare i at time t (really time t_{ij} since the j-th measurement time for the i-th mare varies for each animal), the model is

$$y_{i,t} = (\beta_0 + b_{0,i}) + (\beta_1 + b_{1,i})\sin(2\pi t) + \beta_2 \cos(2\pi t) + \varepsilon_{i,t},$$

where

$$\boldsymbol{\beta} = \begin{bmatrix} \beta_0 \\ \beta_1 \\ \beta_2 \end{bmatrix}, \qquad \boldsymbol{b}_i = \begin{bmatrix} b_{0,i} \\ b_{1,i} \end{bmatrix} \sim \mathrm{NID}(\mathbf{0}, \boldsymbol{G}), \qquad \boldsymbol{G} = \begin{bmatrix} \sigma_0^2 & 0 \\ 0 & \sigma_1^2 \end{bmatrix},$$

and the vector of within-mare errors, ε_i, follows a zero-mean ARMA(1,1) with covariance matrix $\Sigma_i := \Sigma(\phi, \theta, \sigma_a^2)$ for each $i = 1, \ldots, 11$, specified by the recursions $\phi(B)\varepsilon_{i,t} = \theta(B)a_{i,t}$ with $a_{i,t} \sim \mathrm{NID}(0, \sigma_a^2)$. This is easily seen to be a LMM in the form of (11.26), with

$$\boldsymbol{x}_{i,t}^T = [1, \sin(2\pi t), \cos(2\pi t)], \qquad \text{and} \qquad \boldsymbol{z}_{i,t}^T = [1, \sin(2\pi t)].$$

In the first step of a SS formulation we write $\boldsymbol{b}_i := \boldsymbol{b}_{i,t}$ for the state vector, and cast the model in the special case of $\varepsilon_{i,t} = 0$ as

$$y_{i,t} = \boldsymbol{z}_{i,t}^T \boldsymbol{b}_{i,t} + \boldsymbol{x}_{i,t}^T \boldsymbol{\beta}, \qquad \text{and} \qquad \boldsymbol{b}_{i,t} = \boldsymbol{I}_2 \boldsymbol{b}_{i,t-1},$$

with initial condition, $\boldsymbol{b}_{i,0} \sim \mathrm{N}(\mathbf{0}, \boldsymbol{G})$. In the second step, using $\boldsymbol{H}$ and $\boldsymbol{\Phi}$ to denote the observation and state transition matrices (defined below), we appeal to a SS representation for ARIMA models in the special case of $\boldsymbol{b}_{i,t} = 0$, and write

$$y_{i,t} = \underbrace{[\theta, 1]}_{\boldsymbol{H}} \begin{bmatrix} \eta_{i,t-1} \\ \eta_{i,t} \end{bmatrix} + \boldsymbol{x}_{i,t}^T \boldsymbol{\beta}, \qquad \text{and} \qquad \begin{bmatrix} \eta_{i,t-1} \\ \eta_{i,t} \end{bmatrix} = \underbrace{\begin{bmatrix} 0 & 1 \\ 0 & \phi \end{bmatrix}}_{\boldsymbol{\Phi}} \begin{bmatrix} \eta_{i,t-2} \\ \eta_{i,t-1} \end{bmatrix} + \begin{bmatrix} 0 \\ a_{i,t} \end{bmatrix},$$

with initial condition,

$$\begin{bmatrix} \eta_{i,-1} \\ \eta_{i,0} \end{bmatrix} \sim \mathrm{N}(\mathbf{0}, \boldsymbol{V}), \qquad \boldsymbol{V} = \frac{\sigma_a^2}{1-\phi^2}\begin{bmatrix} 1 & \phi \\ \phi & 1 \end{bmatrix}.$$

The SS representation we chose here is outlined in Brockwell and Davis [32, ch. 12], and ensures there is no observational noise process; but in principle any such representation could be used (e.g., the ones discussed in Chapter 2). Putting these two steps together, we create the combined 4-dimensional state vector $\boldsymbol{s}_{i,t} = [\boldsymbol{b}_{i,t}^T, \eta_{i,t-1}, \eta_{i,t}]^T$, leading to the observation and state equations

$$y_{i,t} = \left[\boldsymbol{z}_{i,t}^T, \boldsymbol{H}\right] \boldsymbol{s}_{i,t} + \boldsymbol{x}_{i,t}^T \boldsymbol{\beta}, \qquad \text{and} \qquad \boldsymbol{s}_{i,t} = \begin{bmatrix} \boldsymbol{I}_2 & \mathbf{0} \\ \mathbf{0} & \boldsymbol{\Phi} \end{bmatrix} \boldsymbol{s}_{i,t-1} + \underbrace{\begin{bmatrix} \mathbf{0} \\ a_{i,t} \end{bmatrix}}_{\boldsymbol{w}_{i,t}},$$

where the state noise vector and state initial conditions are given by,

$$\boldsymbol{w}_{i,t} \sim \mathrm{NID}\left(\mathbf{0}, \begin{bmatrix} \mathbf{0} & 0 \\ 0 & \sigma_a^2 \end{bmatrix}\right), \qquad \text{and} \qquad \boldsymbol{s}_{i,0} \sim \mathrm{N}\left(\mathbf{0}, \begin{bmatrix} \boldsymbol{G} & \mathbf{0} \\ \mathbf{0} & \boldsymbol{V} \end{bmatrix}\right).$$

Fitting the LMM via the `lme` function of R [186], yields the maximum likelihood estimates

$$\begin{aligned} &\hat{\beta}_0 = 12.13, && \hat{\sigma}_0^2 = 6.034, && \hat{\phi} = 0.7855, \\ &\hat{\beta}_1 = -2.921, && \hat{\sigma}_1^2 = 0.7263, && \hat{\theta} = -0.2835, \\ &\hat{\beta}_2 = -0.8486, && \hat{\sigma}_\varepsilon^2 = 13.72, && \end{aligned}$$

where $\hat{\sigma}_\varepsilon^2$ is the estimated variance of the ARMA(1,1) model for the (unobserved) $\{\varepsilon_{i,t}\}$ process, from which the estimate of the corresponding white noise variance is obtained from the ARMA(1,1) model variance equation:

$$\hat{\sigma}_a^2 = \hat{\sigma}_\varepsilon^2 \left[1 + \frac{(\hat{\theta}+\hat{\phi})^2}{1-\hat{\phi}^2}\right]^{-1} = 8.7263.$$

Taking these estimates to be the (true) model, and employing the KF and recursions for 12-step ahead forecasting (11.7)–(11.15), results in the predictions and 95% prediction bands for the 11th mare ($i = 11$) displayed in Figure 11.1.

11.6.2 Smoothing and prediction of missing values for the beluga whales data

The data file `whale.dat` in Chatterjee, Handcock and Simonoff [51] gives the values of several time series variables pertaining to the nursing behavior of two newborn captive beluga whale calves, Hudson and Casey. The variables recorded for each animal are as follows.

- `period`: The (consecutive) time period since birth, each period corresponding to a 6 hour time interval. This is the *time* variable, and will be designated by t.

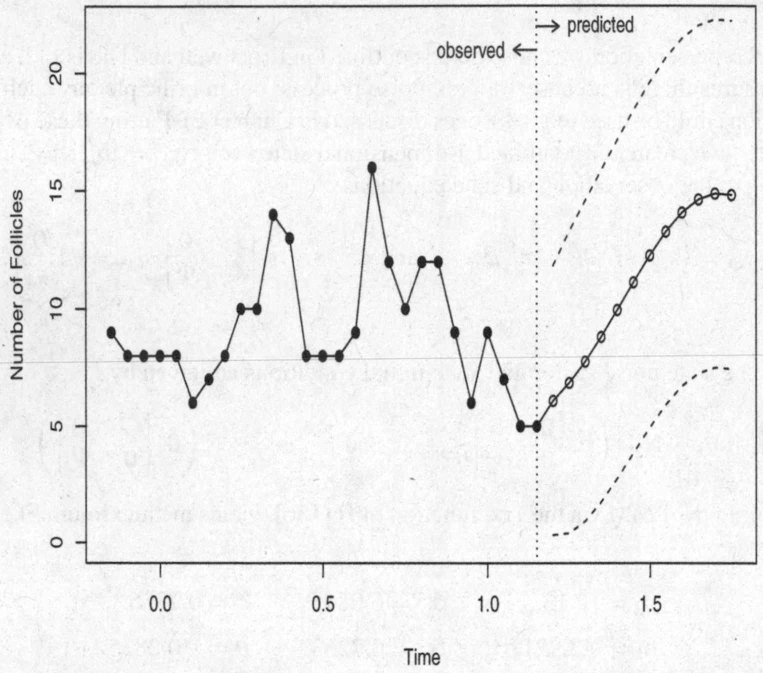

Figure 11.1 *Observed and predicted time series of ovarian follicles for the 11th mare in dataset* `Ovary` *of Section 11.6.1. The 12-step ahead predictions (open circles) and corresponding 95% prediction bands (dashed lines), are obtained by running the Kalman recursions for the SS formulation of the LMM described for the series.*

- `nursing`: Total time (in seconds) spent nursing in the period. This is the *response* variable, and will be designated by y_t.
- `bouts`: Number of nursing bouts that occurred in the period. A nursing bout is defined as a successful nursing episode where milk was obtained from the mother.
- `lockons`: Number of lockons in period. A lockon occurs when the calf attaches itself to the mother's mammary gland while suckling milk.
- `daytime`: Indicator variable indicating whether the period was day (1), or night (0). Day is defined as 8AM-8PM, and night 8PM-8AM.

The calves were observed over $n = 228$ consecutive periods, each period 6 hours long, totaling approximately 57 days. However, Casey was only observed up to period 223. The interesting question then arises of how to appropriately "fill in" the remaining 5 (missing) values for Casey, while "borrowing strength" from a combined analysis of both calves. Since the animals were born to different mothers at different times, it makes sense to treat them as independent subjects.

Chatterjee, Handcock and Simonoff [51] did not attempt a longitudinal analysis, but merely explored an ARMAX model for Hudson in order to predict `nursing` as a function of the remaining variables, eventually arriving at

$$y_t = [\beta_1, \ldots, \beta_4]\boldsymbol{u}_t + \phi y_{t-1} + v_t, \qquad \{v_t\} \sim \text{NID}(0, \sigma^2), \tag{11.55}$$

where the exogenous covariate vector $\boldsymbol{u}_t^T = [1, t, \texttt{lockons}_t, \texttt{lockons}_{t-1}]$ allows for a linear time trend and the instantaneous and lagged effects of the `lockons` variable. The remaining covariates, instantaneous and at lag one, were not found to be statistically significant. Since our goal is to handle the missingness in `nursing`, `bouts`, and `lockons` for Casey, we entertain the following SS representation of the above ARMAX-model-based longitudinal framework for both Hudson ($i = 1$) and Casey ($i = 2$),

$$x_{i,t}^a = \Phi^a x_{i,t-1}^a + \boldsymbol{\Gamma}\boldsymbol{u}_{i,t-1} + \boldsymbol{w}_{i,t}^a, \qquad \{w_{i,t}^a\} \sim \text{NID}(0, Q^a), \tag{11.56}$$

$$y_{i,t} = H_{i,t}^a x_{i,t}^a + \boldsymbol{v}_{i,t}^a, \qquad \{v_{i,t}^a\} \sim \text{NID}(0, R^a), \tag{11.57}$$

with $\boldsymbol{u}_{i,t}^T = [1, t, \texttt{bouts}_{i,t}, \texttt{lockons}_{i,t}]$ as the chosen exogenous covariate vector, and initial state distribution $x_{i,0} \sim \text{N}(\mu^a, P^a)$. Note that both observation and state variables are one-dimensional, and that we have omitted the covariates in the observation equation in order to render the resulting system in the ARMAX form of model (11.55) for each i. We must also allow for correlation between the state and observation error processes, so that we have additionally,

$$w_{i,t}^a = \Phi^a v_{i,t-1}^a, \quad \Longrightarrow \quad S^a := \text{cov}(v_{i,t}^a, w_{i,t+1}^a) = \text{cov}(v_{i,t}^a, \Phi^a v_{i,t}^a) = \Phi^a R^a. \tag{11.58}$$

In order to provide a mechanism for imputing the missing values, we introduce the additional SS model defined by (11.40)–(11.41) as described in Subsection 11.5.2, and note that both observation and state variables in the latter are of dimension 4. Combining the two models, leads to the overall (5-dimensional) state equation,

$$\boldsymbol{x}_{i,t} \equiv \begin{bmatrix} x_{i,t}^a \\ \boldsymbol{x}_{i,t}^b \end{bmatrix} = \underbrace{\begin{bmatrix} \Phi^a & \boldsymbol{\Gamma}H_{i,t}^a \\ \boldsymbol{0} & \boldsymbol{\Phi}^b \end{bmatrix}}_{\boldsymbol{\Phi}} \begin{bmatrix} x_{i,t-1}^a \\ \boldsymbol{x}_{i,t-1}^b \end{bmatrix} + \boldsymbol{w}_{i,t}, \qquad \{\boldsymbol{w}_{i,t}\} \sim \text{NID}(0, Q), \tag{11.59}$$

and the overall (5-dimensional) observation equation,

$$\boldsymbol{z}_{i,t} \equiv \begin{bmatrix} y_{i,t} \\ \boldsymbol{u}_{i,t} \end{bmatrix} = \begin{bmatrix} H_{i,t}^a & \boldsymbol{0} \\ \boldsymbol{0} & \boldsymbol{H}_{i,t}^b \end{bmatrix} \begin{bmatrix} \boldsymbol{x}_{i,t}^a \\ \boldsymbol{x}_{i,t}^b \end{bmatrix} + \boldsymbol{v}_{i,t}, \qquad \{\boldsymbol{v}_{i,t}\} \sim \text{NID}(0, R), \tag{11.60}$$

with the following values for the transition matrices,

$$H_{i,t}^a = \begin{cases} 1, & \text{if } y_{i,t} \text{ is observed,} \\ 0, & \text{if } y_{i,t} \text{ is missing,} \end{cases} \qquad \text{and} \qquad \boldsymbol{H}_{i,t}^b = \begin{cases} I_4, & \text{if } \boldsymbol{u}_{i,t} \text{ is observed,} \\ \boldsymbol{0}, & \text{if } \boldsymbol{u}_{i,t} \text{ is missing,} \end{cases}$$

and initial state distribution, $\boldsymbol{x}_{i,0} \sim \mathrm{N}(\boldsymbol{\mu}, \boldsymbol{P})$.

The EM algorithm of Subsection 11.5.1 can now be used to estimate the entire set of parameters, $\Theta = \{\boldsymbol{\mu}, \boldsymbol{P}, \boldsymbol{\Phi}, \boldsymbol{Q}, \boldsymbol{R}\}$, whence the estimates for ARMAX model (11.56)–(11.57) can be reconstructed, e.g., $\boldsymbol{\Gamma} = \boldsymbol{\Phi}(1, 2:5)$ is obtained by selecting columns 2 through 5 in the first row of $\boldsymbol{\Phi}$, etc. The high-dimensionality of Θ, 75 parameters in total, makes the determination of initial values for this exercise a daunting task! A structured approach was attempted by fitting separate ARMAX models for each whale, estimating parameters via least squares in a regression setting, and then averaging out parameter values for the two models. This did not however, seem to work as well as the simple strategy of starting with values close to zero for all parameters. Table 11.1 displays the final parameter estimates, as well as the values used to initialized the EM algorithm (square brackets). The procedure converged in 45 iterations to within a tolerance of 10^{-4} (as measured by the absolute relative difference in the log-likelihood between successive iterations) with a final log-likelihood value of -4908.3.

Table 11.1 *Final parameter estimates for the SS model fitted to the beluga whale data of Subsection 11.6.2. Estimates were obtained via the EM algorithm initialized with the values in square brackets. The final estimate of the initial state mean was* $\widehat{\boldsymbol{\mu}}^T = [3.296, 1.009, 1.390, -8.848 \times 10^{-3}, 6.653 \times 10^{-1}]$, *starting from the zero vector.*

i	j	$\widehat{\boldsymbol{\Phi}}(i,j)$	$\widehat{\boldsymbol{P}}(i,j)$	$\widehat{\boldsymbol{Q}}(i,j)$	$\widehat{\boldsymbol{R}}(i,j)$
1	1	1.100 [1.0]	8.780×10^{-2} [0.1]	5.881×10^{3} [0.1]	4.156×10^{3} [0.1]
1	2	1.098×10^{2} [0.0]	-7.742×10^{-6} [0.0]	-4.242×10^{-1} [0.0]	0.000 [0.0]
1	3	5.684×10^{-2} [0.0]	1.462×10^{-4} [0.0]	1.301 [0.0]	0.000 [0.0]
1	4	-1.011×10^{1} [0.0]	8.191×10^{-4} [0.0]	1.850×10^{2} [0.0]	0.000 [0.0]
1	5	-1.877 [0.0]	1.599×10^{-3} [0.0]	5.910×10^{2} [0.0]	0.000 [0.0]
2	1	-9.467×10^{-6} [0.0]	=	=	=
2	2	9.912×10^{-1} [1.0]	2.068×10^{-5} [0.1]	1.172×10^{-4} [0.1]	1.149×10^{-2} [0.1]
2	3	-5.914×10^{-6} [0.0]	-2.740×10^{-4} [0.0]	-1.260×10^{-4} [0.0]	0.000 [0.0]
2	4	8.238×10^{-4} [0.0]	-1.810×10^{-5} [0.0]	-1.471×10^{-2} [0.0]	0.000 [0.0]
2	5	1.656×10^{-4} [0.0]	5.752×10^{-6} [0.0]	-4.572×10^{-2} [0.0]	0.000 [0.0]
3	1	-1.149×10^{-4} [0.0]	=	=	=
3	2	8.772×10^{-1} [0.0]	=	=	=
3	3	9.999×10^{-1} [1.0]	2.420×10^{-2} [0.1]	1.985×10^{-2} [0.1]	5.622×10^{2} [0.1]
3	4	1.983×10^{-2} [0.0]	-3.229×10^{-3} [0.0]	5.817×10^{-2} [0.0]	0.000 [0.0]
3	5	-1.986×10^{-4} [0.0]	-8.867×10^{-4} [0.0]	1.739×10^{-1} [0.0]	0.000 [0.0]
4	1	5.695×10^{-3} [0.0]	=	=	=
4	2	4.141 [0.0]	=	=	=
4	3	3.697×10^{-3} [0.0]	=	=	=
4	4	5.820×10^{-1} [1.0]	1.919×10^{-2} [0.1]	6.892 [0.1]	2.539 [0.1]
4	5	-8.297×10^{-2} [0.0]	1.505×10^{-2} [0.0]	2.123×10^{1} [0.0]	0.000 [0.0]
5	1	1.871×10^{-2} [0.0]	=	=	=
5	2	1.172×10^{1} [0.0]	=	=	=
5	3	1.175×10^{-2} [0.0]	=	=	=
5	4	-1.060 [0.0]	=	=	=
5	5	6.726×10^{-1} [1.0]	6.739×10^{-2} [0.1]	6.791×10^{1} [0.1]	1.158×10^{1} [0.1]

Figure 11.2 displays the results of applying Kalman smoothing (period 1 to 223) and prediction (period 224 to 228) for Casey, based on the fitted model. Smoothed

values (solid lines) and predictions (open circles) are shown for nursing time, number of bouts, and number of lockons, along with appropriate 95% confidence bands. In order not to overly compress the display, the plots only span the time frame from period 100 onward. The actual predictions and 95% confidence bounds for Casey's 5 missing values are tabulated in Table 11.2.

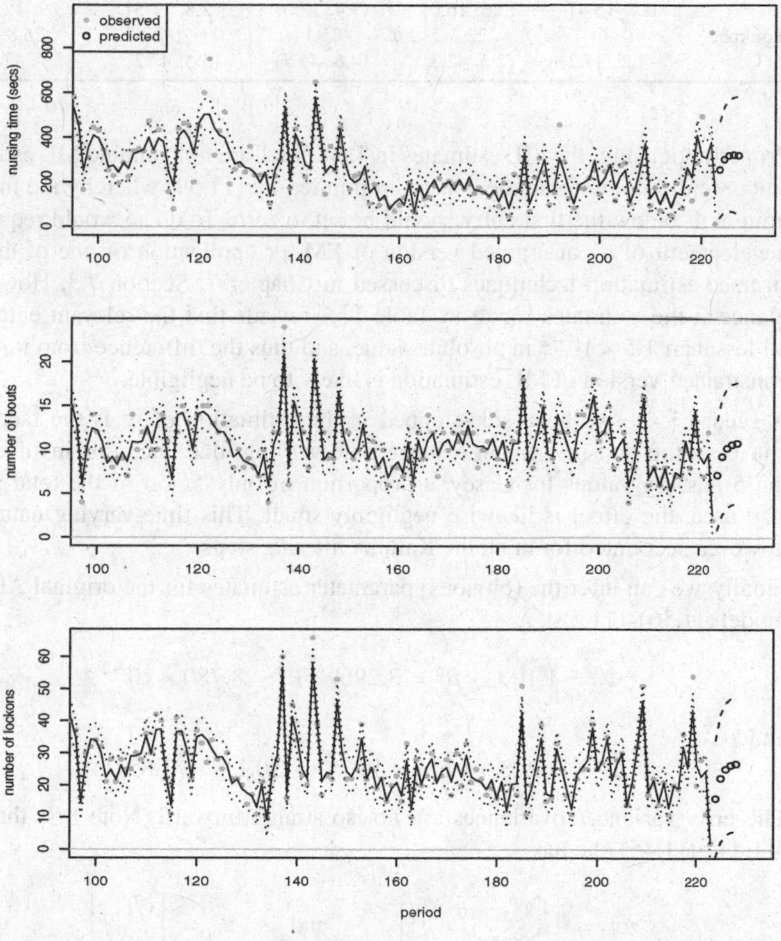

Figure 11.2 *Kalman smoothing and prediction for beluga whale Casey in the* `whale.dat` *dataset of Subsection 11.6.2. Shown are smoothed values (solid lines) and 5-step ahead predictions (open circles) for nursing time (top panel), number of bouts (middle panel), and number of lockons (bottom panel), from period 100 onward. Corresponding 95% confidence bands for the smoothed and predicted values are shown as dotted and dashed lines, respectively.*

In conclusion, a few remarks are in order concerning the complexity of longitudinal modeling via SS models, that this particular example highlights.

Table 11.2 *Five-step ahead predictions and 95% confidence bounds for Casey in the beluga whale data of Subsection 11.6.2.*

Variable	$t = 224$	$t = 225$	$t = 226$	$t = 227$	$t = 228$
`nursing`	257.3	303.1	319.8	323.3	320.6
	(40.9, 473.8)	(65.9, 540.2)	(71.4, 568.2)	(66.7, 579.9)	(57.3, 584.0)
`bouts`	7.2	9.3	10.3	10.6	10.8
	(0.9, 13.4)	(2.6, 16.0)	(3.4, 17.0)	(3.8, 17.5)	(3.9, 17.7)
`lockons`	16.0	22.3	25.1	26.3	26.8
	(−2.2, 34.2)	(2.3, 42.3)	(4.6, 45.7)	(5.5, 47.2)	(5.7, 47.8)

(i) Properly speaking, the ML estimates in Table 11.1 are not genuine ML estimates, since we did not enforce the constraint implied by (11.59) whereby the first column of $\mathbf{\Phi}$ below the first entry should be set to zero. To do so would require the development of a constrained version of EM, or application of one of the constrained estimation techniques discussed in Chapter 7, Section 7.3. However, a glance at the estimates for $\mathbf{\Phi}$ in Table 11.1 reveals that the relevant entries are all less than 1.5×10^{-2} in absolute value, and thus the difference from running a constrained version of ML estimation is likely to be negligible.

(ii) A second issue we have side-stepped in the estimation of $\mathbf{\Phi}$ is the fact that 4 entries in the first row are in essence time-varying, but since this involves only the 5 missing values for Casey, a proportion of only 5/456 of the total sample size used, the effect is likewise negligibly small. This time-varying nature was however, accounted for in all the Kalman filtering steps.

(iii) Finally, we can infer the (obvious) parameter estimates for the original ARMAX model (11.56)–(11.58),

$$\widehat{\Phi}^a = 1.100, \quad \widehat{\mu}^a = 3.296, \quad \widehat{P}^a = 8.780 \times 10^{-2},$$

and

$$\widehat{\mathbf{\Gamma}} = \begin{bmatrix} 109.8 & 5.684 \times 10^{-2} & -10.11 & -1.877 \end{bmatrix}.$$

The error variance/covariances are not so straightforward. Note first that from (11.44)–(11.45) one has

$$\boldsymbol{w}_{i,t} = \begin{bmatrix} \mathbf{\Gamma}\boldsymbol{v}^b_{i,t-1} + w^a_{i,t} \\ \boldsymbol{w}^b_{i,t} \end{bmatrix}, \qquad \text{and} \qquad \boldsymbol{v}_{i,t} = \begin{bmatrix} v^a_{i,t-1} \\ \boldsymbol{v}^b_{i,t} \end{bmatrix},$$

which implies the following covariance structure in the entries of $\boldsymbol{Q} = \text{var}(\boldsymbol{w}_{i,t})$,

$$\begin{aligned} \boldsymbol{Q}(1,1) &= \text{var}(\mathbf{\Gamma}\boldsymbol{v}^b_{i,t-1} + w^a_{i,t}) = \mathbf{\Gamma}\boldsymbol{R}^b\mathbf{\Gamma}^T + Q^a, \\ \boldsymbol{Q}(2:5,2:5) &= \text{var}(\boldsymbol{w}^b_{i,t}) = \boldsymbol{Q}^b, \end{aligned}$$

and

$$\begin{aligned}
\boldsymbol{Q}(1,2:5) &= \mathrm{cov}(\boldsymbol{\Gamma}\boldsymbol{v}_{i,t-1}^{b} + w_{i,t}^{a}, \boldsymbol{w}_{i,t}^{b}) \\
&= \boldsymbol{\Gamma}\mathrm{cov}(\boldsymbol{v}_{i,t-1}^{b}, \boldsymbol{w}_{i,t}^{b}) + \boldsymbol{\Gamma}\underbrace{\mathrm{cov}(w_{i,t}^{a}, \boldsymbol{w}_{i,t}^{b})}_{\boldsymbol{0}} \\
&= \boldsymbol{\Gamma}(\boldsymbol{S}^{b})^{T},
\end{aligned}$$

where $\boldsymbol{S}^{b} := \mathrm{cov}(\boldsymbol{w}_{i,t}^{b}, \boldsymbol{v}_{i,t-1}^{b}) = \mathrm{cov}(\boldsymbol{v}_{i,t-1}^{b}, \boldsymbol{w}_{i,t}^{b})^{T}$ denotes the correlation between the state and observation noise processes in the imputation model. The corresponding result for $\boldsymbol{R} = \mathrm{var}(\boldsymbol{v}_{i,t})$ is simpler,

$$\boldsymbol{R} = \begin{bmatrix} \mathrm{var}(v_{i,t}^{a}) & \mathrm{cov}(v_{i,t}^{a}, \boldsymbol{v}_{i,t}^{b}) \\ = & \mathrm{var}(\boldsymbol{v}_{i,t}^{b}) \end{bmatrix} = \begin{bmatrix} R^{a} & \boldsymbol{0} \\ = & \boldsymbol{R}^{b} \end{bmatrix},$$

whence we obtain $\widehat{R}^{a} = 4.156 \times 10^{3}$ and $\widehat{\boldsymbol{R}}^{b} = \widehat{\boldsymbol{R}}(2:5,2:5)$. Armed with these results we can finally retrieve the estimates for Q^{a},

$$\widehat{Q}^{a} = \widehat{\boldsymbol{Q}}(1,1) - \widehat{\boldsymbol{\Gamma}}\widehat{\boldsymbol{R}}^{b}\widehat{\boldsymbol{\Gamma}}^{T} = 5440.3,$$

and for S^{a},

$$\widehat{S}^{a} = \widehat{\Phi}^{a}\widehat{R}^{a} = 4572.8.$$

The fact that $\widehat{\boldsymbol{Q}}(1,2:5) \neq \boldsymbol{0}$ automatically implies that $\widehat{\boldsymbol{S}}^{b} \neq \boldsymbol{0}$, which suggests the presence of correlation between the state and observation noise processes in the fitted imputation model. However, throughout this analysis we have ignored the fact that the noise and observational error processes may be correlated in the overall model (11.59)–(11.60), and assumed $\boldsymbol{S} = \boldsymbol{0}$ in the relevant estimation and Kalman filtering steps. A more sophisticated analysis should not make this simplification, since the above estimates suggest, in particular, that the relevant correlation between $v_{i,t}^{a}$ and $w_{i,t+1}^{a}$ is of the order of 0.9.

Appendices

Appendix A

Some results in numerical algebra and linear systems

A.1 QR Decomposition 193
A.2 Schur decomposition 195
A.3 The Hessenberg Form 196
A.4 SVD Decomposition 197
A.5 Canonical Correlations 198
A.6 Algebraic Lyapunov and Sylvester equations 198
A.7 Numerical solution of a Sylvester equation 200
A.8 Block-diagonalization of a matrix 201
A.9 Reduced rank least squares 202
A.10 Riccati equations 203
A.10.1 Definition 204
A.10.2 Solving the ARE in the general case 205
A.10.3 Solving the ARE for GARCH models 207
A.11 Kalman filter 208

This Appendix provides a summary of several important results in numerical algebra and linear systems theory which are used liberally throughout the book. While this may be helpful in making the book more "self-contained," it is not meant to provide a complete treatment of this material, so the interested reader may need to consult additional resources.

A.1 QR Decomposition

Any real matrix $\boldsymbol{A}(m \times n)$, with $m \geq n$, can be factored as:

$$\boldsymbol{A} = \boldsymbol{Q}\,\boldsymbol{R}, \tag{A.1}$$

where the matrix $\boldsymbol{Q}(m \times n)$ is orthonormal, and $\boldsymbol{R}(n \times n)$ is an upper triangular matrix. This common factorization is known as the *QR decomposition*.

The numerical determination of $\boldsymbol{Q}$ and $\boldsymbol{R}$ is complicated, and requires additional discussion concerning *triangularization*. Any matrix $\boldsymbol{A}$, with dimension $m \times n$ and $m \leq n$, can be reduced to an upper triangular matrix by applying a sequence of orthogonal transformations, such as the *Givens rotations* or the *Householder reflections* (see [60]). A Householder reflection is an orthonormal and symmetric matrix of the form:

$$\boldsymbol{U} = \boldsymbol{I} - 2\frac{\boldsymbol{u}\boldsymbol{u}^T}{\boldsymbol{u}^T\boldsymbol{u}}. \tag{A.2}$$

Given a vector $\boldsymbol{x}$, by defining $\boldsymbol{u} = \boldsymbol{x} + h\boldsymbol{e}$, with $h = \text{sign}(x_1)\sqrt{\boldsymbol{x}^T\boldsymbol{x}}$ and $\boldsymbol{e} = [1 \quad 0 \quad \ldots \quad 0]^T$, then $\boldsymbol{U}\boldsymbol{x} = -h\boldsymbol{e}$. The algorithm for the triangularization of the matrix $\boldsymbol{A}$ by elemental reflection (known as *Householder reduction*), can be described by the following sequence of steps:

1. Build the reflection matrix $\boldsymbol{U}_1$, such that $\boldsymbol{U}_1\boldsymbol{a}_1 = [r_{11} \quad 0 \quad \ldots \quad 0]^T$, where $\boldsymbol{a}_1$ denotes the first column of $\boldsymbol{A}$. Then, define $\boldsymbol{A}_2$ with the following structure:

$$\boldsymbol{A}_2 = \begin{bmatrix} r_{11} & r_{12} & \cdots & r_{1n} \\ 0 & & & \\ \vdots & & \bar{\boldsymbol{A}}_2 & \\ 0 & & & \end{bmatrix}. \tag{A.3}$$

2. For all $i = 2, \ldots, n$:
   (a) Build the reflection matrix $\bar{\boldsymbol{U}}_i$ corresponding to the first column of the matrix $\bar{\boldsymbol{A}}_i(\bar{\boldsymbol{a}}_{i1})$, such that: $\bar{\boldsymbol{U}}_1\bar{\boldsymbol{a}}_1 = [r_{ii} \quad 0 \quad \ldots \quad 0]^T$.
   (b) Calculate $\boldsymbol{A}_{i+1} = \boldsymbol{U}_i\boldsymbol{A}_i$, where $\boldsymbol{U}_i = \text{diag}(\boldsymbol{I}_{i-1}, \bar{\boldsymbol{U}}_i)$, and define $\bar{\boldsymbol{A}}_{i+1}$ as the block made up of the last $m-i$ rows and $n-i$ columns of $\boldsymbol{A}_{i+1}$.

After n iterations, we obtain the matrix $\boldsymbol{A}_{n+1} = \boldsymbol{U}_n\boldsymbol{U}_{n-1}\ldots\boldsymbol{U}_1\boldsymbol{A}$, which exhibits the structure:

$$\boldsymbol{A}_{n+1} = \begin{bmatrix} \boldsymbol{R} \\ \boldsymbol{0} \end{bmatrix}, \tag{A.4}$$

with $\boldsymbol{R}$ an upper triangular matrix. By defining the matrix $\boldsymbol{Q} = (\boldsymbol{U}_n\boldsymbol{U}_{n-1}\ldots\boldsymbol{U}_1)^T$, it is straightforward to see that $\boldsymbol{Q}^T\boldsymbol{Q} = \boldsymbol{Q}\boldsymbol{Q}^T = \boldsymbol{I}$, and premultiplying (A.4) by $\boldsymbol{Q}$ yields:

$$\boldsymbol{A} = \boldsymbol{Q}\boldsymbol{A}_{n+1} = \boldsymbol{Q}\begin{bmatrix} \boldsymbol{R} \\ \boldsymbol{0} \end{bmatrix}. \tag{A.5}$$

Moreover, if $\boldsymbol{Q}$ is partitioned as $\boldsymbol{Q} = [\boldsymbol{Q}_1 \quad \boldsymbol{Q}_2]$, $\boldsymbol{A}$ can be written as $\boldsymbol{A} = \boldsymbol{Q}_1\boldsymbol{R}$, where $\boldsymbol{Q}_1{}^T\boldsymbol{Q}_1 = \boldsymbol{I}_m$.

The QR decomposition is stable numerically no matter the type of orthogonal transformation employed (*Givens* or *Householder*), although the Householder reflection is more efficient computationally; see [60]. In particular, the algorithm described above has a computational load of $n^2(m - \frac{1}{3}n)$ flops, in contrast with the $n^2(m - \frac{1}{3}n)$ flops needed when using the Givens transformations. The subspace algorithms described in Chapter 7 often use this decomposition.

A.2 Schur decomposition

Let $\mathbb{C}^n$ denote the complex-valued n-space, and $\mathbb{C}^{n\times n} \equiv \mathbb{C}^n \times \mathbb{C}^n$ the crossproduct space of $\mathbb{C}^n$ with itself. Similarly, denote by $\mathbb{R}^{n\times n}$ the corresponding crossproduct space of real numbers. Any matrix $\boldsymbol{A} \in \mathbb{C}^{n\times n}$ can be factored as:

$$\boldsymbol{A} = \boldsymbol{U}\boldsymbol{D}\boldsymbol{U}^{\boldsymbol{H}}, \tag{A.6}$$

where $\boldsymbol{U} \in \mathbb{C}^{n\times n}$ is a unitary matrix, $\boldsymbol{U}^{\boldsymbol{H}}$ denotes the conjugate transpose of $\boldsymbol{U}$, and $\boldsymbol{D}$ is an upper triangular matrix known as the *complex Schur form* of $\boldsymbol{A}$, whose diagonal elements correspond to the eigenvalues of $\boldsymbol{A}$.

The importance of this decomposition is due to the fact that any matrix can be reduced to its Schur form by means of a succession of similar unitary transformations, so this reduction provides a numerically stable method to calculate eigenvalues; see Golub and Van Loan [109] and Datta [60].

If $\boldsymbol{A}$ has complex eigenvalues, matrices $\boldsymbol{U}$ and $\boldsymbol{D}$ will be complex. In this case, an alternative procedure to compute (A.6) is the so-called *real Schur decomposition*, which is noteworthy due to its avoidance of operations with complex numbers. The real Schur decomposition of $\boldsymbol{A} \in \mathbb{R}^{n\times n}$ is given by:

$$\boldsymbol{A} = \boldsymbol{W}\boldsymbol{R}\boldsymbol{W}^T, \tag{A.7}$$

where $\boldsymbol{W} \in \mathbb{R}^{n\times n}$ is a orthonormal matrix and $\boldsymbol{R} \in \mathbb{R}^{n\times n}$ is an upper Hessenberg matrix (called the *real Schur form* of $\boldsymbol{A}$) with the following structure:

$$\boldsymbol{R} = \begin{bmatrix} \boldsymbol{R}_{11} & \boldsymbol{R}_{12} & \cdots & \boldsymbol{R}_{1k} \\ \boldsymbol{0} & \boldsymbol{R}_{22} & \cdots & \boldsymbol{R}_{2k} \\ \vdots & \vdots & \ddots & \vdots \\ \boldsymbol{0} & \boldsymbol{0} & \cdots & \boldsymbol{R}_{kk} \end{bmatrix}, \tag{A.8}$$

where $\boldsymbol{R}_{11}, \boldsymbol{R}_{22}, \ldots, \boldsymbol{R}_{kk}$ are either scalars or (2×2) matrices with complex conjugate eigenvalues. The eigenvalues of $\boldsymbol{A}$ coincide with those of the matrices $\boldsymbol{R}_{11}, \boldsymbol{R}_{22}, \ldots, \boldsymbol{R}_{kk}$, so that k equals the number of real eigenvalues of $\boldsymbol{A}$ plus the number of pairs of complex conjugate eigenvalues of $\boldsymbol{A}$. For a proof of this result, see Golub and Van Loan [109].

From a computational point of view, the reduction of a real Schur form should be performed before obtaining the complex Schur form. The transformation of the real Schur form to the complex representation only requires applying the complex Schur decomposition to each of the $\boldsymbol{R}_{ii}$ blocks in (A.8).

Computing the Schur form of a matrix requires iterations. A detailed description of the available algorithms for real and complex Schur forms can be found in Golub and Van Loan [109] and Datta [60]. In general, these procedures include two steps:

Step 1) Reduce the matrix $\boldsymbol{A}$ to an upper Hessenberg form. This consists in computing the factorization $\boldsymbol{A} = \boldsymbol{U}\boldsymbol{H}\boldsymbol{U}^T$, where $\boldsymbol{H}$ is the upper Hessenberg matrix and $\boldsymbol{U}$ is an orthonormal matrix.

Step 2) QR iterations. With the initial condition $A_0 = \boldsymbol{A}$, the iterations:

$$\boldsymbol{A_i} = \boldsymbol{Q_i}\boldsymbol{R_i}, \tag{A.9}$$

and

$$\boldsymbol{A}_{i+1} = \boldsymbol{R_i}\boldsymbol{Q_i} = \boldsymbol{Q_i^T}\boldsymbol{A_i}\boldsymbol{Q_i}, \tag{A.10}$$

with $i = 0, 1, 2, \ldots$, generate a sequence $\boldsymbol{A_i}$ of similar matrices which converges to a real Schur form under non restrictive conditions. If $\boldsymbol{A_0} = \boldsymbol{H}$ and $\boldsymbol{H}$ is an upper unreduced Hessenberg form, then all the matrices $\boldsymbol{A_i}$ are Hessenberg forms. This has a significant computational advantage, because the computational burden in flops of determining the QR decomposition of a Hessenberg matrix is proportional to n^2, which is smaller than the cost of decomposing an arbitrary matrix (proportional to n^3).

In some cases it may be desirable for the Schur form to be a lower Hessenberg matrix. This requires applying the Schur decomposition only to the transpose of $\boldsymbol{A}$. Then, $\boldsymbol{A} = \boldsymbol{W}\boldsymbol{R}^T\boldsymbol{W}^T$, where $\boldsymbol{R}^T$ is a lower Hessenberg matrix.

A.3 The Hessenberg Form

Any matrix $\boldsymbol{A}(n \times n)$, can be reduced to an upper Schur form by applying a finite sequence of orthogonal transformations such that $\boldsymbol{A}$ is factored as:

$$\boldsymbol{A} = \boldsymbol{U}^T\boldsymbol{H}\boldsymbol{U}, \tag{A.11}$$

where $\boldsymbol{U}(n \times n)$ is an orthonormal matrix, and $\boldsymbol{H}$ is an upper Hessenberg matrix.

The procedure to reduce a Hessenberg matrix can be divided into the following steps:

Step 1) Build a Householder reflection matrix $\bar{\boldsymbol{U}}_1$ such that:

$$\bar{\boldsymbol{U}}_1 \begin{bmatrix} a_{21} \\ a_{31} \\ \vdots \\ a_{n1} \end{bmatrix} = \begin{bmatrix} h_{21} \\ 0 \\ \vdots \\ 0 \end{bmatrix}, \tag{A.12}$$

and define $\boldsymbol{U_1} = \text{diag}(1, \bar{\boldsymbol{U}}_1)$, where $\boldsymbol{U_1}$ is such that $\boldsymbol{A_2} = \boldsymbol{U_1}\boldsymbol{A}\boldsymbol{U_1^T}$, being $\boldsymbol{A_2}$ a matrix with the following structure:

$$\boldsymbol{A_2} = \begin{bmatrix} \times & \times & \ldots & \times \\ \times & \times & \ldots & \times \\ 0 & \times & \ldots & \times \\ \vdots & \vdots & \ddots & \vdots \\ 0 & \times & \ldots & \times \end{bmatrix}, \tag{A.13}$$

where "×" represents potential nonzero elements. Now define $\bar{\boldsymbol{a}}_2 = \boldsymbol{A_2}(3:n, 2)$ to be the vector formed from rows 3 to n of the second column of $\boldsymbol{A_2}$.

Step 2) For all $i = 2, \dots, n-2$:
Build the reflection matrix $\bar{U}_i$ such that $\bar{U}_i \bar{a}_i = [h_{i+1,i} \quad 0 \quad \dots \quad 0]^T$, and then compute $A_{i+1} = U_i A_i U_i^T$, with $U_i = \text{diag}(I_i, \bar{U}_i)$. Now define $\bar{a}_{i+1} = A_{i+1}(i+2:n, i+1)$, similar to what was done at the end of Step 1.
The matrix A_{n-1} has an upper Hessenberg form which is obtained from A as:

$$A_{n-1} = U_{n-2} U_{n-3} \dots U_1 A U_1^T \dots U_{n-3}^T U_{n-2}^T = U A U^T, \tag{A.14}$$

where $U = U_{n-2} U_{n-3} \dots U_1$. Note that U is orthonormal, and therefore $A = U^T A_{n-1} U$.

A.4 SVD Decomposition

The *Singular Value Decomposition* (SVD) is based on the following theorem.

Theorem A.1 *For any real matrix* A$\in \mathbb{R}^{m \times n}$*, there exist two orthogonal matrices,* $U = [u_1, \dots, u_m] \in \mathbb{R}^{m \times m}$ *and* $V = [v_1, \dots, v_n] \in \mathbb{R}^{n \times n}$*, such that* $U^T A V = S =$ *diag*$(\sigma_1, \dots, \sigma_p) \in \mathbb{R}^{m \times n}$*, with* $p = \min\{m, n\}$ *and* $\sigma_1 \geq \sigma_2 \geq \dots \geq \sigma_p \geq 0$.

Proof. See [109], Theorem 2.5.2.

We denote by σ_i, $i = 1, \dots, p$, the singular values of A, while the vectors u_i and v_i are the i left and right singular vectors of A, respectively. We may also define the Euclidean norm of A as $\|A\|_2 = \sigma_1$, and the Frobenius norm of A as $\|A\|_F = \sqrt{\sum_{i=1}^{p} \sigma_i^2}$. Some of the properties of the SVD used in Chapters 7 and 9 are:

1. Rank and singular values. If rank$(A) = k \leq p$ then rank$(A) = $ rank$(USV^T) =$ rank$(S_k) = k$ holds, where $S_k = \text{diag}(\sigma_1, \dots, \sigma_k)$ and $\sigma_1 \geq \sigma_2 \geq \dots \geq \sigma_k > 0$. Therefore, the rank of A coincides with the number of nonzero singular values.
2. Moore–Penrose pseudo-inverse. The matrix $A^+ = V_k S_k^{-1} U_k^T$ is the so-called Moore–Penrose pseudo-inverse of matrix A, where rank$(A) = k \leq p$. The matrix A^+ satisfies the following properties:
   (a) $(A^+)^+ = A$
   (b) $AA^+A = A$
   (c) $A^+AA^+ = A^+$
   (d) $(A^+)^T = (A^T)^+$
   (e) $A^+ = (A^T A)^+ A^T$
   (f) When rank$(A) = n \leq m$, then $A^+ = (A^T A)^{-1}$ and $A^+ A = I_n$
3. Low-rank approximation. Assume that rank$(A) = r$. For any matrix B such that rank$(B) = k$, the matrix $A_k = U_k S_k V_k^T$ with $k < r$ and rank$(A_k) = k$, satisfies the condition:

$$\|A - B\|_2 \geq \|A - A_k\|_2, \qquad \text{and} \qquad \|A - B\|_F \geq \|A - A_k\|_F. \tag{A.15}$$

   This property assures that the SVD is the adequate method to determine the rank of a matrix as the number of singular values that are significantly different from zero.

A.5 Canonical Correlations

Let $\boldsymbol{x}$ and $\boldsymbol{y}$ be two vectors of random variables with dimensions m and n, respectively, and satisfying: $m \geq n$, $\mathrm{var}(\boldsymbol{x}) = \boldsymbol{\Sigma}_{xx} > \mathbf{0}$, $\mathrm{var}(\boldsymbol{y}) = \boldsymbol{\Sigma}_{yy} > \mathbf{0}$, and $\mathrm{cov}(\boldsymbol{x},\boldsymbol{y}) = \boldsymbol{\Sigma}_{xy}$. The n canonical correlations between $\boldsymbol{x}$ and $\boldsymbol{y}$, denoted by ρ_k, $k = 1,\ldots,n$, are defined as:

$$\rho_k(\boldsymbol{x},\boldsymbol{y}) = \max_{\substack{\alpha_k \in \mathbb{R}^m \\ \beta_k \in \mathbb{R}^n}} \mathrm{cov}(\boldsymbol{\alpha}_k^T \boldsymbol{x}, \boldsymbol{\beta}_k^T \boldsymbol{y}), \tag{A.16}$$

subject to the conditions that for all $i = 1,\ldots,k-1$, we have

$$\mathrm{var}(\boldsymbol{\alpha}_k^T \boldsymbol{x}) = 1, \quad \mathrm{var}(\boldsymbol{\beta}_k^T \boldsymbol{y}) = 1, \quad \mathrm{cov}(\boldsymbol{\alpha}_k^T \boldsymbol{x}, \boldsymbol{\alpha}_i^T \boldsymbol{x}) = 0, \quad \mathrm{cov}(\boldsymbol{\beta}_k^T \boldsymbol{y}, \boldsymbol{\beta}_i^T \boldsymbol{y}) = 0.$$

Obviously, we must have $\rho_1 \geq \rho_2 \geq \ldots \geq \rho_n \geq 0$. The first canonical correlation, ρ_1, is the maximum correlation between any linear combination of $\boldsymbol{x}$ and $\boldsymbol{y}$ with unit variance. The vectors $\boldsymbol{\alpha}_1^T \boldsymbol{x}$ and $\boldsymbol{\beta}_1^T \boldsymbol{y}$ are the so-called *canonical variables* corresponding to ρ_1. The second canonical correlation, ρ_2, represents the maximum correlation between linear combinations of $\boldsymbol{x}$ and $\boldsymbol{y}$ that are uncorrelated with the previous canonical variables, and so on.

Proposition A.1 *The canonical correlations between $\boldsymbol{x}$ and $\boldsymbol{y}$ correspond to the singular values of the matrix:*

$$\boldsymbol{\Sigma}_{xx}^{-\frac{1}{2}} \boldsymbol{\Sigma}_{xy} \boldsymbol{\Sigma}_{yy}^{-\frac{1}{2}} \stackrel{svd}{=} \boldsymbol{U}_n \boldsymbol{S}_n \boldsymbol{V}_n^T, \tag{A.17}$$

where $\begin{bmatrix} \boldsymbol{\alpha}_1 & \boldsymbol{\alpha}_2 & \ldots & \boldsymbol{\alpha}_n \end{bmatrix} = \boldsymbol{\Sigma}_{xx}^{-\frac{1}{2}} \boldsymbol{U}_n$, *and* $\begin{bmatrix} \boldsymbol{\beta}_1 & \boldsymbol{\beta}_2 & \ldots & \boldsymbol{\beta}_n \end{bmatrix} = \boldsymbol{\Sigma}_{yy}^{-\frac{1}{2}} \boldsymbol{V}_n$.

Proof. See Casals [38], Proposition A.9.2.

In practice, the information in $\boldsymbol{X} \in \mathbb{R}^{N \times m}$ and $\boldsymbol{Y} \in \mathbb{R}^{N \times n}$ is used to estimate the following sample covariance matrices:

$$\hat{\boldsymbol{\Sigma}}_{xx} = \frac{\boldsymbol{X}^T \boldsymbol{X}}{N} > 0, \qquad \hat{\boldsymbol{\Sigma}}_{yy} = \frac{\boldsymbol{Y}^T \boldsymbol{Y}}{N} > 0, \qquad \hat{\boldsymbol{\Sigma}}_{xy} = \frac{\boldsymbol{X}^T \boldsymbol{Y}}{N}.$$

Therefore, the canonical correlations between $\boldsymbol{X}$ and $\boldsymbol{Y}$ coincide with the singular values of $(\boldsymbol{X}^T \boldsymbol{X})^{-\frac{1}{2}} \boldsymbol{X}^T \boldsymbol{Y} (\boldsymbol{Y}^T \boldsymbol{Y})^{-\frac{1}{2}} \stackrel{svd}{=} \boldsymbol{U}_n \boldsymbol{S}_n \boldsymbol{V}_n^T$.

A.6 Algebraic Lyapunov and Sylvester equations

The so-called algebraic Lyapunov and Sylvester equations often appear in the theory of linear systems and optimal control. The discrete-time Lyapunov equation is given by:

$$\boldsymbol{X} = \boldsymbol{A} \boldsymbol{X} \boldsymbol{A}^T + \boldsymbol{C}, \tag{A.18}$$

where $\boldsymbol{A}$ is a $(n \times n)$ matrix, $\boldsymbol{C}$ is a $(n \times n)$ symmetric matrix and $\boldsymbol{X}$ is an unknown $(n \times n)$ symmetric matrix.

On the other hand, the Sylvester equation is given by:

$$\boldsymbol{A}\boldsymbol{X}+\boldsymbol{X}\boldsymbol{B}=\boldsymbol{C}, \tag{A.19}$$

where the matrices $\boldsymbol{A}$, $\boldsymbol{B}$, $\boldsymbol{X}$, and $\boldsymbol{C}$ are $(n\times n)$, $(m\times m)$, $(n\times m)$, and $(n\times m)$, respectively, with $\boldsymbol{A}$, $\boldsymbol{B}$, and $\boldsymbol{C}$ unknown matrices. Both equations, (A.18) and (A.19), can be represented as linear systems equations. Applying the $vec(\cdot)$ operator to (A.18) and (A.19), yields:

$$(\boldsymbol{I}-\boldsymbol{A}^T\otimes\boldsymbol{A})vec(\boldsymbol{X})=vec(\boldsymbol{C}), \tag{A.20}$$

and

$$(\boldsymbol{I}_m\otimes\boldsymbol{A}+\boldsymbol{B}^T\otimes\boldsymbol{I}_n)vec(\boldsymbol{X})=vec(\boldsymbol{C}). \tag{A.21}$$

The matrix $(\boldsymbol{I}-\boldsymbol{A}^T\otimes\boldsymbol{A})$ has eigenvalues $1-\lambda_i\lambda_j$, for all $i=1,\ldots,n$ and $j=1,\ldots,n$, with λ_i denoting the i-th eigenvalue of $\boldsymbol{A}$. Thus, system (A.20) has a unique solution if matrix $\boldsymbol{A}$ has no reciprocal eigenvalues, that is, if $\lambda_i\lambda_j\neq 1$, for all $i=1,\ldots,n$ and $j=1,\ldots,n$. For example, if $\boldsymbol{A}$ is a transition matrix in a stable dynamic system, then $\lambda_i<1$, for all $i=1,\ldots,n$, and the condition of reciprocity between eigenvalues is satisfied. Moreover, in this case, if $\boldsymbol{C}$ is positive semidefinite, then so is the solution to equation (A.18). Writing (A.18) as:

$$\boldsymbol{X}=\boldsymbol{A}^N\boldsymbol{X}(\boldsymbol{A}^T)^N+\sum_{i=0}^{N-1}\boldsymbol{A}^i\boldsymbol{C}(\boldsymbol{A}^T)^i, \tag{A.22}$$

demonstrates that if the eigenvalues of $\boldsymbol{A}$ are all strictly within the unit circle, it holds that (see Anderson and Moore [6, ch. 4]):

$$\lim_{N\to\infty}\boldsymbol{A}^N\boldsymbol{X}(\boldsymbol{A}^T)^N=\boldsymbol{0}, \tag{A.23}$$

$$\lim_{N\to\infty}\sum_{i=0}^{N}\boldsymbol{A}^i\boldsymbol{C}(\boldsymbol{A}^T)^i<\infty, \tag{A.24}$$

and therefore:

$$\boldsymbol{X}=\sum_{i=0}^{\infty}\boldsymbol{A}^i\boldsymbol{C}(\boldsymbol{A}^T)^i. \tag{A.25}$$

That is, the matrix $\boldsymbol{X}$ is defined as a convergent infinite sum of positive semidefinite matrices. If $\mathrm{rank}([\boldsymbol{A}^{n-1}\boldsymbol{C}^{\frac{1}{2}}\ \boldsymbol{A}^{n-2}\boldsymbol{C}^{\frac{1}{2}}\ \ldots\ \boldsymbol{A}\boldsymbol{C}^{\frac{1}{2}}\ \boldsymbol{C}^{\frac{1}{2}}])=n$, it is easy to see that $\boldsymbol{X}$ is full rank.

In the Sylvester equation, matrix $(\boldsymbol{I}_m\otimes\boldsymbol{A}+\boldsymbol{B}^T\otimes\boldsymbol{I}_n)$ has eigenvalues $\lambda_i+\mu_j$, for all $i=1,\ldots,n$ and $j=1,\ldots,m$, where μ_j denotes the j-th eigenvalue of $\boldsymbol{B}$. Then, the system equations given by (A.21) have a unique solution if matrices $\boldsymbol{A}$ and $-\boldsymbol{B}$ have no common eigenvalues.

Calculating the solutions to (A.18) and (A.19) by solving the systems equations (A.20) and (A.21) is inefficient for large values of n and m. First, because of the computational load involved $(1/6n^6+n^4)$ flops when performing the Cholesky decomposition of $(\boldsymbol{I}-\boldsymbol{A}^T\otimes\boldsymbol{A})$), and second because of the memory storage required. Note that the matrix $(\boldsymbol{I}-\boldsymbol{A}^T\otimes\boldsymbol{A})$ is $(n^2\times n^2)$, while $(\boldsymbol{I}_m\otimes\boldsymbol{A}+\boldsymbol{B}^T\otimes\boldsymbol{I}_n)$ is $(nm\times nm)$. There are, however, computationally more efficient alternative methods based on the reduction of the matrices $\boldsymbol{A}$ and $\boldsymbol{B}$ to simpler similar forms.

A.7 Numerical solution of a Sylvester equation

Solving the Sylvester equation (A.19) is far from trivial. It can be done by an algorithm due to Bartels and Stewart [15], or using the *Hessenberg–Schur* procedure ([181])[1]. The Bartels and Stewart algorithm proceeds as follows:

Step 1) Reduce $\boldsymbol{A}$ and $\boldsymbol{B}$ to lower and upper real Schur forms, respectively, so that $\boldsymbol{U}^T\boldsymbol{A}\boldsymbol{U} = \boldsymbol{A}^*$ and $\boldsymbol{V}^T\boldsymbol{B}\boldsymbol{V} = \boldsymbol{B}^*$.

Step 2) Premultiplying (A.19) by $\boldsymbol{U}^T$ and postmultiplying by $\boldsymbol{V}$, yields:

$$\boldsymbol{U}^T\boldsymbol{A}\boldsymbol{U}\boldsymbol{U}^T\boldsymbol{X}\boldsymbol{V} + \boldsymbol{U}^T\boldsymbol{X}\boldsymbol{V}\boldsymbol{V}^T\boldsymbol{B}\boldsymbol{V} = \boldsymbol{U}^T\boldsymbol{C}\boldsymbol{V}. \tag{A.26}$$

Now, defining $\boldsymbol{X}^* = \boldsymbol{U}^T\boldsymbol{X}\boldsymbol{V}$ and $\boldsymbol{C}^* = \boldsymbol{U}^T\boldsymbol{C}\boldsymbol{V}$, (A.26) can be written as:

$$\boldsymbol{A}^*\boldsymbol{X}^* + \boldsymbol{X}^*\boldsymbol{B}^* = \boldsymbol{C}^*. \tag{A.27}$$

Consider now the following partition of matrices $\boldsymbol{X}^*$ and $\boldsymbol{C}^*$:

$$\boldsymbol{X}^* = \begin{bmatrix} \boldsymbol{X}_{11}^* & \boldsymbol{X}_{1p}^* \\ \vdots & \vdots \\ \boldsymbol{X}_{k1}^* & \boldsymbol{X}_{kp}^* \end{bmatrix}, \tag{A.28}$$

and

$$\boldsymbol{C}^* = \begin{bmatrix} \boldsymbol{C}_{11}^* & \boldsymbol{C}_{1p}^* \\ \vdots & \vdots \\ \boldsymbol{C}_{k1}^* & \boldsymbol{C}_{kp}^* \end{bmatrix}. \tag{A.29}$$

For all $h = 1, \dots, k$ and $l = 1, \dots, k$, the following equations are then immediately derived from the quasi-triangular form of $\boldsymbol{A}^*$ and $\boldsymbol{B}^*$:

$$\sum_{i=1}^{l} \boldsymbol{A}_{li}^*\boldsymbol{X}_{ih}^* + \sum_{i=1}^{h} \boldsymbol{X}_{li}^*\boldsymbol{B}_{ih}^* = \boldsymbol{C}_{lh}^*, \tag{A.30}$$

or

$$\boldsymbol{A}_{ll}^*\boldsymbol{X}_{lh}^* + \boldsymbol{X}_{lh}^*\boldsymbol{B}_{hh}^* = \boldsymbol{C}_{lh}^* - \sum_{i=1}^{l-1} \boldsymbol{A}_{li}^*\boldsymbol{X}_{ih}^* - \sum_{i=1}^{h-1} \boldsymbol{X}_{li}^*\boldsymbol{B}_{ih}^*. \tag{A.31}$$

Equations (A.31) are Sylvester equations with dimension (1×1) or (2×2). If these equations are solved in the order $\boldsymbol{X}_{11}^*, \boldsymbol{X}_{21}^*, \dots, \boldsymbol{X}_{12}^*, \boldsymbol{X}_{22}^*, \dots, \boldsymbol{X}_{kp}^*$, then the right-hand side of expression (A.31) does not contain any element of $\boldsymbol{X}^*$ that has not been previously calculated. The solution of equations (A.31) constitutes Step 2).

Step 3) Calculate the matrix $\boldsymbol{X} = \boldsymbol{U}\boldsymbol{X}^*\boldsymbol{V}^T$.

[1]The main difference between the two methods lies in the fact that Bartels and Stewart [15] reduce the matrix $\boldsymbol{A}$ to a Schur form, while the Hessenberg–Schur algorithm uses the more efficient Hessenberg form.

The computational load of this algorithm is roughly $13(n^3+m^3)+5/2(nm^2+mn^2)$ flops and it is numerically stable, considering the stability properties of the Schur decomposition. A variant of this algorithm reduces matrices $\boldsymbol{A}$ and $\boldsymbol{B}$ to Schur complex forms, such that $\boldsymbol{A}^*$ and $\boldsymbol{B}^*$ are lower and upper triangular matrices, respectively. In this case, Step 2 is simplified as all of the equations in (A.31) are (1×1), but it is necessary to perform complex arithmetic operations.

The procedure to solve the Lyapunov equation (A.18) is very similar to the solution method that we have just described, see Petkov, Christov and Konstantinov [181].

A.8 Block-diagonalization of a matrix

For any real square matrix $\boldsymbol{A}$ there is a transformation matrix $\boldsymbol{T}$, such that $\boldsymbol{T}^{-1}\boldsymbol{A}\boldsymbol{T}=\boldsymbol{R}$, where $\boldsymbol{R}$ is a real block diagonal matrix $\boldsymbol{R}=\text{diag}(\boldsymbol{R}_{11},\boldsymbol{R}_{22},\ldots,\boldsymbol{R}_{kk})$, and blocks $\boldsymbol{R}_{11},\boldsymbol{R}_{22},\ldots,\boldsymbol{R}_{kk}$ have no common eigenvalues.

The block-diagonalization algorithm is described in Petkov, Christov and Konstantinov [181], and is structured in three steps:

Step 1) Obtain the real Schur transformation of matrix $\boldsymbol{A}$:

$$\boldsymbol{U}^T\boldsymbol{A}\boldsymbol{U}=\boldsymbol{R}, \tag{A.32}$$

where

$$\boldsymbol{R}=\begin{bmatrix}\boldsymbol{R}_{11} & \boldsymbol{R}_{12} & \ldots & \boldsymbol{R}_{1k}\\ \boldsymbol{0} & \boldsymbol{R}_{22} & \ldots & \boldsymbol{R}_{2k}\\ \vdots & \vdots & \ddots & \vdots\\ \boldsymbol{0} & \boldsymbol{0} & \ldots & \boldsymbol{R}_{kk}\end{bmatrix}, \tag{A.33}$$

with $\boldsymbol{R}_{11},\boldsymbol{R}_{22},\ldots,\boldsymbol{R}_{kk}$ either scalars or (2×2) matrices with complex conjugate eigenvalues. Transformation matrices $\boldsymbol{T}$ and $\boldsymbol{T}^{-1}$ are initialized to: $\boldsymbol{T}=\boldsymbol{U}$ and $\boldsymbol{T}^{-1}=\boldsymbol{U}^T$.

Step 2) Block diagonalize of the Schur form. We can define a transformation matrix with the following structure:

$$\boldsymbol{X}=\begin{bmatrix}\boldsymbol{I} & \boldsymbol{Z}\\ \boldsymbol{0} & \boldsymbol{I}\end{bmatrix}, \tag{A.34}$$

and given that:

$$\boldsymbol{X}^{-1}=\begin{bmatrix}\boldsymbol{I} & -\boldsymbol{Z}\\ \boldsymbol{0} & \boldsymbol{I}\end{bmatrix}, \tag{A.35}$$

we have:

$$\boldsymbol{X}^{-1}\begin{bmatrix}\boldsymbol{R}_{11} & \boldsymbol{R}_{12}\\ \boldsymbol{0} & \boldsymbol{R}_{22}\end{bmatrix}\boldsymbol{X}=\begin{bmatrix}\boldsymbol{R}_{11} & \boldsymbol{R}_{11}\boldsymbol{Z}-\boldsymbol{Z}\boldsymbol{R}_{22}+\boldsymbol{R}_{12}\\ \boldsymbol{0} & \boldsymbol{R}_{22}\end{bmatrix}. \tag{A.36}$$

The problem then consists of finding the value of $\boldsymbol{Z}$ that solves the following Sylvester equation:

$$\boldsymbol{R}_{11}\boldsymbol{Z}-\boldsymbol{Z}\boldsymbol{R}_{22}+\boldsymbol{R}_{12}=0. \tag{A.37}$$

This equation has a unique solution if $\boldsymbol{R}_{11}$ and $\boldsymbol{R}_{22}$ do not have common eigenvalues. Once we select a block $\boldsymbol{R}_{ij}$ to make it zero, we should take into account that the proposed transformation affects blocks $\boldsymbol{R}_{1:i-1,j}$ and $\boldsymbol{R}_{i,j+1:k}$, as follows:

$$\boldsymbol{R}_{1:i-1,j} = \boldsymbol{R}_{1:i-1,j} + \boldsymbol{R}_{1:i-1,i}\boldsymbol{Z}, \tag{A.38}$$

and

$$\boldsymbol{R}_{i,j+1:k} = \boldsymbol{R}_{i,j+1:k} - \boldsymbol{Z}\boldsymbol{R}_{j,j+1:k}. \tag{A.39}$$

Additionally, the transformation matrix T has to be updated as:

$$\boldsymbol{T}_{1:k,j} = \boldsymbol{T}_{1:k,j} + \boldsymbol{T}_{1:k,i}\boldsymbol{Z}, \tag{A.40}$$

and

$$(\boldsymbol{T}^{-1})_{i,1:k} = (\boldsymbol{T}^{-1})_{i,1:k} - \boldsymbol{Z}(\boldsymbol{T}^{-1})_{j,1:k}. \tag{A.41}$$

Then the algorithm consists of starting bottom-to-top and left-to-right: $\boldsymbol{R}_{k-1,k}, \boldsymbol{R}_{k-2,k-1}, \boldsymbol{R}_{k-2,k}, \ldots, \boldsymbol{R}_{1,2}, \ldots, \boldsymbol{R}_{1,k}$, and solving the Sylvester equations for pairs of blocks with no common eigenvalues.

Step 3) Reorder the matrix $\boldsymbol{R}$ obtained in Step 2. Given that the Schur transformation usually does not provide the eigenvalues ordered in a specific way, we could have common eigenvalues in non-consecutive blocks. For example, if $\boldsymbol{R}_{11}$ and $\boldsymbol{R}_{33}$ share the same eigenvalue, the block $\boldsymbol{R}_{13}$ will not be null. Then, to provide a block diagonal matrix, we have to reorder the rows and columns of $\boldsymbol{R}$, while ensuring that common eigenvalues are placed in consecutive blocks. We apply the same reordering operation to both the columns of $\boldsymbol{T}$ and the rows of $\boldsymbol{T}^{-1}$.

A.9 Reduced rank least squares

Consider the regression model:

$$\boldsymbol{y}_t = \boldsymbol{A}\boldsymbol{B}\boldsymbol{x}_t + \boldsymbol{\varepsilon}_t, \qquad t = 1, 2, \ldots, N, \tag{A.42}$$

where $\boldsymbol{A}$ is $(j \times n)$ and $\boldsymbol{B}$ is $(n \times i)$ such that $n < \min\{i, j\}$, and $\boldsymbol{\varepsilon}_t$ is a zero-mean white noise process with $\mathrm{E}[\boldsymbol{\varepsilon}_t \boldsymbol{\varepsilon}_t^T] = \boldsymbol{\Omega}$. Therefore, $\boldsymbol{C} = \boldsymbol{A}\boldsymbol{B}$ is a reduced-rank matrix. A general representation of (A.42) used in Chapter 7 is:

$$\boldsymbol{y}_t = \boldsymbol{A}\boldsymbol{B}\boldsymbol{x}_t + \boldsymbol{H}\boldsymbol{u}_t + \boldsymbol{\varepsilon}_t, \tag{A.43}$$

where the rank restriction only affects one part of the model parameters. In the following propositions, we will derive least-squares estimators for $\boldsymbol{A}$ and $\boldsymbol{B}$ in model (A.42), and for $\boldsymbol{A}$, $\boldsymbol{B}$, and $\boldsymbol{H}$ in model (A.43).

Proposition A.2 *Let $\boldsymbol{\Omega}$, $\boldsymbol{Y}$, $\boldsymbol{X}$, $\boldsymbol{A}$, and $\boldsymbol{B}$ be matrices with dimensions $(j \times j)$, $(j \times N)$, $(i \times N)$, $(j \times n)$, and $(n \times i)$, respectively, such that rank$(\boldsymbol{\Omega}) = j$, rank$(\boldsymbol{Y}) = j$, and rank$(\boldsymbol{X}) = i$. Then, the problem given by:*

$$\min_{\boldsymbol{A},\boldsymbol{B}} \left\| \boldsymbol{\Omega}^{-1/2}(\boldsymbol{Y} - \boldsymbol{A}\boldsymbol{B}\boldsymbol{X}) \right\|_F^2, \tag{A.44}$$

1. *Has an infinite number of solutions.*
2. *May be solved using the SVD decomposition of the matrix:*

$$\boldsymbol{\Omega}^{-1/2}\boldsymbol{Y}\boldsymbol{X}^T(\boldsymbol{X}\boldsymbol{X}^T)^{-1/2} \overset{svd}{=} \boldsymbol{U}\boldsymbol{S}\boldsymbol{V}^T. \tag{A.45}$$

3. $\hat{\boldsymbol{A}} = \boldsymbol{\Omega}^{1/2}\boldsymbol{U}_n$ *and* $\hat{\boldsymbol{B}} = \boldsymbol{S}_n\boldsymbol{V}_n^T(\boldsymbol{X}\boldsymbol{X}^T)^{-1/2}$ *are optimal solutions, where* $\boldsymbol{U}_n$ *and* $\boldsymbol{V}_n$ *correspond to the matrices made up of the first n columns from* $\boldsymbol{U}$ *and* $\boldsymbol{V}$*, respectively, and* $\boldsymbol{S}_n$ *is the diagonal matrix that contains the first n singular values of* $\boldsymbol{S}$.

Proof. See Casals [38], Proposition A.13.1.

Proposition A.3 *Let us assume now that the rank restriction only affects one part of the model parameters, so that the problem (A.44) can be expressed as:*

$$\min_{\boldsymbol{A},\boldsymbol{B},\boldsymbol{H}} \left\|\boldsymbol{\Omega}^{-1/2}(\boldsymbol{Y}-\boldsymbol{A}\boldsymbol{B}\boldsymbol{X}-\boldsymbol{H}\boldsymbol{U})\right\|_F^2, \tag{A.46}$$

where $\boldsymbol{H}$ *is* $(j\times r)$*,* $\boldsymbol{U}$ *is* $(r\times N)$*, and rank*$(\boldsymbol{U}) = r$*. Then:*

1. *Matrices* $\hat{\boldsymbol{A}}$ *and* $\hat{\boldsymbol{B}}$ *arise from applying Proposition A.2 to:*

$$\min_{\boldsymbol{A},\boldsymbol{B}} \left\|\boldsymbol{\Omega}^{-1/2}(\boldsymbol{Y}\boldsymbol{\Pi}_{\boldsymbol{U}}^{\perp}-\boldsymbol{A}\boldsymbol{B}\boldsymbol{X}\boldsymbol{\Pi}_{\boldsymbol{U}}^{\perp})\right\|_F^2, \tag{A.47}$$

where $\boldsymbol{\Pi}_{\boldsymbol{U}}^{\perp} = \boldsymbol{I}-\boldsymbol{U}^T(\boldsymbol{U}\boldsymbol{U}^T)^{-1}\boldsymbol{U}$ *is the projection onto the orthogonal complement of* $\boldsymbol{U}$.

2. *The optimal estimate for* $\boldsymbol{H}$ *is obtained by solving the unrestricted problem:*

$$\min_{\boldsymbol{H}} \left\|\boldsymbol{\Omega}^{-1/2}(\boldsymbol{Y}-\hat{\boldsymbol{A}}\hat{\boldsymbol{B}}\boldsymbol{X}-\boldsymbol{H}\boldsymbol{U})\right\|_F^2. \tag{A.48}$$

Proof. See Casals [38], Proposition A.13.2.

Under the assumption of Gaussian innovations, setting the weighting matrix $\boldsymbol{\Omega}$ to be equal to the variance of the innovations of models (A.42) and (A.43), leads to ML estimates when solving (A.44) and (A.46). However, as $\boldsymbol{\Omega}$ is generally unknown, it should be replaced by its estimate $\hat{\boldsymbol{\Omega}} = \frac{1}{N}\boldsymbol{Y}\boldsymbol{\Pi}_{\boldsymbol{P}}^{\perp}\boldsymbol{Y}^T$, where $\boldsymbol{P}=\boldsymbol{X}$ in (A.44), or $\boldsymbol{P} = [\boldsymbol{X} \quad \boldsymbol{U}]$ in (A.46), and $\boldsymbol{\Pi}_{\boldsymbol{P}}^{\perp} = \boldsymbol{I}-\boldsymbol{P}^T(\boldsymbol{P}\boldsymbol{P}^T)^{-1}\boldsymbol{P}$ is the projection onto the orthogonal complement of $\boldsymbol{P}$. That is, $\hat{\boldsymbol{\Omega}}$ is obtained as the residual variance ignoring the rank restriction.

The literature also considers $\boldsymbol{\Omega} = \boldsymbol{Y}\boldsymbol{Y}^T$ in model (A.44), and $\boldsymbol{\Omega} = \boldsymbol{Y}\boldsymbol{\Pi}_{\boldsymbol{U}}^{\perp}\boldsymbol{Y}^T$ in model (A.46), as weighting matrices. In such cases, the decomposition (A.45) is equivalent to calculating the canonical correlations; see Section A.5 in this Appendix.

It is noteworthy that there is no asymptotic difference between the two previous approaches, as the optimal solutions obtained with $\boldsymbol{\Omega} = \boldsymbol{Y}\boldsymbol{\Pi}_{\boldsymbol{P}}^{\perp}\boldsymbol{Y}^T$ equal those obtained with $\boldsymbol{\Omega} = \boldsymbol{Y}\boldsymbol{\Pi}_{\boldsymbol{U}}^{\perp}\boldsymbol{Y}^T$.

A.10 Riccati equations

A.10.1 Definition

The so-called Riccati difference equation in discrete time is given by the expression:

$$X_{t+1} = AX_tA^T + M - AX_tH^T(HX_tH^T + N)^{-1}HX_tA^T, \qquad \text{(A.49)}$$

where $\{X_t\}$ is a sequence of symmetric matrices of dimension $n \times n$. If $\{X_t\}$ converges as $t \to \infty$ to an equilibrium solution, $\bar{X}$ (steady-state), then this solution satisfies the so-called *Algebraic Riccati Equation* (ARE):

$$\bar{X} = A\bar{X}A^T + M - A\bar{X}H^T(H\bar{X}H^T + N)^{-1}H\bar{X}A^T. \qquad \text{(A.50)}$$

Equations (A.49)–(A.50) are common in optimal control and signal extraction problems. There is an extensive literature about the convergence of the equation (A.49), and the existence, uniqueness, and computation of the solutions to the ARE (A.50); see Hewer [123], Pappas, Laub, and Sandell [176], Van Dooren [214], Chan, Goodwin, and Sin [49], De Souza, Gevers, and Goodwin [71], or Ionescu, Oara, and Weiss [128]. In particular, the KF equation:

$$\begin{aligned} P_{t+1} &= \Phi P_t \Phi^T + EQE^T \\ &\quad - (\Phi P_t H^T + ESC^T)(HP_tH^T + CRC^T)^{-1}(\Phi P_t H^T + ESC^T)^T, \end{aligned} \qquad \text{(A.51)}$$

is a Riccati difference equation, despite the fact that it looks different from (A.49).

Assuming matrix CRC^T to be nonsingular, the term $(\Phi P_t H^T + ESC^T)$ can be written as:

$$\Phi P_t H^T + ESC^T = \bar{\Phi} P_t H^T + ESC^T(CRC^T)^{-1}(HP_tH^T + CRC^T), \qquad \text{(A.52)}$$

where $\bar{\Phi} = \Phi - ESC^T(CRC^T)^{-1}H$. Substituting (A.52) in (A.51), we obtain:

$$P_{t+1} = \bar{\Phi} P_t \bar{\Phi}^T + \bar{Q} - \bar{\Phi} P_t H^T (HP_tH^T + CRC^T)^{-1} HP_t \bar{\Phi}^T, \qquad \text{(A.53)}$$

with $\bar{Q} = E(Q - SC^T(CRC^T)^{-1}CS^T)E^T$. This last equation corresponds to the definition given by (A.49).

De Souza, Gevers, and Goodwin [71] study the convergence conditions of the difference Riccati equation (A.49), as well as the existence of the solution to the ARE (A.50). Some of their main results are as follows.

Definition A.1. The symmetric, real, and positive semidefinite matrix $\bar{X}$ is a strong solution to the ARE (A.50) if $|\lambda(A - \bar{K}H)| \leq 1$, for all $i = 1, \ldots, n$, where $\bar{K} = A\bar{X}H^T(H\bar{X}H^T + N)^{-1}$ corresponds to the KF steady state gain. If $|\lambda(A - \bar{K}H)| < 1$, for all $i = 1, \ldots, n$, then the strong solution is called "stabilizable."

Theorem A.2 *The strong solution to the ARE exists and is unique if the pair* (H, A) *is detectable. Moreover, the solution coincides with the stabilizing solution if the pair* (H, A) *is detectable, and the pair* $(A, \bar{M})$ *has no unreachable modes on the unit circle, where* $\bar{M}$ *a matrix such that* $M = \bar{M}\bar{M}^T$ *and rank*$(M) =$ *rank*$(\bar{M})$.

Proof. See De Souza, Gevers, and Goodwin [71], Theorem 3.2.

Corollary A.1 *A minimal SEM satisfies that: (a) the ARE has a strong and unique solution; (b) if the eigenvalues of* $\boldsymbol{\Phi} - \boldsymbol{EH}$ *are not outside the unit circle, then the strong solution is* $\boldsymbol{P} = \mathbf{0}$*; and (c) if these eigenvalues are strictly within the unit circle (invertibility), then* $\boldsymbol{P} = \mathbf{0}$ *is a stabilizable solution.*

Proof. See Casals [38], Appendix A.12.

Theorem A.3 *When* $\boldsymbol{X}_0 - \bar{\boldsymbol{X}} \geq \mathbf{0}$ *then* $\lim_{t\to\infty} \boldsymbol{X}_t = \bar{\boldsymbol{X}}$*, where* $\bar{\boldsymbol{X}}$ *is the strong solution for the ARE (A.50), if and only if the pair* $(\boldsymbol{H}, \boldsymbol{A})$ *is detectable.*

Proof. See De Souza, Gevers, and Goodwin [71], Theorem 4.2.

Corollary A.2 *If the eigenvalues of* $\boldsymbol{\Phi} - \boldsymbol{EH}$ *are not outside the unit circle*[2] *in a a minimal SEM, then the difference Riccati equation (A.51) converges to* $\boldsymbol{P} = \mathbf{0}$*. When the eigenvalues of* $\boldsymbol{\Phi} - \boldsymbol{EH}$ *are strictly within the unit circle, this convergence becomes exponentially fast.*

Proof. See Casals [38], Appendix A.12.

A.10.2 Solving the ARE in the general case

The procedures to solve the ARE (A.50) can be divided into iterative and non-iterative methods. The first are based on the convergence properties of the difference Riccati equation to achieve an equilibrium solution by propagating this equation. Among them, it is important to mention Hewer's method [123], which is based on the following result.

Suppose that $\{\boldsymbol{X}_k\}$ is the sequence of Lyapunov equations:

$$\boldsymbol{X}_k = \boldsymbol{A}_k \boldsymbol{X}_k \boldsymbol{A}_k^T + \boldsymbol{M} + \boldsymbol{K}_k \boldsymbol{N} \boldsymbol{K}_k^T, \qquad k = 0, 1, 2, \ldots, \tag{A.54}$$

with $\boldsymbol{A}_k = \boldsymbol{A} - \boldsymbol{K}_k \boldsymbol{H}$, and:

$$\boldsymbol{K}_k = \boldsymbol{A}_k \boldsymbol{X}_{k-1} \boldsymbol{H}^T (\boldsymbol{H} \boldsymbol{X}_{k-1} \boldsymbol{H}^T + \boldsymbol{N})^{-1}, \qquad k = 1, 2, \ldots, \tag{A.55}$$

where $\boldsymbol{K}_0$ is an arbitrary matrix such that $\boldsymbol{A}_0 = \boldsymbol{A} - \boldsymbol{K}_0 \boldsymbol{H}$ is stable. Then:

$$\boldsymbol{X}_0 \geq \boldsymbol{X}_1 \geq \cdots \geq \bar{\boldsymbol{X}}, \tag{A.56}$$

and

$$\lim_{k\to\infty} \boldsymbol{X}_k = \bar{\boldsymbol{X}}. \tag{A.57}$$

The Hewer [123] algorithm can be summarized by the following iterative scheme:

[2] This condition is equivalent to requiring that the MA roots of the model lie on or within the unit circle, thus excluding strict noninvertibility.

Step 1) Select $\boldsymbol{K}_0$ so that $\boldsymbol{A}_0$ is stable. Note that if $\boldsymbol{A}$ is stable, the procedure can be initialized with the condition $\boldsymbol{K}_0 = \mathbf{0}$.

Step 2) Calculate $\boldsymbol{X}_k$, for all $k = 0, 1, 2, \ldots$, by solving Lyapunov equation (A.54).

Step 3) Calculate $\boldsymbol{K}_{k+1}$, for all $k = 0, 1, 2, \ldots$, from (A.55).

Step 4) If $\|\boldsymbol{X}_k - \boldsymbol{X}_{k-1}\| / \|\boldsymbol{X}_{k-1}\| > \tau$, where τ is a scalar that sets a convergence criterion, go back to Step 2). Otherwise, stop the iteration.

The value for τ will be set according to the ε-machine and the condition number of the Riccati equation; see Petkov, Christov, and Konstantinov [181]. Obviously, the computational load of this algorithm depends on the number of iterations until convergence.

Non-iterative methods are based on solving a problem of generalized eigenvalues; see Pappas, Laub, and Sandell [176], Van Dooren [214], Chan, Goodwin, and Sin [49], and Ionescu, Oara, and Weiss [128]. Suppose that the pair $(\boldsymbol{C}, \boldsymbol{A})$ is detectable, meaning that there is a strong and unique solution to the difference Riccati equation. The equation (A.50) can then be solved by computing the generalized eigenvalue matrix in:

$$\boldsymbol{T}\boldsymbol{v} = \lambda \boldsymbol{Y}\boldsymbol{v}, \tag{A.58}$$

where:

$$\boldsymbol{T} = \begin{bmatrix} \boldsymbol{A}^T & \mathbf{0} \\ -\boldsymbol{M} & \boldsymbol{I}_n \end{bmatrix}, \tag{A.59}$$

and

$$\boldsymbol{Y} = \begin{bmatrix} \boldsymbol{I}_n & (\boldsymbol{H}^T \boldsymbol{N}^{-1} \boldsymbol{H}) \\ \mathbf{0} & \boldsymbol{A} \end{bmatrix}. \tag{A.60}$$

If matrix $\boldsymbol{A}$ is nonsingular, then $\boldsymbol{Y}$ is also nonsingular, and solving (A.58) then reduces to solving the following eigenvalue problem[3]:

$$(\boldsymbol{Y}^{-1}\boldsymbol{T})\boldsymbol{v} = \lambda \boldsymbol{v}. \tag{A.61}$$

The problem defined in (A.61) has the following properties (see De Souza, Gevers, and Goodwin [71]):

1. If λ is an eigenvalue of the pencil $\boldsymbol{T} - \lambda \boldsymbol{Y}$, then $1/\lambda$ is an eigenvalue of the same pencil, with the same multiplicity.
2. If $\boldsymbol{A}$ is nonsingular, then all its generalized eigenvalues are nonzero.
3. If $\boldsymbol{A}$ is singular, then the pencil $\boldsymbol{T} - \lambda \boldsymbol{Y}$ has at least one null generalized eigenvalue.
4. If $\lambda = 0$ is a generalized eigenvalue of $\boldsymbol{A}$ with multiplicity r, then the pencil $\boldsymbol{T} - \lambda \boldsymbol{Y}$ has only $2n - r$ finite generalized eigenvalues.

---

[3]Petkov, Christov, and Konstantinov [181] provide an alternative form to this generalized eigenvalue problem that avoids the inversion of the matrix $\boldsymbol{N}$ in (A.60). This is the solution implemented in E^4; see Appendix C

The eigenvalues of the problem (A.58) can be sorted as:

$$\overbrace{0,\ldots,0}^{r},\lambda_{r+1},\ldots,\lambda_n,\frac{1}{\lambda_n},\ldots,\frac{1}{\lambda_{r+1}},\overbrace{\infty,\ldots,\infty}^{r} \tag{A.62}$$

where $0<|\lambda_i|\leq 1$, for all $i=r+1,\ldots,n$. If $\boldsymbol{V}$ is the matrix of ordered generalized eigenvalues according to (A.62), partitioning $\boldsymbol{V}$ into four blocks of dimension $(n\times n)$ yields:

$$\boldsymbol{V}=\begin{bmatrix}\boldsymbol{V}_{11} & \boldsymbol{V}_{12}\\ \boldsymbol{V}_{21} & \boldsymbol{V}_{22}\end{bmatrix}. \tag{A.63}$$

Then:

1. $\boldsymbol{V}_{11}$ and $\boldsymbol{V}_{12}$ are nonsingular.
2. $\bar{\boldsymbol{X}}=\boldsymbol{V}_{21}\boldsymbol{V}_{11}^{-1}$ and $\tilde{\boldsymbol{X}}=\boldsymbol{V}_{22}\boldsymbol{V}_{12}^{-1}$ are solutions to the ARE (A.50).
3. The eigenvalues of the matrix $\boldsymbol{A}-\bar{\boldsymbol{K}}\boldsymbol{H}$, where $\bar{\boldsymbol{K}}=\boldsymbol{A}\bar{\boldsymbol{X}}\boldsymbol{H}^T(\boldsymbol{H}\bar{\boldsymbol{X}}\boldsymbol{H}^T+\boldsymbol{N})^{-1}$, coincide with the n eigenvalues of the problem given by (A.58) with modulus less than or equal to unity.
4. $\bar{\boldsymbol{X}}=\boldsymbol{V}_{21}\boldsymbol{V}_{11}^{-1}$ is the strong solution to (A.50).

Petkov, Christov, and Konstantinov [181] recommend first using the non-iterative procedure explained above, and then refining the solution $\bar{\boldsymbol{X}}$ by means of Hewer's algorithm with $\boldsymbol{X}_0=\bar{\boldsymbol{X}}$.

A.10.3 Solving the ARE for GARCH models

In Section 6.3 we stated that in a representation such as (6.56)–(6.57), with GARCH errors, $\bar{\boldsymbol{P}}=\boldsymbol{0}$ is the solution of the algebraic Riccati equation:

$$\begin{aligned}\bar{\boldsymbol{P}}=&\,\boldsymbol{\Phi}\bar{\boldsymbol{P}}\boldsymbol{\Phi}^T+\boldsymbol{E}\boldsymbol{\Sigma}_t\boldsymbol{E}^T\\ &-(\boldsymbol{\Phi}\bar{\boldsymbol{P}}\boldsymbol{H}^T+\boldsymbol{E}\boldsymbol{\Sigma}_t)(\boldsymbol{H}\bar{\boldsymbol{P}}\boldsymbol{H}^T+\boldsymbol{\Sigma}_t)^{-1}(\boldsymbol{\Phi}\bar{\boldsymbol{P}}\boldsymbol{H}^T+\boldsymbol{E}\boldsymbol{\Sigma}_t)^T.\end{aligned} \tag{A.64}$$

This statement can be proved as follows. In models with GARCH errors the conditional covariances $\boldsymbol{\Sigma}_t$ are a source of parametric variation. However, it holds that:

1. $\bar{\boldsymbol{P}}=\boldsymbol{0}$ is the solution to the algebraic Riccati equation (A.64), associated with the Riccati difference equation (6.62) in Chapter 6, for any value of $\boldsymbol{\Sigma}_t$.
2. Following De Souza, Gevers, and Goodwin [71] and denoting $\bar{\boldsymbol{\Sigma}}=\sup\{\boldsymbol{\Sigma}_t\}$, it can be shown that:
   (a) $\boldsymbol{P}_{t+1|t}=\boldsymbol{\Phi}\boldsymbol{P}_{t|t-1}\boldsymbol{\Phi}^T+\boldsymbol{E}\bar{\boldsymbol{\Sigma}}\boldsymbol{E}^T-\boldsymbol{K}_t\boldsymbol{B}_t\boldsymbol{K}_t^T$, where $\boldsymbol{K}_t=(\boldsymbol{\Phi}\boldsymbol{P}_{t/t-1}\boldsymbol{H}^T+\boldsymbol{E}\bar{\boldsymbol{\Sigma}})\boldsymbol{B}_t^{-1}$ and $\boldsymbol{B}_t=\boldsymbol{H}\boldsymbol{P}_{t/t-1}\boldsymbol{H}^T+\bar{\boldsymbol{\Sigma}}$, converges to its strong solution if and only if the model (6.56)–(6.57) is detectable.
   (b) The strong solution is $\bar{\boldsymbol{P}}=\boldsymbol{0}$ if the system (6.56)–(6.57) does not have moving average roots strictly inside the unit circle.
3. Following Bitmead *et al.* [23] it can be shown that, if $\boldsymbol{P}^*_{t|t-q}\leq\boldsymbol{P}_{t|t-1}$ then $\boldsymbol{P}^*_{t+1|t}\leq\boldsymbol{P}_{t+1|t}$, where $\boldsymbol{P}^*_{t+1|t}$ is propagated with $\boldsymbol{\Sigma}_t$ instead of $\bar{\boldsymbol{\Sigma}}$. Because of

this, the convergence of $P_{t+1|t}$ to zero implies that $P^*_{t+1|t}$ will also converge to zero. ■

A.11 Kalman filter

The Kalman filter (KF) is a minimum-variance recursive algorithm which predicts the state vector x_{t+1} of a linear dynamic system from the information available up to time t, as well as the covariance matrix of the prediction error. The Kalman filter equations are easily derived from the following results:

Result 1) Given the random vectors x and y, with $\mathrm{E}(x) = \bar{x}$, $\mathrm{E}(y) = \bar{y}$, $\tilde{x} = x - \bar{x}$, $\tilde{y} = y - \bar{y}$, $\mathrm{E}(\tilde{x}\tilde{x}^T) = \Sigma_{xx}$, $\mathrm{E}(\tilde{y}\tilde{y}^T) = \Sigma_{yy}$, and $\mathrm{E}(\tilde{y}\tilde{x}^T) = \Sigma_{yx}$, consider linear estimators of x of the form $F(y) = Ay + b$. Then, the minimum-variance (linear) estimator of x conditional on y can be shown to be[4]:

$$\mathrm{E}(x|y) = \bar{x} + \Sigma_{xy}\Sigma_{yy}^{-1}\tilde{y}, \tag{A.65}$$

where the prediction error variance is given by:

$$\mathrm{E}[(x - \mathrm{E}(x|y))(x - \mathrm{E}(x|y))^T] = \Sigma_{xx} - \Sigma_{xy}\Sigma_{yy}^{-1}\Sigma_{xy}^T. \tag{A.66}$$

This estimator satisfies,

$$\mathrm{E}[\|x - \mathrm{E}(x|y)\|_F^2] \leq \mathrm{E}[\|x - F(y)\|_F^2],$$

among all linear estimators of the form $F(y) = Ay + b$. If x and y are jointly normally distributed, then the linear minimum-variance estimator (A.65) coincides with the minimum-variance estimator of x conditional on y, where the form of the estimator is any function $F(y)$, not necessarily linear in y.

Result 2) If x, y, and z are random vectors, then:

$$\mathrm{E}(x|y,z) = \mathrm{E}(x|z) - \bar{\Sigma}_{xx}\bar{\Sigma}_{yy}^{-1}[y - \mathrm{E}(y|z)], \tag{A.67}$$

and

$$\mathrm{E}[(x - \mathrm{E}(x|y,z))(x - \mathrm{E}(x|y,z))^T] = \bar{\Sigma}_{xx} - \bar{\Sigma}_{xy}\bar{\Sigma}_{yy}^{-1}\bar{\Sigma}_{xy}^T, \tag{A.68}$$

with

$$\bar{\Sigma}_{xx} = \mathrm{E}[(x - \mathrm{E}[x|z])(x - \mathrm{E}[x|z])^T] \tag{A.69}$$

$$\bar{\Sigma}_{xy} = \mathrm{E}[(x - \mathrm{E}[x|z])(y - \mathrm{E}[y|z])^T] \tag{A.70}$$

$$\bar{\Sigma}_{yy} = \mathrm{E}[(y - \mathrm{E}[y|z])(y - \mathrm{E}[y|z])^T]. \tag{A.71}$$

[4]Although we follow the frequently used convention of denoting this estimator by the conditional expectation operator, $\mathrm{E}(x|y)$, it does not coincide with the actual conditional expectation of x given y, unless x and y are jointly normally distributed. Thus, in general $\mathrm{E}(x|y)$ is used here in the sense of optimal (minimum error variance) projection operator of x onto the space of all linear combinations of $(1, y)$.

Result 3) Assume that the vector $\boldsymbol{x}$ can be defined as $\boldsymbol{x} = \boldsymbol{A}\boldsymbol{y} + \boldsymbol{b} + \boldsymbol{v}$, where $\boldsymbol{v}$ is a zero-mean random vector with $\mathrm{E}(\boldsymbol{v}\boldsymbol{v}^T) = \boldsymbol{Q}$ which is uncorrelated with random vectors $\boldsymbol{y}$ and $\boldsymbol{z}$. Then:

$$\mathrm{E}(\boldsymbol{x}|\boldsymbol{y}) = \boldsymbol{A}\boldsymbol{y} + \boldsymbol{b} \tag{A.72}$$

$$\mathrm{E}[(\boldsymbol{x} - \mathrm{E}(\boldsymbol{x}|\boldsymbol{y}))(\boldsymbol{x} - \mathrm{E}(\boldsymbol{x}|\boldsymbol{y}))^T] = \boldsymbol{Q} \tag{A.73}$$

$$\mathrm{E}(\boldsymbol{x}|\boldsymbol{z}) = \boldsymbol{A}\mathrm{E}(\boldsymbol{y}|\boldsymbol{z}) + \boldsymbol{b} \tag{A.74}$$

$$\bar{\boldsymbol{\Sigma}}_{xx} = \boldsymbol{A}\bar{\boldsymbol{\Sigma}}_{yy}\boldsymbol{A}^T + \boldsymbol{Q}. \tag{A.75}$$

Anderson and Moore [6, ch. 5] provide a detailed treatment of the linear minimum-variance estimator, from which it is straightforward to derive Results (1)–(3).

In deriving the KF recursions below, we will use the following notation:

$$\hat{\boldsymbol{x}}_{t|t-j} = \mathrm{E}(\boldsymbol{x}_t|\boldsymbol{z}_0, \ldots, \boldsymbol{z}_{t-j}, \boldsymbol{u}_0, \ldots, \boldsymbol{u}_N) \tag{A.76}$$

$$\tilde{\boldsymbol{x}}_{t|t-j} = \boldsymbol{x}_t - \hat{\boldsymbol{x}}_{t|t-j} \tag{A.77}$$

$$\hat{\boldsymbol{z}}_{t|t-j} = \mathrm{E}(\boldsymbol{z}_t|\boldsymbol{z}_0, \ldots, \boldsymbol{z}_{t-j}, \boldsymbol{u}_0, \ldots, \boldsymbol{u}_N) \tag{A.78}$$

$$\tilde{\boldsymbol{z}}_{t|t-j} = \boldsymbol{z}_t - \hat{\boldsymbol{z}}_{t|t-j} \tag{A.79}$$

$$\boldsymbol{P}_{t|t-j} = \mathrm{E}(\tilde{\boldsymbol{x}}_{t|t-j}\tilde{\boldsymbol{x}}_{t|t-j}^T) \tag{A.80}$$

$$\boldsymbol{B}_t = \mathrm{E}(\tilde{\boldsymbol{z}}_t\tilde{\boldsymbol{z}}_t^T). \tag{A.81}$$

Now, for a MEM in state space and according to Result 3), we have:

$$\begin{bmatrix} \hat{\boldsymbol{x}}_{t+1|t-1} \\ \hat{\boldsymbol{z}}_{t|t-1} \end{bmatrix} = \begin{bmatrix} \boldsymbol{\Phi} \\ \boldsymbol{H} \end{bmatrix} \hat{\boldsymbol{x}}_{t|t-1} + \begin{bmatrix} \boldsymbol{\Gamma} \\ \boldsymbol{D} \end{bmatrix} \boldsymbol{u}_t, \tag{A.82}$$

$$\begin{bmatrix} \boldsymbol{P}_{t+1|t-1} \\ \tilde{\boldsymbol{B}}_t \end{bmatrix} = \begin{bmatrix} \boldsymbol{\Phi} \\ \boldsymbol{H} \end{bmatrix} \boldsymbol{P}_{t|t-1} \begin{bmatrix} \boldsymbol{\Phi}^T & \boldsymbol{H}^T \end{bmatrix} + \begin{bmatrix} \boldsymbol{E} & \boldsymbol{0} \\ \boldsymbol{0} & \boldsymbol{C} \end{bmatrix} \begin{bmatrix} \boldsymbol{Q} & \boldsymbol{S} \\ \boldsymbol{S}^T & \boldsymbol{R} \end{bmatrix} \begin{bmatrix} \boldsymbol{E}^T & \boldsymbol{0} \\ \boldsymbol{0} & \boldsymbol{C}^T \end{bmatrix}, \tag{A.83}$$

and applying Result 2):

$$\hat{\boldsymbol{x}}_{t+1|t} = \hat{\boldsymbol{x}}_{t+1|t-1} + (\boldsymbol{\Phi}\boldsymbol{P}_{t|t-1}\boldsymbol{H}^T + \boldsymbol{E}\boldsymbol{S}\boldsymbol{C}^T)\boldsymbol{B}_t^{-1}(\boldsymbol{z}_t - \hat{\boldsymbol{z}}_{t|t-1}), \tag{A.84}$$

$$\boldsymbol{P}_{t+1|t} = \boldsymbol{P}_{t+1|t-1} - (\boldsymbol{\Phi}\boldsymbol{P}_{t|t-1}\boldsymbol{H}^T + \boldsymbol{E}\boldsymbol{S}\boldsymbol{C}^T)\boldsymbol{B}_t^{-1}(\boldsymbol{\Phi}\boldsymbol{P}_{t|t-1}\boldsymbol{H}^T + \boldsymbol{E}\boldsymbol{S}\boldsymbol{C}^T)^T. \tag{A.85}$$

Then, the Kalman filter is given by the following expressions:

$$\hat{\boldsymbol{z}}_{t|t-1} = \boldsymbol{H}\hat{\boldsymbol{x}}_{t|t-1} + \boldsymbol{D}\boldsymbol{u}_t \tag{A.86}$$

$$\hat{\boldsymbol{x}}_{t+1|t} = \boldsymbol{\Phi}\hat{\boldsymbol{x}}_{t|t-1} + \boldsymbol{\Gamma}\boldsymbol{u}_t + \boldsymbol{K}_t(\boldsymbol{z}_t - \hat{\boldsymbol{z}}_{t|t-1}) \tag{A.87}$$

$$\boldsymbol{K}_t = (\boldsymbol{\Phi}\boldsymbol{P}_{t|t-1}\boldsymbol{H}^T + \boldsymbol{E}\boldsymbol{S}\boldsymbol{C}^T)\boldsymbol{B}_t^{-1} \tag{A.88}$$

$$\boldsymbol{B}_t = \boldsymbol{H}\boldsymbol{P}_{t|t-1}\boldsymbol{H}^T + \boldsymbol{C}\boldsymbol{R}\boldsymbol{C}^T \tag{A.89}$$

$$P_{t+1|t} = \Phi P_{t|t-1} \Phi^T + EQE^T - K_t B_t K_t^T, \tag{A.90}$$

where the initial conditions for the filter are:

$$\hat{x}_{0|-1} = \mathrm{E}(x_0 | u_0, \ldots, u_N), \tag{A.91}$$

and

$$P_{0|-1} = \mathrm{E}[(x_0 - \hat{x}_{0|-1})(x_0 - \hat{x}_{0|-1})^T]. \tag{A.92}$$

Matrix K_t is the Kalman filter gain, and B_t corresponds to the prediction error variance of z_t obtained from past values of the input and output information.

Equations (A.86)–(A.92) correspond to the general case, where the input of the system is stochastic. The analytical expression for the moments (A.91) and (A.92) was derived in Chapter 5.

Appendix B

Asymptotic properties of maximum likelihood estimates

B.1	Preliminaries	211
B.2	Basic likelihood results for the State-Space model	214
B.2.1	The information matrix	214
B.2.2	Regularity conditions	216
B.2.3	Choice of estimation method	217
B.3	The State-Space model with cross-sectional extension	218

The aim of this Appendix is to provide a brief and accessible discussion of the asymptotic theory underlying the ML estimates for simplified versions of the SS model with time-invariant parameters. These include the MEM of Chapter 5, and its cross-sectionally extended counterpart in Chapter 11 without inputs. The hope is that this will provide the reader with enough background to understand something of what goes on "behind the scenes" as software packages produce statistical inference output in the form of standard errors and p-values.

B.1 Preliminaries

Assume that the observation equation sample is $\boldsymbol{z} = \text{vec}(\boldsymbol{z_1}, \boldsymbol{z_2}, ..., \boldsymbol{z_N})$, with N being the sample size and $\boldsymbol{z} \in \mathbb{R}^{mN}$ a vectorization of all the observed values. Denote by $\boldsymbol{\theta}_0 \in \mathbb{D}$ the *true* value of the d-dimensional parameter vector $\boldsymbol{\theta}$ in the model under consideration, which is simply a chosen vectorization of the entire set of model coefficients, Θ, and where $\mathbb{D} \subset \mathbb{R}^d$ is the parameter space. In addition, let $F(\boldsymbol{z};\boldsymbol{\theta})$ denote the cumulative distribution function of $\boldsymbol{z}$. In this context, $\boldsymbol{\theta}_0$ is said to be *identifiable* in the sense already defined in Chapter 9, i.e., if and only if

$$\boldsymbol{\theta}_1 \neq \boldsymbol{\theta}_0 \quad \Rightarrow \quad F(\boldsymbol{z};\boldsymbol{\theta}_1) \neq F(\boldsymbol{z};\boldsymbol{\theta}_0), \qquad \text{for at least one } \boldsymbol{z}. \tag{B.1}$$

One can then define two types of identifiability for $\boldsymbol{\theta}_0$ (or the model): *global identifiability* if (B.1) holds for all $\boldsymbol{\theta}_1 \in \mathbb{D}$, and *local identifiability* if (B.1) holds for all $\boldsymbol{\theta}_1$ only in a neighborhood of $\boldsymbol{\theta}_0$. For the asymptotic properties discussed here, it suffices to have local identifiability.

Throughout the discussion, we will let $\ell(\boldsymbol{\theta})$ denote the *negative of the log-likelihood* function. The ML estimate, which we denote generically by $\hat{\boldsymbol{\theta}}_N$, is then the minimizer of $\ell(\boldsymbol{\theta})$. We will assume that the reader of this more technical material is familiar with basic notions of large sample theory for estimators in the context of parametric models, namely *consistency* and *asymptotic normality*, a very accessible introduction to which can be found in, e.g., Davidson and MacKinnon [61].

The ML approach was first proposed by Fisher [96]. Cramer [58] proved its consistency in the case of IID observations, under the assumption that the likelihood function is differentiable at least three times. Many relevant estimation problems, including those treated in this book, do not correspond to this classical formulation because the observations are neither identically nor independently distributed. Extensions concerning the consistency of the ML estimator when the observations are dependent, were first proposed by Cramer [58] and Wald [218], among others.

Nowadays, a standard approach for establishing asymptotic normality (and hence consistency) uses a Taylor series expansion of the *score* equations,

$$\frac{\partial \ell(\boldsymbol{\theta})}{\partial \boldsymbol{\theta}} = \mathbf{0},$$

in conjunction with a central limit theorem on the (uncorrelated) elements of the score vector, and involves the associated *Hessian* matrix,

$$\mathcal{H}_N(\boldsymbol{\theta}) = -\frac{\partial^2 \ell(\boldsymbol{\theta})}{\partial \boldsymbol{\theta} \partial \boldsymbol{\theta}^T}.$$

From this, three versions of the (Fisher) *information* matrix can be defined:

1. the *expected* information matrix,

$$\boldsymbol{J}_N(\boldsymbol{\theta}) = -\mathrm{E}[\mathcal{H}_N(\boldsymbol{\theta})],$$

2. the *observed* (or *empirical*) information matrix,

$$\hat{\boldsymbol{J}}_N(\hat{\boldsymbol{\theta}}) = -\left.\mathcal{H}_N(\boldsymbol{\theta})\right|_{\boldsymbol{\theta}=\hat{\boldsymbol{\theta}}},$$

3. and the *asymptotic* information matrix,

$$\mathcal{I}(\boldsymbol{\theta}) = \operatorname*{plim}_{N\to\infty} \frac{1}{N} \boldsymbol{J}_N(\boldsymbol{\theta}),$$

where the "plim" in the last expression denotes "limit in probability."

For time series data, and under appropriate regularity conditions, the same (familiar) results are obtained as in the IID case.

Consistency:

$$\operatorname*{plim}_{N\to\infty} \hat{\boldsymbol{\theta}}_N = \boldsymbol{\theta}_0. \tag{B.2}$$

Asymptotic unbiasedness: as a consequence of consistency,

$$\lim_{N\to\infty} \mathrm{E}(\hat{\boldsymbol{\theta}}_N) = \boldsymbol{\theta}_0. \tag{B.3}$$

Asymptotic normality. As $N \to \infty$, a scaled and centered version of $\hat{\boldsymbol{\theta}}_N$ converges in distribution to a Gaussian random vector with mean $\mathbf{0}$ and covariance matrix given by the inverse of the asymptotic information matrix,

$$\sqrt{N}\left(\hat{\boldsymbol{\theta}}_N - \boldsymbol{\theta}_0\right) \xrightarrow{d} \mathrm{N}\left(\mathbf{0}, \mathcal{I}(\boldsymbol{\theta}_0)^{-1}\right). \tag{B.4}$$

Asymptotic efficiency: as a consequence of the above asymptotic normality result,

$$\lim_{N\to\infty} \mathrm{var}\left(\sqrt{N}(\hat{\boldsymbol{\theta}}_N - \boldsymbol{\theta}_0)\right) = \mathcal{I}(\boldsymbol{\theta}_0)^{-1}, \tag{B.5}$$

so that the variance of the ML estimator attains the *Cramer-Rao lower bound*.

These results rely on invoking the *information matrix equality*,

$$\mathcal{I}(\boldsymbol{\theta}) = -\operatorname*{plim}_{N\to\infty} \frac{1}{N}\mathcal{H}_N(\boldsymbol{\theta}),$$

which holds under correct model specification; with correspondingly more complex results otherwise; see Davidson and MacKinnon [62]. The primary usefulness of (B.4) lies in the fact that it can be exploited for the construction of a $(1-\alpha)100\%$ *confidence region* for $\boldsymbol{\theta}_0$ defined by the set:

$$CR(\boldsymbol{\theta}) = \left\{ \boldsymbol{\theta} \in \mathbb{R}^d : \left(\boldsymbol{\theta} - \hat{\boldsymbol{\theta}}_N\right)^T \mathcal{I}(\boldsymbol{\theta}_0)\left(\boldsymbol{\theta} - \hat{\boldsymbol{\theta}}_N\right) \leq \frac{1}{N}\chi^2_{1-\alpha}(d) \right\}, \tag{B.6}$$

where $\chi^2_{1-\alpha}(d)$ is the $1-\alpha$ quantile from a chi-square distribution with d degress of freedom. From a practical point of view, one replaces the (unknown) asymptotic information matrix in (B.6) with either the expected or observed information matrices normalized by N, that is,

$$\mathcal{I}(\boldsymbol{\theta}_0) \approx \frac{\boldsymbol{J}_N(\hat{\boldsymbol{\theta}}_N)}{N}, \qquad \text{or} \qquad \mathcal{I}(\boldsymbol{\theta}_0) \approx \frac{\hat{\boldsymbol{J}}_N(\hat{\boldsymbol{\theta}}_N)}{N}.$$

The latter is often the obvious choice in a software package implementation seeking automation in the calculation of standard errors, since it requires only numerical evaluation of the Hessian. However, the results can be very inaccurate and highly unstable if (as is often the case) nothing is assumed about the form of the derivatives; see Jamshidian and Jennrich ([129]). Improvements can therefore be obtained by understanding something about the structure of the Hessian.

Estimates of the Hessian are useful not only in inference, but also in diagnosing model identifiability. It can be shown that (except for possibly a set of Lebesgue measure zero) local identifiability and asymptotic nonsingularity of the Hessian are equivalent conditions; see Caines [37].

B.2 Basic likelihood results for the State-Space model

Consider the MEM with exogenous inputs and time-invariant transition matrices:

$$\boldsymbol{x}_{t+1} = \boldsymbol{\Phi}\boldsymbol{x}_t + \boldsymbol{\Gamma}\boldsymbol{u}_t + \boldsymbol{E}\boldsymbol{w}_t, \tag{B.7}$$

$$\boldsymbol{z}_t = \boldsymbol{H}\boldsymbol{x}_t + \boldsymbol{D}\boldsymbol{u}_t + \boldsymbol{C}\boldsymbol{v}_t, \tag{B.8}$$

observed for $t = 1, \ldots, N$, with corresponding error covariances

$$\mathrm{E}\left(\begin{bmatrix}\boldsymbol{w}_t \\ \boldsymbol{v}_t\end{bmatrix}\begin{bmatrix}\boldsymbol{w}_t^T & \boldsymbol{v}_t^T\end{bmatrix}\right) = \begin{bmatrix}\boldsymbol{Q} & \boldsymbol{S} \\ \boldsymbol{S}^T & \boldsymbol{R}\end{bmatrix}, \tag{B.9}$$

and initial condition $\boldsymbol{x}_1 \sim \mathrm{N}(\boldsymbol{\mu}, \boldsymbol{P})$. As seen in Chapter 5, the prediction error decomposition (also called innovations) provides a very concise expression for the log-likelihood. Dropping constants and making explicit the dependence on $\boldsymbol{\theta}$, we have the negative log-likelihood expression

$$\begin{aligned}\ell(\boldsymbol{\theta}) &:= -\log L(\boldsymbol{\theta}) = \frac{1}{2}\sum_{t=1}^{N}\left\{\log|\boldsymbol{B}_t(\boldsymbol{\theta})| + \tilde{\boldsymbol{z}}_t(\boldsymbol{\theta})^T \boldsymbol{B}_t(\boldsymbol{\theta})^{-1}\tilde{\boldsymbol{z}}_t(\boldsymbol{\theta})\right\} \\ &:= \frac{1}{2}\sum_{t=1}^{N} \nu_t(\boldsymbol{\theta}),\end{aligned} \tag{B.10}$$

where the innovations, $\tilde{\boldsymbol{z}}_t(\boldsymbol{\theta}) = \boldsymbol{z}_t - \boldsymbol{H}_t\boldsymbol{x}_{t|t-1} - \boldsymbol{D}\boldsymbol{u}_t$, are as defined in Chapter 5, and we have $\tilde{\boldsymbol{z}}_t(\boldsymbol{\theta}) \sim \mathrm{NID}(\boldsymbol{0}, \boldsymbol{B}_t(\boldsymbol{\theta}))$ under the Gaussian assumption. The ML estimate $\hat{\boldsymbol{\theta}}_N$ is therefore the minimizer of (B.10).

In the context of linear SS models, Pagan [173] obtained consistency and asymptotic normality for $\hat{\boldsymbol{\theta}}_N$ by applying the conditions developed by Crowder [59] to a univariate model without inputs; a result which generalizes immediately to the multivariate case. The conditions required to assure properties (B.2)–(B.5) are: (a) model identifiability, (b) some regularity conditions on $F(\boldsymbol{z};\boldsymbol{\theta})$, and (c) some conditions on the convergence properties of the Hessian evaluated in a neighborhood of $\boldsymbol{\theta}_0$.

B.2.1 The information matrix

Calculation of the Hessian for $\ell(\boldsymbol{\theta})$ relies on evaluating the matrix of second partial derivatives. Caines [37] provides this in element-wise form, while Lutkepohl [158] opts for a matrix representation, but both omit some details which we present in the following lemma.

Lemma 1 *For the expression $\nu_t(\boldsymbol{\theta})$ as defined in (B.10), and omitting explicit dependence on t and $\boldsymbol{\theta}$ for notational expedience, the $(d \times d)$ Hessian matrix with respect to the vector $\boldsymbol{\theta}$ is:*

(i) given in matrix form as

$$\frac{\partial^2 v(\boldsymbol{\theta})}{\partial\boldsymbol{\theta}\partial\boldsymbol{\theta}^T} = 2\boldsymbol{J}_{\tilde{\boldsymbol{z}}}^T\boldsymbol{B}^{-1}\boldsymbol{J}_{\tilde{\boldsymbol{z}}} - \boldsymbol{J}_{\boldsymbol{B}}^T\left(\boldsymbol{B}^{-1}\otimes\boldsymbol{B}^{-1}\right)\boldsymbol{J}_{\boldsymbol{B}} \tag{B.11}$$

$$+\boldsymbol{J}_{\boldsymbol{B}}^T\left(\boldsymbol{B}^{-1}\tilde{\boldsymbol{z}}\tilde{\boldsymbol{z}}^T\boldsymbol{B}^{-1}\otimes\boldsymbol{B}^{-1} + \boldsymbol{B}^{-1}\otimes\boldsymbol{B}^{-1}\tilde{\boldsymbol{z}}\tilde{\boldsymbol{z}}^T\boldsymbol{B}^{-1}\right)\boldsymbol{J}_{\boldsymbol{B}} \tag{B.12}$$

$$-2\boldsymbol{J}_{\boldsymbol{B}}^T\left(\boldsymbol{B}^{-1}\boldsymbol{J}_{\tilde{\boldsymbol{z}}}\otimes\boldsymbol{B}^{-1}\tilde{\boldsymbol{z}}\right) + 2\boldsymbol{H}_{\tilde{\boldsymbol{z}}}^T\left(\boldsymbol{I}_d\otimes\boldsymbol{B}^{-1}\tilde{\boldsymbol{z}}\right) \tag{B.13}$$

$$-\boldsymbol{J}_{\tilde{\boldsymbol{z}}}^T\left(\boldsymbol{I}_m\otimes\tilde{\boldsymbol{z}}^T + \tilde{\boldsymbol{z}}^T\otimes\boldsymbol{I}_m\right)\left(\boldsymbol{B}^{-1}\otimes\boldsymbol{B}^{-1}\right)\boldsymbol{J}_{\boldsymbol{B}} \tag{B.14}$$

$$+\boldsymbol{H}_{\boldsymbol{B}}^T\left\{\boldsymbol{I}_d\otimes vec\left[\boldsymbol{B}^{-1}\left(\boldsymbol{I}_d - \tilde{\boldsymbol{z}}\tilde{\boldsymbol{z}}^T\boldsymbol{B}^{-1}\right)\right]\right\} \tag{B.15}$$

(ii) whose (p,q) element is given by

$$\begin{aligned}\frac{\partial^2 v(\boldsymbol{\theta})}{\partial\theta_p\partial\theta_q} =& 2\tilde{\boldsymbol{z}}^T\boldsymbol{B}^{-1}\frac{\partial^2\tilde{\boldsymbol{z}}}{\partial\theta_p\partial\theta_q} - 2\tilde{\boldsymbol{z}}^T\boldsymbol{B}^{-1}\frac{\partial\boldsymbol{B}}{\partial\theta_q}\boldsymbol{B}^{-1}\frac{\partial\tilde{\boldsymbol{z}}}{\partial\theta_p} + 2\frac{\partial\tilde{\boldsymbol{z}}^T}{\partial\theta_p}\boldsymbol{B}^{-1}\frac{\partial\tilde{\boldsymbol{z}}}{\partial\theta_q}\\ &-2\tilde{\boldsymbol{z}}^T\boldsymbol{B}^{-1}\frac{\partial\boldsymbol{B}}{\partial\theta_p}\boldsymbol{B}^{-1}\frac{\partial\tilde{\boldsymbol{z}}}{\partial\theta_q} + 2\tilde{\boldsymbol{z}}^T\boldsymbol{B}^{-1}\frac{\partial\boldsymbol{B}}{\partial\theta_q}\boldsymbol{B}^{-1}\frac{\partial\boldsymbol{B}}{\partial\theta_p}\boldsymbol{B}^{-1}\tilde{\boldsymbol{z}}\\ &-\tilde{\boldsymbol{z}}^T\boldsymbol{B}^{-1}\frac{\partial^2\boldsymbol{B}}{\partial\theta_p\partial\theta_q}\boldsymbol{B}^{-1}\tilde{\boldsymbol{z}} + \mathrm{tr}\left(\boldsymbol{B}^{-1}\frac{\partial^2\boldsymbol{B}}{\partial\theta_p\partial\theta_q} - \boldsymbol{B}^{-1}\frac{\partial\boldsymbol{B}}{\partial\theta_p}\boldsymbol{B}^{-1}\frac{\partial\boldsymbol{B}}{\partial\theta_q}\right).\end{aligned}$$

Both expressions depend on Jacobians and Hessians of $\tilde{\boldsymbol{z}}$ and $\boldsymbol{B}$. While this is explicit in (ii), the notation employed in (i) is as follows:

- *$m\times d$ Jacobian matrix of $\tilde{\boldsymbol{z}}$, and $m^2\times d$ Jacobian matrix of $\boldsymbol{B}$,*

$$\boldsymbol{J}_{\tilde{\boldsymbol{z}}} = \frac{\partial\tilde{\boldsymbol{z}}}{\partial\boldsymbol{\theta}^T}, \quad \text{and} \quad \boldsymbol{J}_{\boldsymbol{B}} = \frac{\partial vec(\boldsymbol{B})}{\partial\boldsymbol{\theta}^T}$$

- *$md\times d$ Hessian matrix of $\tilde{\boldsymbol{z}}$, and $m^2d\times d$ Hessian matrix of $\boldsymbol{B}$,*

$$\boldsymbol{H}_{\tilde{\boldsymbol{z}}} = \frac{\partial vec(\boldsymbol{J}_{\tilde{\boldsymbol{z}}})}{\partial\boldsymbol{\theta}^T}, \quad \text{and} \quad \boldsymbol{H}_{\boldsymbol{B}} = \frac{\partial vec(\boldsymbol{J}_{\boldsymbol{B}})}{\partial\boldsymbol{\theta}^T}.$$

Proof for Lemma 1 These results are tedious to derive, but follow by straightforward application of elementary matrix calculus rules, which can be found in, e.g., Lutkepohl [157] and Harville [121]. ■

Now, using the fact that $\mathrm{E}(\tilde{\boldsymbol{z}}_t) = \boldsymbol{0}$ and $\mathrm{E}(\tilde{\boldsymbol{z}}_t\tilde{\boldsymbol{z}}_t^T) = \boldsymbol{B}_t$, we obtain the expected information matrix,

$$\boldsymbol{J}_N(\boldsymbol{\theta}) = \frac{1}{2}\sum_{t=1}^{N}\mathrm{E}\left[\frac{\partial^2 v_t(\boldsymbol{\theta})}{\partial\boldsymbol{\theta}\partial\boldsymbol{\theta}^T}\right] \tag{B.16}$$

$$= \frac{1}{2}\sum_{t=1}^{N}\left[\frac{\partial\mathrm{vec}(\boldsymbol{B}_t)^T}{\partial\boldsymbol{\theta}}\left(\boldsymbol{B}_t^{-1}\otimes\boldsymbol{B}_t^{-1}\right)\frac{\partial\mathrm{vec}(\boldsymbol{B}_t)}{\partial\boldsymbol{\theta}^T} + 2\mathrm{E}\left(\frac{\partial\tilde{\boldsymbol{z}}_t^T}{\partial\boldsymbol{\theta}}\boldsymbol{B}_t^{-1}\frac{\partial\tilde{\boldsymbol{z}}_t}{\partial\boldsymbol{\theta}^T}\right)\right] \tag{B.17}$$

$$:= \frac{1}{2}\boldsymbol{\Sigma}_N(\boldsymbol{\theta}). \tag{B.18}$$

This follows straightforwardly by taking the expectation of terms (B.11) and (B.12), and noting that the expectation of (B.15) is zero. Each of the remaining terms contains both $\tilde{z}_t$ and its Jacobian $\boldsymbol{J}_{\tilde{z}}$; consequently, the fact that these two terms are statistically independent implies that the expectation of terms (B.13) and (B.14) is also null. Using similar arguments, the (p,q) element of the expected information matrix is:

$$J_N(\boldsymbol{\theta})_{p,q} = \frac{1}{2}\sum_{t=1}^{N}\left[\operatorname{tr}\left(\boldsymbol{B}_t^{-1}\frac{\partial \boldsymbol{B}_t}{\partial \theta_p}\boldsymbol{B}_t^{-1}\frac{\partial \boldsymbol{B}_t}{\partial \theta_q}\right) + 2\mathrm{E}\left(\frac{\partial \tilde{z}_t^T}{\partial \theta_p}\boldsymbol{B}_t^{-1}\frac{\partial \tilde{z}_t}{\partial \theta_q}\right)\right]. \quad \text{(B.19)}$$

B.2.2 Regularity conditions

Under appropriate regularity conditions, (B.2)–(B.5) will hold, and in particular,

$$\sqrt{N}\,(\hat{\boldsymbol{\theta}}_N - \boldsymbol{\theta}_0) \xrightarrow{d} \mathrm{N}\left(\mathbf{0}, \left[\frac{\boldsymbol{\Sigma}_N(\boldsymbol{\theta}_0)}{2N}\right]^{-1}\right), \quad \text{(B.20)}$$

with $\boldsymbol{\Sigma}_N(\boldsymbol{\theta})$ defined by (B.18). To give some idea of the nature of the assumptions involved we adapt the conditions given by Lutkepohl [158] to the situation of SS model (B.7)–(B.9).

(i) The model is *stable* in the sense of the Definition 2.1, given in Chapter 2.

(ii) The state and observational white noise errors are Gaussian, $\boldsymbol{w}_t \sim N(\mathbf{0},\boldsymbol{Q})$, $\boldsymbol{v}_t \sim N(\mathbf{0},\boldsymbol{R})$, with $\operatorname{cov}(\boldsymbol{w}_t,\boldsymbol{v}_t) = \boldsymbol{S} \equiv \mathbf{0}$.

(iii) The initial state vector is Gaussian, $\boldsymbol{x}_1 \sim N(\boldsymbol{\mu},\boldsymbol{P})$, and is independent of $\{\boldsymbol{w}_t, \boldsymbol{v}_{t+1}\}$, for $t \geq 1$.

(iv) $\boldsymbol{H}, \boldsymbol{\Gamma}$, and $\boldsymbol{D}$ are known.

(v) The parameter set Θ is therefore comprised of $\{\boldsymbol{\Phi},\boldsymbol{E},\boldsymbol{C},\boldsymbol{Q},\boldsymbol{R},\boldsymbol{\mu},\boldsymbol{P}\}$, with $\boldsymbol{\theta}_0$ belonging to the interior of Θ, assumed to be compact.

(vi) The inputs $\{\boldsymbol{u}_t\}$ are non-stochastic and uniformly bounded, meaning that there exist real numbers c_1 and c_2 such that $c_1 \leq \boldsymbol{u}_t^T\boldsymbol{u}_t \leq c_2$, for all $t \geq 0$.

(vii) There exists a constant c_0 such that $N^{-1}\boldsymbol{J}_N(\boldsymbol{\theta}_0) - c_0\boldsymbol{I}_N$ is positive definite in the limit as $N \to \infty$, where $\boldsymbol{I}_N$ denotes the identity matrix of dimension N.

This is obviously a fairly restricted situation, particularly (iv), but it is important to realize that these are a set of *sufficient* (not *necessary*) conditions; they do not exclude other less restrictive sets of conditions. Condition (ii), independence of the errors in state and observation equations, seems quite restrictive, but it really is not when one realizes that any model formulation with $\boldsymbol{S} \neq \mathbf{0}$ can be transformed into an observationally equivalent formulation with $\boldsymbol{S} = \mathbf{0}$; see Anderson and Moore [6, ch. 5]. Condition (vi) excludes the possibility of having time trends; while (vii) insists that the sequence of normalized information matrices be bounded from below by a positive definite matrix.

Generalizing the above results to a model where the input matrices $\boldsymbol{\Gamma}$ and $\boldsymbol{D}$ are unknown, is analytically complex and requires further assumptions on the properties of the inputs. In particular, Caines [37] discussed the asymptotic properties of $\hat{\boldsymbol{\theta}}_N$ under the assumptions of deterministic and stochastic inputs. Under general conditions, he proves consistency and asymptotic normality for these estimates. An analogous analysis was done by Hannan and Deistler [116, ch. 4], in a slightly more general framework, as they avoided the hypothesis of compactness.

A bootstrap procedure can be implemented in situations where: (a) the asymptotics may not hold, (b) specific (non-Gaussian) distributions for the noise terms are entertained, or (c) small sample sizes deliver poor asymptotic performance. An instance of the latter is detailed by Stoffer and Wall [201].

B.2.3 *Choice of estimation method*

A natural generalization of ML is to use the ML criterion as the objective function to maximize, but not assume a distributional structure for the model noise terms. This is what Caines [37, ch. 8] calls the "minimum prediction error" identification method, where the term is broadly applied in reference to a particular *objective function* (computing terminology) or *loss function* (statistical terminology), $\ell(\boldsymbol{\theta})$, that one chooses to optimize (minimize) in order to produce parameter estimates. In this regard, the innovations (prediction errors) $\tilde{\boldsymbol{z}}_t(\boldsymbol{\theta}) \sim (\boldsymbol{0}, \boldsymbol{B}_t(\boldsymbol{\theta}))$ are a natural starting point, since they measure the discrepancies between the realized values of the process and its optimal (mean squared error sense) one-step ahead predictions.

"Good" loss functions might then seek to minimize the sum of these discrepancies, leading to the ubiquitous *least squares* criterion,

$$\ell_1(\boldsymbol{\theta}) = \sum_{t=1}^{N} \tilde{\boldsymbol{z}}_t(\boldsymbol{\theta})^T \tilde{\boldsymbol{z}}_t(\boldsymbol{\theta}).$$

Since the innovations have different variances in general, it may be more appealing to minimize the *generalized least squares*, by standardizing the innovations to have unit variance,

$$\ell_2(\boldsymbol{\theta}) = \sum_{t=1}^{N} \tilde{\boldsymbol{z}}_t(\boldsymbol{\theta})^T \boldsymbol{B}_t(\boldsymbol{\theta})^{-1} \tilde{\boldsymbol{z}}_t(\boldsymbol{\theta}).$$

Employing the ML principle with $\tilde{\boldsymbol{z}}_t$ Gaussian, adds additionally a log determinant term in ℓ_2, so that one is lead to the objective function,

$$\ell_3(\boldsymbol{\theta}) = \sum_{t=1}^{N} \left\{ \log |\boldsymbol{B}_t(\boldsymbol{\theta})| + \tilde{\boldsymbol{z}}_t(\boldsymbol{\theta})^T \boldsymbol{B}_t(\boldsymbol{\theta})^{-1} \tilde{\boldsymbol{z}}_t(\boldsymbol{\theta}) \right\},$$

given in (B.10). However, note that any one of $\{\ell_1, \ell_2, \ell_3\}$ is a sensible loss function, leading to different estimators with (possibly) distinct asymptotic properties. In a quest for "robustness," one then tries to establish the relevant asymptotics under more lenient assumptions that do **not** include specifying a distributional structure for $\tilde{\boldsymbol{z}}_t$ (e.g., Gaussian), thereby freeing one from making unrealistic assumptions[1].

[1] The resulting estimates under this paradigm where one does not fully specify the likelihood, are sometimes known as *pseudo*-ML or *quasi*-ML estimates.

Several such results for ARMA models are well-known in the literature; e.g., Davis and Dunsmuir [64] established consistency and asymptotic normality under minimization of a least absolute deviations criterion, Brockwell and Davis [32, ch. 8] do so for the ML criterion ℓ_3 above, while Barnard, Trindade, and Wickramasinghe [14] shed light on the robustness of ℓ_3 with regard to the exponential power family of distributions.

For SS models, under this paradigm (and appropriate regularity conditions) it is often still possible to recover asymptotic properties (B.2)–(B.4). The situation when ℓ_1 and ℓ_2 are used as loss functions is discussed in Caines [37, ch. 6], while that for ℓ_3 is detailed in Caines [37, ch. 7] (an overview of which we have strived to give in this section); see also Ljung [152].

In essence, the quest for robustness involves a trade-off in relinquishing a particular distributional assumption (e.g., normality) at the expense of asymptotic efficiency, as the variance of the resulting estimator is no longer guaranteed to attain the Cramer-Rao lower bound. For example, the asymptotic variance of the estimator under ℓ_3 loss and Gaussian errors is known to be the smallest it can possibly be (in the class of all asymptotically normal estimators), if the errors are *truly* Gaussian. However, if the errors do not follow a Gaussian distribution, then the asymptotic variance of the estimator under ℓ_3 loss may well be larger.

B.3 The State-Space model with cross-sectional extension

Unlike Chapter 5 where there is only one asymptotic regime as the longitudinal sample size $N = n_1 \to \infty$, here we may also take into account the limiting behavior of the ML estimators as the cross-sectional sample size $k \to \infty$. (The most general situation where both the longitudinal and cross-sectional sample sizes grow will not be discussed here, as it does not appear to have yet been fully investigated.) We will also simplify the panel structure in order to have a "balanced" layout: $n_i = n$ for each $i = 1, \ldots, k$. Under this paradigm we then have that the effective sample size, denoted generically by N, will be $N = n$ under the increasing longitudinal sample size situation, and $N = k$ under the increasing cross-sectional sample size regime. As we shall see shortly, and for the simplified model (B.21)–(B.22) below, the same asymptotic behavior is obtained in either regime.

To avoid overly technical regularity conditions we will restrict attention to the SS model without exogenous inputs and constant transition matrices:

$$\boldsymbol{x}_{i,t+1} = \boldsymbol{\Phi}\boldsymbol{x}_{i,t} + \boldsymbol{w}_{i,t+1}, \tag{B.21}$$

$$\boldsymbol{z}_{i,t} = \boldsymbol{H}\boldsymbol{x}_{i,t} + \boldsymbol{v}_{i,t}, \tag{B.22}$$

for $i = 1, \ldots, k$ and $t = 1, \ldots, n$, with corresponding error covariances:

$$\mathrm{E}\left(\begin{bmatrix} \boldsymbol{w}_{i,t+1} \\ \boldsymbol{v}_{i,t} \end{bmatrix} \begin{bmatrix} \boldsymbol{w}_{i,t+1}^T & \boldsymbol{v}_{i,t}^T \end{bmatrix}\right) = \begin{bmatrix} \boldsymbol{Q} & \boldsymbol{S} \\ \boldsymbol{S}^T & \boldsymbol{R} \end{bmatrix},$$

and initial condition $\boldsymbol{x}_{i,0} \sim \mathrm{N}(\boldsymbol{\mu}, \boldsymbol{P})$. The estimator for the set of parameters $\Theta = \{\boldsymbol{\mu}, \boldsymbol{P}, \boldsymbol{\Phi}, \boldsymbol{Q}, \boldsymbol{R}, \boldsymbol{S}\}$ we will be concerned with is the ML estimator $\hat{\theta}_N$ discussed in

Chapter 11, obtained under the usual Gaussian optimality criterion. If $\boldsymbol{\theta} \in \Theta$ is a vectorization of the parameters (d-dimensional), the negative log-likelihood expression from (11.32) is given by

$$\begin{aligned} \ell(\boldsymbol{\theta}) := -\log L(\boldsymbol{\theta}) &= -\frac{1}{2}\sum_{i=1}^{k}\sum_{t=1}^{n}\{\log|\boldsymbol{B}_{i,t}(\boldsymbol{\theta})^{-1}| - \tilde{z}_{i,t}(\boldsymbol{\theta})^T \boldsymbol{B}_{i,t}(\boldsymbol{\theta})^{-1}\tilde{z}_{i,t}(\boldsymbol{\theta})\} \\ &= \frac{1}{2}\sum_{i=1}^{k}\sum_{t=1}^{n}\{\log|\boldsymbol{B}_t(\boldsymbol{\theta})| + \tilde{z}_{i,t}(\boldsymbol{\theta})^T \boldsymbol{B}_t(\boldsymbol{\theta})^{-1}\tilde{z}_{i,t}(\boldsymbol{\theta})\} \qquad \text{(B.23)} \\ &:= \frac{1}{2}\sum_{i=1}^{k}\sum_{t=1}^{n} v_{i,t}(\boldsymbol{\theta}), \qquad \text{(B.24)} \end{aligned}$$

where the innovations are as defined in (11.33). Note that the structure of the model (B.21)–(B.22) implies the simpler structure for the distribution of the innovations, $\tilde{z}_{i,t}(\boldsymbol{\theta}) \sim \text{NID}(\mathbf{0}, \boldsymbol{B}_t(\boldsymbol{\theta}))$, so that we have made the substitution $\boldsymbol{B}_{i,t}(\boldsymbol{\theta}) \equiv \boldsymbol{B}_t(\boldsymbol{\theta})$ in (B.23).

Calculation of the expected Hessian for $\ell(\boldsymbol{\theta})$ now follows, similar to the arguments given in Section B.2:

$$\boldsymbol{J}_N(\boldsymbol{\theta}) = \frac{1}{2}\text{E}\left[\sum_{i=1}^{k}\sum_{t=1}^{n}\frac{\partial^2 v_{i,t}(\boldsymbol{\theta})}{\partial\boldsymbol{\theta}\partial\boldsymbol{\theta}^T}\right] = \frac{k}{2}\sum_{t=1}^{n}\text{E}\left[\frac{\partial^2 v_t(\boldsymbol{\theta})}{\partial\boldsymbol{\theta}\partial\boldsymbol{\theta}^T}\right],$$

and by using the fact that the moments of the innovations vectors are the same over the index i. Now note that the summand in the above is identical to that of (B.16), and thus we obtain a correspondingly similar form,

$$\begin{aligned} \boldsymbol{J}_N(\boldsymbol{\theta}) &= \frac{k}{2}\sum_{t=1}^{n}\left[\frac{\partial \text{vec}(\boldsymbol{B}_t)^T}{\partial\boldsymbol{\theta}}\left(\boldsymbol{B}_t^{-1}\otimes\boldsymbol{B}_t^{-1}\right)\frac{\partial\text{vec}(\boldsymbol{B}_t)}{\partial\boldsymbol{\theta}^T} + 2\text{E}\left(\frac{\partial\tilde{z}_t^T}{\partial\boldsymbol{\theta}}\boldsymbol{B}_t^{-1}\frac{\partial\tilde{z}_t}{\partial\boldsymbol{\theta}^T}\right)\right] \\ &= \frac{k}{2}\boldsymbol{\Sigma}_n(\boldsymbol{\theta}). \end{aligned} \qquad \text{(B.25)}$$

Similar regularity conditions to those given in Section B.2 can now be imposed to guarantee that the asymptotic results (B.2)–(B.5) will hold. Among these, Caines [37, ch. 9] cites observability, controllability (and thus minimality), and identifiability as key. Gaussian innovations can again be used as a proxy for more stringent requirements relating to minimum prediction error estimation. Under additionally nonsingularity of the expected Hessian (B.25), we obtain in particular,

1. for the increasing longitudinal sample size $N = n \to \infty$ situation:

$$\sqrt{n}\,(\hat{\boldsymbol{\theta}}_N - \boldsymbol{\theta}_0) \xrightarrow{d} \text{N}\left(\mathbf{0}, \left[\frac{k\boldsymbol{\Sigma}_n(\boldsymbol{\theta}_0)}{2n}\right]^{-1}\right), \qquad \text{(B.26)}$$

2. and for the increasing cross-sectional sample size $N = k \to \infty$ situation:

$$\sqrt{k}\,(\hat{\boldsymbol{\theta}}_N - \boldsymbol{\theta}_0) \xrightarrow{d} \text{N}\left(\mathbf{0}, \left[\frac{\boldsymbol{\Sigma}_n(\boldsymbol{\theta}_0)}{2}\right]^{-1}\right). \qquad \text{(B.27)}$$

Note therefore that, under either regime, we obtain the same approximate large sample behavior for $\hat{\theta}_N$:

$$\hat{\theta}_N \approx \mathrm{N}\left(\theta_0, \frac{2\Sigma_n(\theta_0)^{-1}}{k}\right).$$

An equivalent expression for the Hessian that emphasizes dependence on second order moments, is to vectorize the data longitudinally by concatenating the series cross-sectionally as $z_i = (z_{i,1}, \ldots, z_{i,n})^T$, $i = 1, \ldots, k$. If $z_i \sim \mathrm{N}(\mu_z, M_z)$ for every $i = 1, \ldots, k$, and we take the usual ML estimates for μ_z and M_z[2] as

$$\bar{z} = \frac{1}{k}\sum_{i=1}^{k} z_i, \qquad \text{and} \qquad \widehat{M_z} = \frac{1}{k}\sum_{i=1}^{k}(z_i - \bar{z})(z_i - \bar{z})^T,$$

we can write a scaled version of the negative log-likelihood as

$$\frac{2\ell(\theta)}{k} = \log|M_z| + \frac{1}{k}\sum_{i=1}^{k}(z_i - \mu_z)^T M_z^{-1}(z_i - \mu_z).$$

By "completing the square" on the quadratic form above we obtain,

$$\begin{aligned}
&\frac{1}{k}\sum_{i=1}^{k}(z_i - \mu_z)^T M_z^{-1}(z_i - \mu_z) \\
&= \frac{1}{k}\sum_{i=1}^{k}[(z_i - \bar{z}) + (\bar{z} - \mu_z)]^T M_z^{-1}[(z_i - \bar{z}) + (\bar{z} - \mu_z)] \\
&= \frac{1}{k}\sum_{i=1}^{k}(z_i - \bar{z})^T M_z^{-1}(z_i - \bar{z}) + (\bar{z} - \mu_z)^T M_z^{-1}(\bar{z} - \mu_z),
\end{aligned}$$

since the cross terms cancel. Finally, noting that each term in the above summand is a scalar (and therefore trivially equal to its trace), we have from basic properties of the trace operator that,

$$\begin{aligned}
\frac{1}{k}\sum_{i=1}^{k}\mathrm{tr}\left[(z_i - \bar{z})^T M_z^{-1}(z_i - \bar{z})\right] &= \frac{1}{k}\sum_{i=1}^{k}\mathrm{tr}\left[M_z^{-1}(z_i - \bar{z})(z_i - \bar{z})^T\right] \\
&= \mathrm{tr}\left[M_z^{-1}\frac{1}{k}\sum_{i=1}^{k}(z_i - \bar{z})(z_i - \bar{z})^T\right] \\
&= \mathrm{tr}\left(M_z^{-1}\widehat{M_z}\right),
\end{aligned}$$

whence we obtain

$$\frac{2\ell(\theta)}{k} = \log|M_z| + \mathrm{tr}\left(M_z^{-1}\widehat{M_z}\right) + (\bar{z} - \mu_z)^T M_z^{-1}(\bar{z} - \mu_z) := \eta(\theta). \quad \text{(B.28)}$$

[2]Noting the dependence on the parameters one should properly write $\mu_z(\theta)$ and $M_z(\theta)$, but this detail will be sacrificed in favor of notational expedience.

Calculation of the Hessian for $\ell(\boldsymbol{\theta})$ now relies on derivatives for $\eta(\boldsymbol{\theta})$, given in Lemma 2 below. Under the increasing cross-sectional sample size regime, consistency of the sample mean and covariance estimates implies that

$$\operatorname*{plim}_{k\to\infty} \bar{\boldsymbol{z}} = \boldsymbol{\mu}_{\boldsymbol{z}}, \quad \text{and} \quad \operatorname*{plim}_{k\to\infty} \widehat{\boldsymbol{M}_{\boldsymbol{z}}} = \boldsymbol{M}_{\boldsymbol{z}}.$$

Taking the limits in k of the Hessian results in substantial simplification, and thus this is the preferred route in obtaining a simple expression for the asymptotic information matrix, i.e.

$$\mathcal{I}(\boldsymbol{\theta}) = \operatorname*{plim}_{k\to\infty} \frac{1}{2}\frac{\partial^2 \eta(\boldsymbol{\theta})}{\partial\boldsymbol{\theta}\partial\boldsymbol{\theta}^T}.$$

Lemma 2 *For the expression $\eta(\boldsymbol{\theta})$ as defined in (B.28), the (p,q) element of the limiting Hessian matrix is given by*

$$\operatorname*{plim}_{k\to\infty} \frac{\partial^2 \eta(\boldsymbol{\theta})}{\partial\theta_p\partial\theta_q} = 2\frac{\partial\boldsymbol{\mu}_{\boldsymbol{z}}^T}{\partial\theta_p}\boldsymbol{M}_{\boldsymbol{z}}^{-1}\frac{\partial\boldsymbol{\mu}_{\boldsymbol{z}}}{\partial\theta_q} + \operatorname{tr}\left(\boldsymbol{M}_{\boldsymbol{z}}^{-1}\frac{\partial\boldsymbol{M}_{\boldsymbol{z}}}{\partial\theta_q}\boldsymbol{M}_{\boldsymbol{z}}^{-1}\frac{\partial\boldsymbol{M}_{\boldsymbol{z}}}{\partial\theta_p}\right).$$

Proof for Lemma 2

The result follows by routine matrix calculus similar to those used in the proof of Lemma 1. To simplify notation, we drop the $\boldsymbol{z}$ subscripts on $\boldsymbol{\mu}_{\boldsymbol{z}}$ and $\boldsymbol{M}_{\boldsymbol{z}}$, but introduce p and q subscripts to denote partial derivatives with respect to θ_p and θ_q. Thus, for example, $\boldsymbol{\mu}_p = \partial\boldsymbol{\mu}/\partial\theta_p$, $\boldsymbol{\mu}_{p,q} = \partial^2\boldsymbol{\mu}/(\partial\theta_p\partial\theta_q)$, etc., and similarly for $\boldsymbol{M}$. With this, we obtain the following second derivatives term by term:

$$\begin{aligned}
\frac{\partial^2 \eta(\boldsymbol{\theta})}{\partial\theta_p\partial\theta_q} &= \operatorname{tr}\left[\left(\boldsymbol{M}^{-1} - \boldsymbol{M}^{-1}\widehat{\boldsymbol{M}}\boldsymbol{M}^{-1}\right)\boldsymbol{M}_{p,q} - \boldsymbol{M}^{-1}\boldsymbol{M}_p\boldsymbol{M}^{-1}\boldsymbol{M}_q\right] \\
&\quad + \operatorname{tr}\left(\boldsymbol{M}^{-1}\boldsymbol{M}_q\boldsymbol{M}^{-1}\widehat{\boldsymbol{M}}\boldsymbol{M}^{-1}\boldsymbol{M}_p + \boldsymbol{M}^{-1}\widehat{\boldsymbol{M}}\boldsymbol{M}^{-1}\boldsymbol{M}_q\boldsymbol{M}^{-1}\boldsymbol{M}_p\right) \\
&\quad + 2\boldsymbol{\mu}_p^T\boldsymbol{M}^{-1}\boldsymbol{\mu}_q - \boldsymbol{\mu}^T\boldsymbol{M}^{-1}\boldsymbol{M}_{p,q}\boldsymbol{M}^{-1}\boldsymbol{\mu} - \bar{\boldsymbol{z}}^T\boldsymbol{M}^{-1}\boldsymbol{M}_{p,q}\boldsymbol{M}^{-1}\bar{\boldsymbol{z}} \\
&\quad + 2(\boldsymbol{\mu}^T + \bar{\boldsymbol{z}}^T)\boldsymbol{g}_1 + 2\bar{\boldsymbol{z}}^T\boldsymbol{g}_2,
\end{aligned}$$

where the vectors $\boldsymbol{g}_1$ and $\boldsymbol{g}_2$ are given by:

$$\begin{aligned}
\boldsymbol{g}_1 = \boldsymbol{M}^{-1}\boldsymbol{\mu}_{p,q} + \boldsymbol{M}^{-1}\boldsymbol{M}_q\boldsymbol{M}^{-1}\boldsymbol{\mu}_p + \boldsymbol{M}^{-1}\boldsymbol{M}_p\boldsymbol{M}^{-1}\boldsymbol{\mu}_q \\
- \boldsymbol{M}^{-1}\boldsymbol{M}_q\boldsymbol{M}^{-1}\boldsymbol{M}_p\boldsymbol{M}^{-1}\boldsymbol{\mu},
\end{aligned}$$

and

$$\begin{aligned}
\boldsymbol{g}_2 = \boldsymbol{M}^{-1}\boldsymbol{M}_{p,q}\boldsymbol{M}^{-1}\boldsymbol{\mu} + \boldsymbol{M}^{-1}\boldsymbol{M}_q\boldsymbol{M}^{-1}\boldsymbol{M}_p\boldsymbol{M}^{-1}\bar{\boldsymbol{z}} \\
- \boldsymbol{M}^{-1}\boldsymbol{M}_p\boldsymbol{M}^{-1}\boldsymbol{M}_q\boldsymbol{M}^{-1}\boldsymbol{\mu}.
\end{aligned}$$

Taking limits as $k \to \infty$ and noting the consistency of the sample mean and covariance, leads to much cancellation of terms and the simple expression stated in the lemma. ∎

With this setup we then obtain, under the $k \to \infty$ regime,

$$\sqrt{k}\left(\hat{\boldsymbol{\theta}}_N - \boldsymbol{\theta}_0\right) \xrightarrow{d} \mathrm{N}\left(\mathbf{0}, \mathcal{I}(\boldsymbol{\theta})^{-1}\right),$$

where the (p,q) element of the asymptotic information matrix is

$$\mathcal{I}(\boldsymbol{\theta})_{p,q} = \frac{\partial \boldsymbol{\mu}_{\boldsymbol{z}}(\boldsymbol{\theta})^T}{\partial \theta_p} \boldsymbol{M}_{\boldsymbol{z}}(\boldsymbol{\theta})^{-1} \frac{\partial \boldsymbol{\mu}_{\boldsymbol{z}}(\boldsymbol{\theta})}{\partial \theta_q} + \frac{1}{2} \operatorname{tr}\left[\boldsymbol{M}_{\boldsymbol{z}}(\boldsymbol{\theta})^{-1} \frac{\partial \boldsymbol{M}_{\boldsymbol{z}}(\boldsymbol{\theta})}{\partial \theta_q} \boldsymbol{M}_{\boldsymbol{z}}(\boldsymbol{\theta})^{-1} \frac{\partial \boldsymbol{M}_{\boldsymbol{z}}(\boldsymbol{\theta})}{\partial \theta_p}\right].$$

Appendix C

Software (E^4)

C.1 Models supported in E^4 224
C.1.1 State-Space model 224
C.1.2 The THD format 224
C.1.3 Basic models 226
C.1.3.1 Mathematical definition 226
C.1.3.2 Definition in THD format 227
C.1.4 Models with GARCH errors 228
C.1.4.1 Mathematical definition of the GARCH process 228
C.1.4.2 Defining a model with GARCH errors in THD format 229
C.1.5 Nested models 230
C.2 Overview of computational procedures 233
C.2.1 Standard procedures for time series analysis 233
C.2.2 Signal extraction methods 235
C.2.3 Likelihood and model estimation 238
C.3 Who can benefit from E^4? 241

This Appendix describes a MATLAB toolbox for time series modeling. Its name, E^4, refers to the Spanish: *Estimación de modelos Econométricos en Espacio de los Estados*, meaning "State-Space Estimation of Econometric Models." It includes ready-to-use functions for model specification, parameter estimation, model forecasting, simulation, and signal extraction. In particular, all the examples in this book (except those of Chapter 11) have been computed with E^4.

The structure of this Appendix is as follows. Section C.1 reviews the standard models supported (including e.g., SS, VARMAX or transfer functions), describes how to define them, and shows how a model can be created by joining several building blocks through the model combination procedures that described in subsections 3.2.1 and 3.2.2. Section C.2 provides an overview of the analytical and computational procedures employed by E^4. They fall into three broad categories: standard time series methods, signal extraction procedures, and model estimation algorithms. Last,

Section C.3 provides some remarks about the design priorities of E^4, and discusses what type of user might benefit from this toolbox.

The source code for all the functions in E^4 is freely provided under the terms of the GNU General Public License. Appendix D indicates how to download E^4 and its related materials, as well as the source code and data required to replicate the examples in this book.

C.1 Models supported in E^4

C.1.1 State-Space model

The most important formulation supported by E^4 is the SS model; the standard representation for which most of its core computational algorithms are devised. The basic SS formulations supported are the MEM and SEM forms defined in Sections 2.1 and 2.2.

On the other hand, SS models are difficult to store and manipulate through software. For this reason, E^4 manages all the models supported using a convenient format called THD (THeta-Din).

C.1.2 The THD format

Defining a model in THD format requires creating two numeric matrices: `theta` and `din`, which contain respectively, the parameter values and a description of the model dynamic structure. Additionally, the model can be documented through an optional character matrix, `lab`, which contains names for the parameters in `theta`. While some analyses may require editing `theta` and `lab`, most users will never need to touch the matrix `din`.

The THD representation of an SS model can be obtained via the function `ss2thd` (SS-to-THD), which translates the matrices characterizing any MEM or SEM form into the THD format. Its syntax is:

```
[theta, din, lab] = ss2thd(Phi, Gam, E, H, D, C, Q, S, R);
```

where the input arguments are the parameter matrices in the SS model, and the outputs are the `theta`, `din`, and `lab` matrices previously described. To illustrate the use of this function, the following Example shows how to define the SS model implied by a Hodrick–Prescott filter (see Harvey and Trimbur [120]):

Example C.1.1.1 Defining a SS model in THD format

```
e4init  % Initializes E4 default options and tolerances
% *** Defines the model matrices
Phi= [1 1; NaN 1];
% The Not-A-Number value (NaN) indicates a null parameter
E= [1;NaN];
H=[1 0]; C=[1]; Q=[1/1600]; S=[0]; R=[1];
% *** Obtains the THD representation and displays
```

```
%    the resulting model
[theta, din, lab] = ss2thd(Phi, [], E, H, [], C, Q, S, R);
% The only free parameters in this model are the variances.
% Accordingly, nonzero values in the second column of theta
% indicate which parameters should remain at its current value
theta(1,2)=1; theta(2,2)=1; theta(3,2)=1;
theta(4,2)=1; theta(5,2)=1; theta(6,2)=1;
theta(7,2)=1; theta(9,2)=1;
prtmod(theta, din, lab);
```

and the resulting output is:

```
******************** Options set by user ****************
Filter. . . . . . . . . . . . . : KALMAN
Scaled B and M matrices . . . . : NO
Initial state vector. . . . . . : AUTOMATIC SELECTION
Initial covariance of state v.  : IDEJONG
Variance or Cholesky factor? . : VARIANCE
Optimization algorithm. . . . . : BFGS
Maximum step length . . . . . . : 0.100000
Stop tolerance. . . . . . . . . : 0.000010
Max. number of iterations . . . :       75
Verbose iterations. . . . . . . : YES
*********************************************************

************************** Model ************************
Native SS model
1 endogenous v., 0 exogenous v.
Seasonality: 1
SS vector dimension: 2
Parameters (* denotes constrained parameter):
PHI(1,1)      *     1.0000
PHI(1,2)      *     1.0000
PHI(2,2)      *     1.0000
E(1,1)        *     1.0000
H(1,1)        *     1.0000
H(1,2)        *     0.0000
C(1,1)        *     1.0000
Q(1,1)              0.0006
S(1,1)        *     0.0000
R(1,1)              1.0000
*********************************************************
```

Note that asterisks identify which parameters are constrained to remain at their current values in any subsequent estimation.

C.1.3 Basic models

C.1.3.1 Mathematical definition

E^4 supports three basic models: VARMAX, structural econometric models, and single-output transfer functions. The VARMAX model is given by:

$$\boldsymbol{\phi_p}(\mathrm{B})\,\boldsymbol{\Phi_P}(\mathrm{B}^S)\,\boldsymbol{z_t} = \boldsymbol{G_n}(\mathrm{B})\,\boldsymbol{u_t} + \boldsymbol{\theta_q}(\mathrm{B})\,\boldsymbol{\Theta_Q}(\mathrm{B}^S)\,\boldsymbol{a_t}, \tag{C.1}$$

where S denotes the period of the seasonal cycle, B is the (already defined) backshift operator, the vectors $\boldsymbol{z_t}$, $\boldsymbol{u_t}$, and $\boldsymbol{a_t}$ are the model outputs, inputs, and errors, respectively, and the polynomial matrices in (C.1) are given by:

$$\boldsymbol{\phi_p}(\mathrm{B}) = \boldsymbol{I} + \boldsymbol{\phi_1}\mathrm{B} + \boldsymbol{\phi_2}\mathrm{B}^2 + \ldots + \boldsymbol{\phi_p}\mathrm{B}^p \tag{C.2}$$

$$\boldsymbol{\Phi_P}(\mathrm{B}^S) = \boldsymbol{I} + \boldsymbol{\Phi_1}\mathrm{B}^S + \boldsymbol{\Phi_2}\mathrm{B}^{2\cdot S} + \ldots + \boldsymbol{\Phi_P}\mathrm{B}^{P\cdot S} \tag{C.3}$$

$$\boldsymbol{G_n}(\mathrm{B}) = \boldsymbol{G_1}\mathrm{B} + \boldsymbol{G_2}\mathrm{B}^2 + \ldots + \boldsymbol{G_n}\mathrm{B}^n \tag{C.4}$$

$$\boldsymbol{\theta_q}(\mathrm{B}) = \boldsymbol{I} + \boldsymbol{\theta_1}\mathrm{B} + \boldsymbol{\theta_2}\mathrm{B}^2 + \ldots + \boldsymbol{\theta_q}\mathrm{B}^q \tag{C.5}$$

$$\boldsymbol{\Theta_Q}(\mathrm{B}^S) = \boldsymbol{I} + \boldsymbol{\Theta_1}\mathrm{B}^S + \boldsymbol{\Theta_2}\mathrm{B}^{2\cdot S} + \ldots + \boldsymbol{\Theta_Q}\mathrm{B}^{Q\cdot S}. \tag{C.6}$$

The structural econometric model is similar to the VARMAX specification (C.1), but allows for contemporary relationships between the endogenous variables and their errors. In econometric terms, it is a "structural form" while the VARMAX is a "reduced form." It is defined by:

$$\boldsymbol{\bar{\phi}_p}(\mathrm{B})\boldsymbol{\Phi_P}(\mathrm{B}^S)\boldsymbol{z_t} = \boldsymbol{G_n}(\mathrm{B})\boldsymbol{u_t} + \boldsymbol{\bar{\theta}_q}(\mathrm{B})\boldsymbol{\Theta_Q}(\mathrm{B}^S)\boldsymbol{a_t}, \tag{C.7}$$

with:

$$\boldsymbol{\bar{\phi}_p}(\mathrm{B}) = \boldsymbol{\bar{\phi}_0} + \boldsymbol{\bar{\phi}_1}\mathrm{B} + \bar{\phi}_2\mathrm{B}^2 + \ldots + \boldsymbol{\bar{\phi}_p}\mathrm{B}^p \tag{C.8}$$

$$\boldsymbol{\bar{\theta}_q}(\mathrm{B}) = \boldsymbol{\bar{\theta}_0} + \bar{\theta}_1\mathrm{B} + \boldsymbol{\bar{\theta}_2}\mathrm{B}^2 + \ldots + \boldsymbol{\bar{\theta}_q}\mathrm{B}^q. \tag{C.9}$$

(Note that the polynomial matrices $\boldsymbol{\Phi_P}(\mathrm{B}^S)$, $\boldsymbol{G_n}(\mathrm{B})$, and $\boldsymbol{\Theta_Q}(\mathrm{B}^S)$ in (C.7) were already defined in (C.1).)

Last, the single-output transfer function with r inputs is given by:

$$z_t = \frac{\omega_1(\mathrm{B})}{\delta_1(\mathrm{B})}u_{1,t} + \ldots + \frac{\omega_r(\mathrm{B})}{\delta_r(\mathrm{B})}u_{r,t} + \frac{\theta_q(\mathrm{B})\Theta_Q(\mathrm{B}^S)}{\phi_p(\mathrm{B})\Phi_P(\mathrm{B}^S)}a_t, \tag{C.10}$$

where, for $i = 1, 2, \ldots, r$, we have:

$$\omega_i(\mathrm{B}) = \omega_0 + \omega_1\mathrm{B} + \ldots + \omega_{m_i}\mathrm{B}^{m_i} \tag{C.11}$$

$$\delta_i(\mathrm{B}) = 1 + \delta_1\mathrm{B} + \ldots + \delta_{k_i}\mathrm{B}^{k_i}, \tag{C.12}$$

and the polynomials $\theta_q(\mathrm{B})$, $\Theta_Q(\mathrm{B}^S)$, $\phi_p(\mathrm{B})$, and $\Phi_P(\mathrm{B}^S)$ in (C.10) are scalar versions of the autoregressive and moving average matrix factors in (C.1).

C.1.3.2 Definition in THD format

The E^4 functions arma2thd, str2thd and tf2thd generate the THD formulation for VARMAX, structural econometric models and transfer functions, respectively. The general syntax of these functions is:

```
[theta, din, lab] = arma2thd(FR, FS, AR, AS, V, s, G, r)
[theta, din, lab] = str2thd(FR, FS, AR, AS, V, s, G, r)
[theta, din, lab] = tf2thd(FR, FS, AR, AS, V, s, W, D)
```

where the inputs to these functions are matrices of real numbers containing: (a) the initial values for the regular and seasonal autoregressive factors, FR and FS, (b) the regular and seasonal moving average factors, AR and AS, (c) the terms related to the exogenous variables in the transfer function model, W and D, and in the VARMAX and structural econometric model, G, (d) the noise covariance, V, as well as (e) scalar integers defining the seasonal period, s, and the number of inputs in the VARMAX and structural econometric models, r.

The following Example illustrates the use of arma2thd by defining a quarterly airline model in THD form, displaying its structure and obtaining the matrices in its SS representation:

Example C.1.1.2 ARIMA specification in THD and SS format

```
e4init % Initializes E4 default options and tolerances
% Defines and displays the non-stationary airline model
%     (1-B)(1-B^4) y_t = (1 - .6 B)(1 - .5 B^4) a_t
[theta,din,lab] = arma2thd([-1],[-1],[-.6],[-.5],[.1],4);
theta(1,2)=1; theta(2,2)=1; % Constrains the unit roots
prtmod(theta,din,lab);
% Now obtains the matrices of the equivalent SS representation
[Phi, Gam, E, H, D, C, Q, S, R] = thd2ss(theta, din)
```

and the resulting output is:

```
*************************** Model ***************************
VARMAX model (innovations model)
1 endogenous v., 0 exogenous v.
Seasonality: 4
SS vector dimension: 5
Parameters (* denotes constrained parameter):
FR1(1,1)       *     -1.0000
FS1(1,1)       *     -1.0000
AR1(1,1)             -0.6000
AS1(1,1)             -0.5000
V(1,1)                0.1000
*************************************************************
Phi =
     1     1     0     0     0
```

```
      0     0     1     0     0
      0     0     0     1     0
      1     0     0     0     1
     -1     0     0     0     0

Gam =
     []

E =
    0.4000
         0
         0
    0.5000
   -0.7000

H =
      1     0     0     0     0

D =
     []

C =
      1

Q =
    0.1000

S =
    0.1000

R =
    0.1000
```

C.1.4 Models with GARCH errors

C.1.4.1 Mathematical definition of the GARCH process

E^4 allows one to combine any SEM, VARMAX, structural model or transfer function with a vector GARCH process, for a survey on this important issue, see Bollerslev, Engle, and Nelson [25].

The general conditional heteroscedastic process is defined by the distributions $\boldsymbol{a_t} \sim \mathrm{IID}(\boldsymbol{0}, \boldsymbol{Q})$ and $\boldsymbol{a_t}|\boldsymbol{\Omega}_{t-1} \sim \mathrm{IID}(\boldsymbol{0}, \boldsymbol{\Sigma_t})$ where, according to the notation already defined in Chapter 4, $\boldsymbol{\Omega}_{t-1}$ denotes the information available up to $t-1$. In a GARCH model, the conditional covariances, $\boldsymbol{\Sigma_t}$, are such that:

$$[\boldsymbol{I} - \boldsymbol{B_m}(\mathrm{B})]\, vech\,(\boldsymbol{\Sigma_t}) = \boldsymbol{w} + \boldsymbol{A_k}(\mathrm{B}) vech\left(\boldsymbol{a_t}\boldsymbol{a_t^T}\right), \tag{C.13}$$

where $vech()$ vectorizes the upper triangle and main diagonal of its argument and the polynomial matrices are given by:

$$\boldsymbol{B}_m(\mathrm{B}) = \boldsymbol{B}_1\mathrm{B} + \boldsymbol{B}_2\mathrm{B}^2 + \ldots + \boldsymbol{B}_m\mathrm{B}^m \tag{C.14}$$

$$\boldsymbol{A}_k(\mathrm{B}) = \boldsymbol{A}_1\mathrm{B} + \boldsymbol{A}_2\mathrm{B}^2 + \ldots + \boldsymbol{A}_k\mathrm{B}^k. \tag{C.15}$$

E^4 manages model (C.13) in its alternative VARMAX representation. To derive it, consider the white noise process $\boldsymbol{v}_t = vech\left(\boldsymbol{a}_t\boldsymbol{a}_t^T\right) - vech\left(\boldsymbol{\Sigma}_t\right)$, which has a null conditional mean and a very complex heteroscedastic structure. Then $vech\left(\boldsymbol{\Sigma}_t\right) = vech\left(\boldsymbol{a}_t\boldsymbol{a}_t^T\right) - \boldsymbol{v}_t$, and substituting this expression in (C.13) and rearranging some terms, yields:

$$[\boldsymbol{I} - \boldsymbol{A}_k(\mathrm{B}) - \boldsymbol{B}_m(\mathrm{B})]\, vech\left(\boldsymbol{a}_t\boldsymbol{a}_t^T\right) = \boldsymbol{w} + [\boldsymbol{I} - \boldsymbol{B}_m(\mathrm{B})]\,\boldsymbol{v}_t. \tag{C.16}$$

Equation (C.16) defines a VARMAX model for $vech\left(\boldsymbol{a}_t\boldsymbol{a}_t^T\right)$, which can be written in the alternative form:

$$vech\left(\boldsymbol{a}_t\boldsymbol{a}_t^T\right) = vech(\boldsymbol{Q}) + \boldsymbol{N}_t, \tag{C.17}$$

$$\left[\boldsymbol{I} + \bar{\boldsymbol{A}}_{max(k,m)}(\mathrm{B})\right]\boldsymbol{N}_t = \left[\boldsymbol{I} + \bar{\boldsymbol{B}}_m(\mathrm{B})\right]\boldsymbol{v}_t, \tag{C.18}$$

where $\bar{\boldsymbol{A}}_{max(k,m)}(\mathrm{B}) = -\boldsymbol{A}_k(\mathrm{B}) - \boldsymbol{B}_m(\mathrm{B})$ and $\bar{\boldsymbol{B}}_m(\mathrm{B}) = -\boldsymbol{B}_m(\mathrm{B})$. This is the formulation supported by E^4. Note that this representation allows for IGARCH (Integrated-GARCH) components, which would require specifying some unit roots in the AR factor of (C.18).

C.1.4.2 Defining a model with GARCH errors in THD format

To define a model with GARCH errors in THD format it is necessary to obtain: (a) the THD formulation of the model for the mean, given by `t1`, `d1` and `lab1`, (b) the THD model corresponding to the model (ARCH, GARCH, or IGARCH) for the variance, given by `t2`, `d2` and `lab2`, and (c) link both formulations using the `garc2thd` function, which has the following syntax:

```
[theta, din, lab] = garc2thd(t1, d1, t2, d2, lab1, lab2);
```

Example C.1.1.3 Defining a regression model with GARCH errors

The following code generates and displays the THD representation of a regression model with GARCH errors:

```
e4init % Initializes E4 default options and tolerances

% Defines and displays a regression model with GARCH(1,1) errors:
%   y_t = .7 x_t1 + 1 x_t2 + a_t
%   a_t^2 = .1 + N_t
%   (1 - .8 B) N_t =  (1 - .7 B) v_t
```

```
[t1, d1, lab1] = arma2thd([], [], [], [], [.1], 1,[.7 1],2);
[t2, d2, lab2] = arma2thd([-.8], [], [-.7], [], [.1], 1);
[theta, din, lab] = garc2thd(t1, d1, t2, d2, lab1, lab2);
prtmod(theta, din, lab);
```

and the resulting output is:

```
************************* Model *************************
GARCH model (innovations model)
1 endogenous v., 2 exogenous v.
Seasonality: 1
SS vector dimension: 0
Endogenous variables model:
  White noise model (innovations model)
  1 endogenous v., 2 exogenous v.
  Seasonality: 1
  SS vector dimension: 0
  Parameters (* denotes constrained parameter):
  G0(1,1)            0.7000
  G0(1,2)            1.0000
  V(1,1)             0.1000
  --------------
GARCH model of noise:
  VARMAX model (innovations model)
  1 endogenous v., 0 exogenous v.
  Seasonality: 1
  SS vector dimension: 1
  Parameters (* denotes constrained parameter):
  FR1(1,1)          -0.8000
  AR1(1,1)          -0.7000
  --------------
*********************************************************
```

C.1.5 *Nested models*

A "nested" model is built by combining several simple models. E^4 allows for two types of model combination: nesting in inputs and nesting in errors.

Nesting in inputs, or "endogeneization" was defined mathematically in Subsection 3.2.1. It consists in augmenting a SS model with the models for its exogenous variables which, after doing so, become endogenous variables. This operation is useful, for example, when one wants to combine a transfer function and a VARMAX model for its inputs to compute forecasts for all these variables in a single operation. Another situation where nesting is useful is when there are missing values in the inputs. "Endogeneizing" is then required because SS methods only allow for missing values in the endogenous variables; see Section 11.5.

On the other hand, nesting in errors was defined in Subsection 3.3.1. It consists in augmenting a model in SEM form with another model for its disturbances. Intuitively, it can be seen as defining a model by the product of different polynomials. This is handy, e.g., when one wants to separate the factors containing different types of roots, such as unit, complex, or real roots. It is also useful when modeling a series with several seasonal factors, corresponding to different periods.

The following examples illustrate how E^4 implements the definition of nested models.

Example C.1.1.4 Nesting in inputs

Given the models:

$$z_t = \frac{0.3+0.6\mathrm{B}}{1-0.5\mathrm{B}} u_t + \frac{1-0.8\mathrm{B}}{1-0.6\mathrm{B}+0.4\mathrm{B}^2} a_{1,t}, \qquad \sigma_{a1}^2 = 1, \tag{C.19}$$

$$(1-0.7\mathrm{B})\, u_t = a_{2,t}, \qquad \sigma_{a2}^2 = 0.3, \tag{C.20}$$

the code required to define and combine both models in inputs is:

```
e4init

% Obtains the THD representation for both models:
[t1,d1,l1]=tf2thd([-.6 .4],[],[-.8],[],[1.0],1,[ .3 .6],[-.5]);
[t2,d2,l2]=arma2thd(-.7,[],[],[],.3,1);

% Combines the models for the endogenous variable and the input
[theta, din, lab] = stackthd(t1, d1, t2, d2, l1, l2);
[theta, din, lab] = nest2thd(theta, din, 1, lab);
% The 3rd input argument to nest2thd determines the combination:
%   0 : nesting in errors,  1 : nesting in inputs
%   and displays the resulting model structure
prtmod(theta,din,lab);
```

and the output from the prtmod command is:

```
*************************** Model ***************************
Nested model in inputs (innovations model)
2 endogenous v., 0 exogenous v.
Seasonality: 1
SS vector dimension: 4
Submodels:
{
  Transfer function model (innovations model)
  1 endogenous v., 1 exogenous v.
  Seasonality: 1
  SS vector dimension: 3
  Parameters (* denotes constrained parameter):
```

```
  FR(1,1)          -0.6000
  FR(1,2)           0.4000
  AR(1,1)          -0.8000
  W1(1,1)           0.3000
  W1(2,1)           0.6000
  D1(1,1)          -0.5000
  V(1,1)            1.0000
  --------------
  VARMAX model (innovations model)
  1 endogenous v., 0 exogenous v.
  Seasonality: 1
  SS vector dimension: 1
  Parameters (* denotes constrained parameter):
  FR1(1,1)          -0.7000
  V(1,1)             0.3000
  --------------
}
*************************************************************
```

Note that the final model has two endogenous variables, z_t and $u_{1,t}$, and no inputs.

Example C.1.1.5 Nesting in errors

Consider now the model:

$$\left(1-0.5\mathrm{B}+0.3\mathrm{B}^2\right)\left(1-\mathrm{B}\right)\left(1-\mathrm{B}^{12}\right)z_t = a_t, \qquad \sigma_a^2 = 0.2, \tag{C.21}$$

which has nonstationary

regular and seasonal AR factors. The following code defines and displays model (C.21) through nesting in errors.

```
e4init
% First obtains the THD representation of the factors
[t1, d1, l1] = arma2thd([-1], [-1], [], [], [0], 12);
[t2, d2, l2] = arma2thd([-.5 .3], [], [], [],[.2], 12);

%  and then the stacked and nested models
[ts, ds, ls] = stackthd(t1, d1, t2, d2, l1, l2);
[theta, din, lab] = nest2thd(ts, ds, 0, ls);
% The 3rd input argument to nest2thd determines the combination:
%   0 : nesting in errors,  1 : nesting in inputs

% Constrains the regular and seasonal unit roots
theta(1,2)=1; theta(2,2)=1;
prtmod(theta, din, lab);
% The variance of the model t1-d1 is ignored
%  so we can specify any arbitrary value
```

and the output from prtmod is:

```
************************** Model ***************************
Nested model in errors (innovations model)
1 endogenous v., 0 exogenous v.
Seasonality: 12
SS vector dimension: 15
Submodels:
{
  VARMAX model (innovations model)
  1 endogenous v., 0 exogenous v.
  Seasonality: 12
  SS vector dimension: 13
  Parameters (* denotes constrained parameter):
  FR1(1,1)       *  -1.0000
  FS1(1,1)       *  -1.0000
  --------------
  VARMAX model (innovations model)
  1 endogenous v., 0 exogenous v.
  Seasonality: 12
  SS vector dimension: 2
  Parameters (* denotes constrained parameter):
  FR1(1,1)          -0.5000
  FR2(1,1)           0.3000
  V(1,1)             0.2000
  --------------
}
************************************************************
```

C.2 Overview of computational procedures

E^4 includes three basic groups of functions: (a) standard procedures for time series analysis, (b) signal extraction methods and (c) model estimation algorithms.

C.2.1 Standard procedures for time series analysis

The main standard procedures included are detailed in Table C.1. These functions cover basic needs such as plotting a time series, computing its autocorrelations or its descriptive statistics. They are included to make the Toolbox self-contained, but its functionality is not innovative.

The following example illustrates the use of these functions with simulated data. We will not show the outputs resulting from this and other simulation-based examples, because the exact results obtained will differ in each run.

Example C.2.1 ARIMA identification

Table C.1 *Standard statistics procedures to specify and simulate a time series model.*

Function	**Description**	**References**:
augdft	Computes the augmented Dickey-Fuller test	Dickey and Fuller [78]
descser	Descriptive statistics for a vector of time series	
histsers	Computes and displays histograms for a vector of time series	
lagser	Generates lags and leads from a data matrix	
midents	Multiple sample autocorrelation and partial autoregression functions	Wei [221]
plotsers	Displays standardized plots for a vector of time series	
rmedser	Displays the mean/std. deviation plot of a vector of time series	Box and Cox [28]
transdif	Computes differences and Box-Cox transforms	Box and Cox [28]
uidents	Sample autocorrelation and partial autocorrelation functions	Box, Jenkins, and Reinsel [30]
simmod	Simulates a homoscedastic model	
simgarch	Simulates a model with GARCH errors	Bollerslev, Engle, and Nelson [25]

Consider the ARMA(2,1) model:

$$\left(1-0.5\mathrm{B}+0.3\mathrm{B}^2\right)z_t = (1-0.7\mathrm{B})\,a_t, \qquad \sigma_a^2 = 0.2. \tag{C.22}$$

The following code defines it, simulates a sample, and performs a standard univariate identification process:

```
e4init
% Obtains the THD definition of the model
[theta, din, lab] = arma2thd([-.5 .3], [], [-.7], [], [.2], 1);
prtmod(theta,din,lab)

% Simulates the sample and omits the first 50 observations
z=simmod(theta,din,250); z=z(51:250,1);

% Applies several standard time series analysis tools
% Standard-deviation mean plot
rmedser(z);
% Time series plot
plotsers(z);
```

```
% Augmented Dickey-Fuller test. The second argument (2)
%  specifies the number of lags for the dynamic regression
augdft(z,2);
% Histogram
histsers(z);
% Descriptive statistics
descser(z);
% Sample autocorrelation and partial autocorrelation functions
% The second argument specifies the number of lags
uidents(z,10);
```

Example C.2.2 VARMA identification

The following code illustrates the use of the multiple time series identification functions included in E^4 by simulating a VAR(1) process and then identifying it:

```
e4init
% Obtains the THD definition of the model
Phi=[-.5 .3;NaN -.6];
Sigma=[1 .5;.5 1];
[theta,din,lab]=arma2thd([Phi],[],[],[],[Sigma],1);
prtmod(theta,din,lab)

% Simulates the sample and omits the first 50 observations
y=simmod(theta,din,250); y=y(51:250,:);

% Descriptive statistics for multiple time series
descser(y);

% Computes 10 lags of the multiple autocorrelation, partial
%  autocorrelation and cross-correlation functions
midents(y,10);
```

C.2.2 *Signal extraction methods*

As we saw in Chapter 4, the SS literature considers two main signal extraction problems: filtering and smoothing. Both have as their focus the estimation of the sequence of states, but based on different information sets. E^4 computes these sequences for different purposes. Filtered states are typically used to compute the Gaussian likelihood in prediction-error decomposition form (see Chapters 5 and 6), so that filtering algorithms are embedded in the likelihood evaluation functions of E^4; see Section C.2.3.

On the other hand, smoothed estimates have many uses, such as interpolating missing in-sample values (Kohn and Ansley [137]), calculating the residuals of a model allowing for missing values (Kohn and Ansley [138]), cleaning noise-contaminated samples (Kohn and Ansley [136]), decomposing a time series into meaningful unobserved components (Harvey, [118]; Casals, Jerez, and Sotoca [41]), and detecting outliers (De Jong and Penzer [70]). Table C.2 summarizes the main E^4 functions implementing signal-extraction algorithms, with corresponding literature references where relevant.

Table C.2 *Signal extraction functions. These are the main procedures offered for time series disaggregation, structural decomposition, smoothing and forecasting.*

Function	Description	References
`aggrmod`	Computes smoothed estimates for a high-frequency time series from: (a) a sample with low and high-frequency data, and (b) the high-frequency model for the endogenous variable(s)	Casals, Sotoca and Jerez [46]
`e4trend`	Decomposes a vector of time series into the trend, seasonal, cycle and irregular components implied by an econometric model in THD format	Casals, Jerez, and Sotoca [41]
`fismod`, `fismiss`	Computes smoothed estimates of the state and observable variables of a model in THD form. The function `fismiss` allows for in-sample missing values	Casals, Jerez, and Sotoca [40]
`foremod`	Computes forecasts for a homoscedastic model in THD format	
`foregarc`	Computes forecasts for a model in THD format, allowing for GARCH errors	
`residual`	Computes the residuals for a model in THD format	

Basic Fixed-Interval Smoothing is addressed by the E^4 function `fismod`, which implements the algorithm of Casals, Jerez, and Sotoca [40]. Its syntax is:

```
[xhat, Px, e] = fismod(theta, din, z);
```

The input arguments of these functions are a THD format specification (`theta`, `din`) and the data matrix (`z`) where the missing values, if any, are coded as `NaN` (the "not-a-number" MATLAB special symbol). The output arguments of `fismod` are: `xhat`, the sequence of expected values of the state vector conditional on the sample; `Px`, a matrix with the corresponding covariances; and `e`, a matrix of fixed-interval smoothed errors. There is another smoother function, `fismiss`, which allows for missing values in z.

Smoothing is often used to decompose a time series into the sum of trend, cycle, seasonal, and irregular components. The function `e4trend` implements the exact decomposition described in Casals, Jerez, and Sotoca [41]. Its simplified syntax is:

```
[trend,season,cycle,irreg] = e4trend(theta,din,z)
```

where the input arguments theta, din, and z are identical to those of fismod, and the output arguments are fixed-interval smoothed estimates of the corresponding structural components.

The function foremod predicts future values of the endogenous variables for a given model. Its syntax is:

```
[zf, Bf] = foremod(theta, din, z, k, u)
```

where the input arguments theta, din, and z are identical to those of fismod, k is the number of forecasts to be computed, and u is a matrix with the out-of-sample values of the exogenous variables, if any. The forecasts and the corresponding sequence of covariances are returned in the output arguments zf and Bf. The analogous functions, foremiss and foregarc, compute forecasts for samples with missing values and for models with conditional heteroscedastic errors, respectively. Last, the residuals for a model are calculated through the function residual whose syntax is:

```
[z1, vT, wT, vz1, vvT, vwT] = residual(theta, din, z)
```

where the input arguments theta, din and z are identical to those of fismod. On the other hand, the output arguments are z1, vT, and wT which are, respectively, the model residuals, the fixed-interval smoothed estimates of the observation errors, and the smoothed estimates of the state errors. The arguments vz1, vvT, and vwT store the corresponding sequences of covariances.

Example C.2.3 HP filtering

The following code simulates the HP filter model defined in Example C.1.1.1 and extracts the HP trend using different procedures:

```
e4init
% Working with standard deviations improves the model scaling
sete4opt('var','fac');

Phi= [1 1; NaN 1];
E= [1;NaN];
H=[1 0]; C=[1]; Q=[sqrt(1/1600)]; S=[0]; R=[1];
% Obtains the THD representation and displays the model
[theta, din, lab] = ss2thd(Phi, [], E, H, [], C, Q, S, R);
prtmod(theta, din, lab);
% Simulates the data and omits the first 50 observations
y=simmod(theta,din,250); y=y(51:250,:);

% Extracts and plots the trend and error components
[trend,season,cycle,irreg] = e4trend(theta,din,y);
plotsers([trend,irreg]);

% e4trend does not provide variances for the trend component
```

```
% The following commands do this using fismod
[xhat, px, e] = fismod(theta, din, y); varhptrend=px(1:2:400,1);
figure; hold on
plot(varhptrend.^.5)
hold off
```

C.2.3 Likelihood and model estimation

Table C.3 summarizes the main functions in connection with model estimation. These functions specialize in the computation of likelihood values, information matrices, and parameter estimates.

E^4 includes several functions to compute the Gaussian log-likelihood. The most advanced one is `lfsd`, whose general syntax is:

```
[f, innov, ssvect] = lfsd(theta, din, z)
```

where the input arguments are identical to those defined for previous functions, and the output arguments are: (a) `f`, the value of the log-likelihood, (b) `innov`, a matrix of one-step-ahead forecast errors, and (c) `ssvect`, a matrix of estimates of the state variables organized so that its t-th row contains the filtered estimate of the state vector at time t, conditional on sample values up to $t-1$. The technical details of this function can be found in Casals, Sotoca and Jerez [47]. Other functions to compute the log-likelihood of a homoscedastic model are `lfmod` (Terceiro [206]), `lffast` (Casals, Sotoca and Jerez [44]), and `lfcd` (Casals, Sotoca and Jerez [47]). They differ only in minor details, and all have the same inputs and outputs. Last, `lfmiss` and `lfgarch` are two specialized variants of `lfsd`, allowing for missing values in the endogenous variables and GARCH errors, respectively.

As in the case of the likelihood, there are three functions dealing with computation of the information matrix: `imod`, `imiss`, and `igarch`. They all have the same syntax. For example, any call to `imod` has the following structure:

```
[std, corrm, varm, Im] = imod(theta, din, z)
```

This function computes the information matrix of `lfmod` and `lffast` (`Im`) as well as the parameter analytical standard errors (`std`), the correlation matrix of the estimates (`corrm`), and the corresponding covariance matrix (`varm`). The functions `imiss` and `igarch` do the same for `lfmiss` and `lfgarch`, respectively. If the model is misspecified or is non-Gaussian, the function `imodg` computes a robust information matrix, alternative to `imod`; see Ljung and Caines [153] and White [226].

Being able to compute the Gaussian log-likelihood is not enough to estimate the parameters of a model: one needs an iterative procedure to compute the optimal value of the parameters. The E^4 function `e4min` implements two main optimization algorithms, BFGS and Newton–Raphson (Dennis and Schnabel [74]). Its general syntax is:

```
[pnew,iter,fnew,gnew,hessin]= ...
       e4min(func,theta,dfunc,P1,P2,P3,P4,P5)
```

Table C.3 *Model estimation functions. These are the main procedures which can be applied to compute the Gaussian likelihood, its first-order derivatives, and information matrix. These functions can be fed to the standard* E^4 *optimizer (*`e4min`*) to compute ML estimates. Initial estimates can be computed using a specialized function (*`e4preest`*).*

Function	**Description**	**References**
Likelihood computation and derivatives		
`lfmod`, `lffast`, `lfsd`, `lfcd`, `lfper`, `lfgarch`, `lfmiss`	Compute the log-likelihood function for a model in THD form. `lfmod`, `lffast`, `lfsd` and `lfcd` support models with homoscedastic errors. `lfper` supports periodic models. Last, `lfgarch` admits for GARCH errors and `lfmiss` allows for in-sample missing values	Terceiro [206], Casals, Sotoca and Jerez [44], Casals, Sotoca and Jerez [47]
`gmod`, `gmiss`, `ggarch`	Computes the analytical gradient for `lfmod` and `lffast` (`gmod`), `lfmiss` (`gmiss`) and `lfgarch` (`ggarch`)	Casals, Sotoca and Jerez [45], Casals, and Sotoca [43]
`imod`, `imiss`, `igarch`, `imodg`	Compute the Gaussian analytical information matrix for `lfmod`, `lffast`, `lfsd` and `lfcd` (`imod`), `lfmiss` (`imiss`) and `lfgarch` (`igarch`). The function `imodg` computes the robustified information matrix for `lfmod` and `lffast`	Terceiro [206], Ljung and Caines [153], White [226]
Model estimation		
`e4min`	Computes the unconstrained numerical minimum of a nonlinear function using the BFGS or the Newton-Raphson algorithms	Dennis and Schnabel [74]
`e4preest`	Computes a fast estimate of the parameters for a model in THD format	Garcia–Hiernaux, Jerez, and Casals [104]
`prtest`	Displays estimation results	
Toolbox defaults and options		
`e4init`	Initializes the global variables and defaults	
`sete4opt`	Modifies the toolbox options	

The operation of `e4min` is as follows. Starting from the initial estimate of the parameters in `theta`, it iterates on the objective function whose name is stored in the argument `func`. The iteration can be based on the analytical gradient or on a numerical approximation, depending on whether `dfunc` contains the name of the analytical gradient function or is an empty string, `''`. Finally, the stopping criterion takes into account the relative changes in the values of the parameters and/or the size of the gradient vector. The parameters `P1, ..., P5` are optional. If they are specified, its values are fed as additional input arguments to the objective function `func`.

Once the iterative process is stopped, the function returns: `pnew`, the value of the parameters; `iter`, the number of iterations performed; `fnew`, the value of the objective function at `pnew`; `gnew`, the analytical or numerical gradient, depending on the contents of `dfunc`; and finally `hessin`, a numerical approximation to the Hessian of the objective function.

The auxiliary function `sete4opt` manages different options controlling the performance of the optimizer, including type of algorithm, tolerance for the stopping criteria, or the maximum number of iterations allowed.

Another important function for model estimation is `e4preest`. It computes fast and consistent estimates of the parameters in `theta` which, in most cases, are adequate starting values for likelihood optimization. The syntax of this function is:

```
thetanew = e4preest(theta, din, z)
```

where the input arguments are identical to those of `lffast`, and the estimates are returned in `thetanew`. The algorithm implemented in `e4preest` is described by Garcia–Hiernaux, Jerez, and Casals [104].

The following code simulates, estimates and applies different signal-extraction procedures to the nonstationary

airline model structure defined in Example C.1.1.2:

Example C.2.4 Simulation, estimation and signal extraction

```
e4init
% Defines and displays the non-stationary airline model
%    (1-B)(1-B^12) y_t = (1 - .6 B)(1 - .5 B^12) a_t
[theta,din,lab] = arma2thd([-1],[-1],[-.6],[-.5],.1,12);
theta(1,2)=1; theta(2,2)=1;  % Constrains the unit roots

% Simulates the sample and omits the first 50 observations
z=simmod(theta,din,250); z=z(51:250,1);

% Computes preliminary estimates
theta=e4preest(theta,din,z);
% ... and then ML estimates
[thopt, it, lval, g, h] = e4min('lffast', theta,'', din, z);
% Computes the analytical standard errors
[std, corrm, varm, Im] = imod(thopt, din, z);
% Summary of estimation results
prtest(thopt, din, lab, z, it, lval, g, h, std, corrm)

% Residual diagnostics
ehat=residual(thopt, din, z);
descser(ehat);
plotsers(ehat);
uidents(ehat,39);
```

```
% Structural decomposition
[trend,season,cycle,irreg] = e4trend(thopt,din,z);
plotsers([trend,season,irreg]);

% Computes and displays 10 forecasts and their standard errors
[zfor, Bfor] = foremod(thopt, din, z, 10);
[zfor Bfor.^.5]

% Marks obs. 10 and 40 as missing and estimates them
z1=z; z1(10)=NaN; z1(40)=NaN;
[zhat, pz] = fismiss(thopt, din, z1);
[z(1:50) z1(1:50) zhat(1:50)]
```

C.3 Who can benefit from E^4?

E^4 combines the intellectual input from many people who faced special needs that commercial software could not satisfy. If these needs coincide with yours, perhaps you should consider using E^4 for your work.

First of all, we needed **transparency**. Statistical software often hides relevant computational outputs, such as the likelihood, gradient, or the information matrix conditioning number. This happens because these results convey one of two messages: either that everything worked as expected, or that there is a problem with the model, the data, or the software. Providing this information may not be the best way to do "business" but, for some users at least, they are important pieces of the puzzle.

Second, we needed **reliability**. Mainstream software tends to prioritize computational speed at the expense of numerical accuracy. This can be done in different ways such as, e.g., choosing fast but sloppy computational algorithms or easy-to-satisfy default convergence criteria. Speed is crucial for a real-time computational system. However, for academic work and many professional applications, speed is attractive but not so valuable when compared to precision or computational reliability.

Third, we needed **consistency**. Common time series packages implement different procedures to perform the same task for different models, and this may produce substantial inconsistencies. These inconsistencies are often due to the intrinsic difficulty of making a clean distinction between models and procedures. For example, a standard time series package may have a specific procedure to forecast an ARIMA model and another one for VARMA, without any guarantee that their outputs will be mutually consistent. E^4 avoids this shortcoming by translating internally all the models supported to the SS form, so that, for example, in the ARIMA and VARMA case, a unique forecasting procedure is employed.

Our fourth and last need was strictly personal: we wanted to exploit the software generated by research. Academic work often produces neat pieces of code which do a specific task well. This software could be useful to other people in an adequate framework, but if this framework does not exist, it will probably vanish forever into the "limbo" reserved for neat pieces of code after the paper is published. On the other hand, a coordinated effort to put together, organize, and document these scientific by-products, may be academically profitable for both authors and end-users.

Our personal answer to these questions was E^4, a MATLAB toolbox intended to harbor the collection of results from a research line in SS methods, whose origins extend back into the early 1980s.

E^4 is our main research tool and, accordingly, it prioritizes academic needs. As a consequence, its functions provide a transparent, well-documented, and user-configurable suite of outputs that provides many clues about the problem such as lack of convergence or ill-conditioning. Also, the code has been carefully optimized for numerical accuracy and robustness, making an intensive use of stability-inducing methods, such as factorization and scaling. Finally, computational tolerances are transparent, user-configurable, and have strict default values.

E^4 does all the important calculations using the equivalent SS form of the models employed. As noted before, this feature is important in enforcing consistency, since all the core algorithms are developed for a SS model. Therefore, when an E^4 function receives an input model, the first thing it does is obtain the equivalent SS representation, and only then does it apply the corresponding procedure. This approach cleanly separates models from procedures, assures a high degree of coherence in the results, simplifies code development, and, last but not least, makes the toolbox very easy to extend, because implementing new models with equivalent SS representations only requires coding new user interface functions. Conversely, a new procedure only needs to support the SS model, because, thanks to the bridge functions provided, it can be immediately applied to all the models supported.

E^4 has been an effective repository to organize and distribute the code resulting from our research. Every new research project we undertake adds new functionality to E^4, which is documented not only in the user manual, but also, with exceptional depth, in the corresponding papers. In 2000 we launched a website (`www.ucm.es/e-4/`), containing the source code for the Toolbox as well as complete documentation. Appendix D provides more information about these issues.

Appendix D

Downloading E^4 and the examples in this book

D.1 The E^4 website 243

D.2 Downloading and installing E^4 243

D.3 Downloading the code for the examples in this book 245

D.1 The E^4 website

You can find the E^4 website at the URL: `https://www.ucm.es/e-4/`. This website distributes many E^4-related materials, including the complete toolbox code, a user manual, as well as all the code and data required to replicate the examples in this book.

D.2 Downloading and installing E^4

To download E^4 you need to:

1. Access `https://www.ucm.es/e-4` using your WEB browser.
2. Click the "Downloads" item in the left-hand-side menu (alternatively, you can type the address `https://www.ucm.es/e-4/downloads` in the navigation bar of your browser).
3. Download the file `E4.zip` in a folder of your choice and extract the files contained in it. By doing so, you will obtain 150+ text files, most of them with the MATLAB `.m` extension, which constitute the core of E^4 code.
4. Optionally, download other optional materials such as: (a) a user manual for E^4, (b) source code and data for the examples in the user manual, (c) source code and data for the examples in this book, as well as (d) the documentation for several groups of specialized functions which are distributed with E^4 but not covered by the user manual, and (e) beta-versions of some functions which are not yet included in the core of E^4.

The only installation required consists of including the folder(s) with E^4 code in the MATLAB search path[1]. To this end, you will need to select "Set Path" from the File menu in the initial MATLAB window and the add the E^4 folder using the menu options. Alternatively, you can access the same menu by typing `pathtool` at the MATLAB Command Window prompt. To edit this search path you may need to execute MATLAB with full administrative rights. Refer to your Operating System documentation, or to specialized Internet resources, to see how to do this.

To use the E^4 functions, you will also need to type the command:

```
e4init;
```

at least once at the beginning of each E^4 session. This command initializes the toolbox options and parameters, and its output is similar to:

```
                         EEEEEEEEE   444  444
                        EEEEEEEEEEE  444  444
                        EEE          44444444
                        EEE           4444444
                        EEEEEEE           444
                        EEEEEEE           444
                        EEE
                        EEE
                        EEEEEEEEEE
                         EEEEEEEE


Toolbox for State Space Estimation of Econometric Models
                  Version  APR-2013

Web: www.ucm.es/e-4/

*********************** Options set by user ********************
Filter. . . . . . . . . . . . . . : KALMAN
Scaled B and M matrices . . . . : NO
Initial state vector. . . . . . : AUTOMATIC SELECTION
Initial covariance of state v.  : IDEJONG
Variance or Cholesky factor? .  : VARIANCE
Optimization algorithm. . . . . : BFGS
Maximum step length . . . . . . : 0.100000
Stop tolerance. . . . . . . . . : 0.000010
Max. number of iterations . . . :        75
Verbose iterations. . . . . . . : YES
****************************************************************
```

Bear in mind that E^4 will not work properly if this `e4init` is not run.

[1]This search path is a list of folders where MATLAB will look for the code of any function which is not part of its core.

After these installation and initialization steps have been completed, MATLAB is able to use the E^4 library.

D.3 Downloading the code for the examples in this book

In the Downloads section of [`https://www.ucm.es/e-4/downloads`] you will find an item titled "Source code and data for the examples in the book: State-Space Methods for Time Series Analysis: Theory, Applications and Software." By clicking on this item you will download a file titled `SSBookExamples.zip`. You should extract its content to any folder of your choice. By doing this, you will create the folders: `ExamplesChapter3`, `ExamplesChapter4`,... and so on until `ExamplesChapter11`. These folders contain the source code and data required to replicate the examples in the corresponding Chapter.

In each of these folders you will find either code files or new folders. In both cases, the name of the file/folder refers to a specific example, e.g. the files contained in `ExamplesChapter3` are:

```
Example0311.m
Example0312.m
Example0313.m
Example0321.m
Example0322.m
Example0331.m
```

where the first nine characters, `Example03`, refer to the Chapter, the tenth character refers to the Section containing the example, and the last character is the sequential number of the example. Therefore, the name `Example0311.m` refers to the first example in Chapter 3, Section 1, `Example0312.m` corresponds to the second example in the same Chapter and Section, and so on.

In other cases the main folder will contain several sub-folders. For example, `ExamplesChapter8` includes the sub-folders:

```
Example0851 (Simulated data)
Example0852 (Wheat price)
Example0853 (Lydia Pinkham data)
```

where the first part of the folder name follows the naming convention described above, and the free text within parentheses refers to the specific dataset employed. Each of these folder contains MATLAB code (in `.m` files) and data (in text files with the extension `.dat`). In some cases you will also find and Excel worksheets with additional information about the dataset and the results.

The `ExamplesChapter11` folder is a little different since the examples were coded in `R`, and it contains just three files:

```
Example1161.R
Example1162.R
whale.txt
```

where `whale.txt` is the dataset required by the second example in Section 6 (`Example1162.R`). The first two files are chunks of `R` code meant to be executed sequentially, with accompanying comments to indicate what exactly is being computed. Each of these may require the user to load specific R libraries[2]. In particular, both examples make liberal use of the "astsa" library that accompanies the book by Shumway and Stoffer [196]

[2]See the website `https://www.r-project.org/` for detailed information on R.

Bibliography

[1] S. Ahn and G. Reinsel. Estimation for partially nonstationary multivariate autoregressive models. *Journal of the American Statistical Association*, 85:813–823, 1990.

[2] H. Akaike. *Canonical Correlation Analysis of Time Series and the Use of an Information Criterion*. New York: Academic Press, 1976.

[3] H. Akaike. Covariance matrix computation of the state variable of a stationary gaussian process. *Annals of the Institute of Statistics and Mathematics*, Series B, 30:499–504, 1978.

[4] G. Alexander and P. Benson. More on beta as a random coefficient. *Journal of Financial and Quantitative Analysis*, XVII, 1:27–36, 1982.

[5] T. Amemiya and R. Wu. The effect of aggregation on prediction in the autoregressive model. *Journal of the American Statistical Association*, 339:628–632, 1972.

[6] B. Anderson and J. Moore. *Optimal Filtering*. Englewood Cliffs (N.J.): Prentice Hall, 1979.

[7] C. Ansley and R. Kohn. Exact likelihood of vector autoregressive-moving average process with missing or aggregated data. *Biometrika*, 70:275–278, 1983.

[8] C. Ansley and R. Kohn. Estimation, filtering and smoothing in state space models with incompletely specified initial conditions. *The Annals of Statistics*, 13:1286–1316, 1985.

[9] C. Ansley and R. Kohn. Filtering and smoothing in state space models with partially diffuse initial conditions. *Journal Time Series Analysis*, 11:275–293, 1990.

[10] C. Ansley and P. Newbold. Finite sample properties of estimators for autoregression moving average models. *Journal of Econometrics*, 13:159–183, 1980.

[11] M. Aoki. *State Space Modelling of Time Series*. Springer Verlag, New York, 1990.

[12] K. Arun and S. Kung. Balanced approximation of stochastic systems. *SIAM Journal of Matrix Analysis and Applications*, 11(1):42–68, 1990.

[13] G. Baiocchi and W. Distaso. GRETL: Econometric software for the GNU generation. *Journal of Applied Econometrics*, 18:105–110, 2003.

[14] R. Barnard, A. Trindade, and R. Wickramasinghe. Autoregressive moving average models under exponential power distributions. *ProbStat Forum*, 7:65–77, 2014.

[15] R. Bartels and G. Stewart. Solution of the equation ax + xb = c. *Communications ACM*, 15:820–826, 1972.

[16] D. Bauer. Order estimation for subspace methods. *Automatica*, 37:1561–1573, 2001.

[17] D. Bauer and L. Ljung. Some facts about the choice of the weighting matrices in Larimore type of subspace algorithms. *Automatica*, 38:763–773, 2002.

[18] D. Bauer and M. Wagner. Estimating cointegrated systems using subspace algorithms. *Journal of Econometrics*, 111:47–84, 2002.

[19] W. Bell and S. Hillmer. Initializing the Kalman filter for nonstationary time series models. *Journal of Time Series Analysis*, 12:283–300, 1991.

[20] T. Bengtsson and J. Cavanaugh. An improved Akaike information criterion for state-space model selection. *Computational Statistics and Data Analysis*, 50(10):2635–2654, 2006.

[21] M. Bentarzi and M. Hallin. On the invertiibility of periodic moving-average models. *Journal of Time Series Analysis*, 15, 3:263–268, 1994.

[22] A. Bera, M. Higgins, and S. Lee. Interaction between autocorrelation and conditional heterocedasticity: a random-coefficient approach. *Journal of Business and Economic Statistics*, 10, 2:133–142, 1992.

[23] R. Bitmead, M. Gevers, I. Petersen, and R. Kaye. Monotonicity and stabilizability properties of the solutions of the Riccati difference equation: Propositions, lemmas, theorems, fallacious conjectures and counterexamples. *System Control Letters*, 5:309–315, 1985.

[24] K. Bollen. *Structural Equations With Latent Variables*. Wiley, New York, 1989.

[25] T. Bollerslev, R. Engle, and D. Nelson. ARCH models. In R. Engle and D. McFadden, editors, *Handbook of Econometrics*, volume IV. North-Holland, 1994.

[26] J. Boot, W. Feibes, and J. Lisman. Further methods of derivation of quarterly figures from annual data. *Applied Statistics*, 16:65–75, 1967.

[27] P. Boswijk and P. Franses. Unit roots in periodic autoregressions. *Journal of Time Series Analysis*, 17, 3:221–245, 1995.

[28] G. Box and D. Cox. An analysis of transformations. *Journal of the Royal Statistical Society, Series B*, 26:211–243, 1964.

[29] G. Box and G. Jenkins. *Time Series Analysis, Forecasting and Control.* Holden-Day, San Francisco, 1970.

[30] G. Box, G. Jenkins, and G. Reinsel. *Time Series Analysis, Forecasting and Control*. Prentice Hall, Englewood Cliffs, 1994.

[31] G. Box and G. Tiao. Intervention analysis with applications to economic and environmental problems. *Journal of the American Statistical Association*, 70(349):70–79, 1975.

[32] P. Brockwell and R. Davis. *Time Series: Theory and Methods*. Springer-Verlag, New York, 2nd edition, 1991.

[33] M. Bujosa, A. Garcia-Ferrer, and P. Young. Linear dynamic harmonic regression. *Computational Statistics and Data Analysis*, 52(2):999–1024, 2007.

[34] J. Burman. Seasonal adjustment by signal extraction. *Journal of the Royal Statistical Society, Series A*, 143:321–337, 1980.

[35] I. Burr. *Statistical Quality Control Methods*. Marcel Dekker, New York, 1976.

[36] Z. Cai, J. Fan, and Q. Yao. Functional-coefficient regression models for nonlinear time series. *Journal of the American Statistical Association*, 95, 451:941–956, 2000.

[37] P. Caines. *Linear Stochastic Systems*. John Wiley, New York, 1988.

[38] J. Casals. *Métodos de Subespacios en Econometría*. PhD thesis, Universidad Complutense de Madrid, 1997.

[39] J. Casals, A. Garcia-Hiernaux, and M. Jerez. From general State-Space to VARMAX models. *Mathematics and Computers in Simulation*, 80(5):924–936, 2012.

[40] J. Casals, M. Jerez, and S. Sotoca. Exact smoothing for stationary and nonstationary time series. *International Journal of Forecasting*, 16(1):199–209, 2000.

[41] J. Casals, M. Jerez, and S. Sotoca. An exact multivariate model-based structural decomposition. *Journal of the American Statistical Association*, 97(458):553–564, 2002.

[42] J. Casals, M. Jerez, and S. Sotoca. Decomposition of a state-space model with inputs. *Journal of Statistical Computation and Simulation*, 80, 9:979–992, 2010.

[43] J. Casals and S. Sotoca. The exact likelihood for a state space model with stochastic inputs. *Computers and Mathematics with Applications*, 42:199–209, 2001.

[44] J. Casals, S. Sotoca, and M. Jerez. Un algoritmo rápido para evaluar la verosimilitud exacta de modelos VARMAX periódicos. *Estadística Española*, 40, 143:269–291, 1998.

[45] J. Casals, S. Sotoca, and M. Jerez. A fast and stable method to compute the likelihood of time invariant state space models. *Economics Letters*, 65(3):329–337, 1999.

[46] J. Casals, S. Sotoca, and M. Jerez. Modeling and forecasting time series sampled at different frequencies. *Journal of Forecasting*, 28, 4:316–342, 2009.

[47] J. Casals, S. Sotoca, and M. Jerez. Minimally conditioned likelihood for a nonstationary state space model. *Mathematics and Computers in Simulation*, 100C:24–40, 2014.

[48] G. Casella and R. Berger. *Statistical Inference*. Duxbury, Pacific Grove, USA., 2nd edition, 2002.

[49] S. Chan, G. Goodwin, and K. Sin. Convergence properties of the Riccati difference equation in optimal filtering of nonstabilizable systems. *IEEE Transactions on Automatic Control*, AC-29, 2:110–118, 1984.

[50] H. Chang and Y. Hsing. The demand for residential electricity: New evidence on time-varying elasticities. *Applied Economics*, 23, 7:1251–56, 1991.

[51] S. Chatterjee, M. Handcock, and J. Simonoff. *A Casebook for a First Course in Statistics and Data Analysis*. Wiley, New York, 1995.

[52] G. Chow and A. Lin. Best linear unbiased interpolation, distribution and extrapolation of time series by related series. *The Review of Economics and Statistics*, 53:372–375, 1971.

[53] S. Chow, M. Ho, E. Hamaker, and C. Dolan. Equivalence and differences between structural equation modeling and state-space modeling techniques. *Structural Equation Modeling*, 17:303—332, 2010.

[54] W. Cleveland and S. Devlin. Calendar effects in monthly time series: Modeling and adjustment. *Journal of the American Statistical Association*, 77, 379:520–528, 1982.

[55] J. Commandeur and S. Koopman. *An Introduction to State Space Time Series Analysis*. Oxford University Press, 2007.

[56] T. Cooley and E. Prescott. An adaptive regression model. *International Economic Review*, 14, 2:364–371, 1973.

[57] D. M. Cooper and E. F. Wood. Identifying multivariate time series models. *Journal of Time Series Analysis*, 3:277–293, 1982.

[58] H. Cramer. *Mathematical Methods of Statistics*. Princeton University Press, Princeton (N.J.), 1946.

[59] M. J. Crowder. Maximum likelihood estimation for dependent observations. *Journal of the Royal Statistical Society. Series B (Methodological)*, 38, 1:45–53, 1976.

[60] B. Datta. *Numerical Linear Algebra and Applications*. Brooks/ Cole, 1995.

[61] R. Davidson and J. MacKinnon. *Econometric Theory and Methods*. Oxford University Press, New York, 2004.

[62] R. Davidson and J. G. MacKinnon. *Estimation and Inference in Econometrics*. Oxford University Press, New York, 1993.

[63] R. Davis, M. Chen, and W. Dunsmuir. Inference for MA(1) processes with a unit root on or near the unit circle. *Probability and Mathematical Statistics*, 15:227–242, 1995.

[64] R. Davis and W. Dunsmuir. Least absolute deviation estimation for regression with ARMA errors. *Journal of Theoretical Probability*, 10(2):481–497, 1997.

[65] R. Davis and L. Song. Unit roots in moving averages beyond first order. *The Annals of Statistics*, 39(6):3062–3091, 2011.

[66] P. de Jong. The likelihood for a state space model. *Biometrika*, 75, 1:165–169, 1988.

[67] P. de Jong. Smoothing and interpolation with the state-space model. *Journal of the American Statistical Association*, 84:1085–1088, 1989.

[68] P. de Jong. The diffuse Kalman filter. *Annals of Statistics*, 2:1073–1083, 1991.

[69] P. de Jong and S. Chu-Chun-Lin. Fast likelihood evaluation and prediction for nonstationary state space models. *Biometrika*, 81:133–142, 1994.

[70] P. de Jong and J. Penzer. Diagnosing shocks in time series. *Journal of the American Statistical Association*, 93, 442:796–806, 1998.

[71] C. De Souza, M. Gevers, and G. Goodwin. Riccati equations in optimal filtering of nonstabilizable systems having singular state transition matrices. *IEEE Transactions on Automatic Control*, 31:831–838, 1986.

[72] J. S. Demetry. A note on the nature of optimality in the discrete Kalman filter. *IEEE Transactions on Automatic Control*, AC-15:603–604, 1970.

[73] A. P. Dempster, N. M. Laird, and D. B. Rubin. Maximum likelihood from incomplete data via the EM algorithm. *Journal of the Royal Statistical Society B*, 39:1–38, 1977.

[74] J. E. Dennis and R. B. Schnabel. *Numerical methods for unconstrained optimization and nonlinear equations*. Prentice Hall, Englewood Cliffs, NJ, 1996.

[75] F. Denton. Adjustment of monthly or quarterly series to annual totals: An approach based on quadratic minimization. *Journal of the American Statistical Association*, 66:99–102, 1971.

[76] T. Di Fonzo. The estimation of m disaggregate time series when contemporaneous and temporal aggregates are known. *The Review of Economics and Statistics*, 72:178–182, 1990.

[77] D. A. Dickey and W. A. Fuller. Distribution of the estimators for autoregressive time series with unit root. *Journal of the American Statistical Association*, 74:427–431, 1979.

[78] D. A. Dickey and W. A. Fuller. Likelihood ratio statistics for autoregressive time series with a unit root. *Econometrica*, 49:1057–1072, 1981.

[79] B. Dickinson, M. Morf, and T. Kailath. Canonical matrix fraction and state space descriptions for deterministic and stochastic linear systems. *IEEE Transactions on Automatic Control*, AC-19:656–667, 1974.

[80] P. J. Diggle, P. Heagerty, K. Liang, and S. L. Zeger. *Analysis of Longitudinal Data*. Oxford University Press, New York, 2nd edition, 2002.

[81] J. M. Dufour and D. Pelletier. Linear estimation of weak VARMA models with macroeconomic applications. Technical report, Centre Interuniversitaire

de Recherche en Economie Quantitative (CIREQ), Université de Montréal, CANADA, 2004.

[82] J. Durbin and S. Koopman. *Time Series Analysis by State Space Methods*. Oxford University Press, 2012.

[83] J. Durbin and B. Quenneville. Benchmarking by state space models. *International Statistical Review / Revue Internationale de Statistique*, 65:23–48, 1997.

[84] R. Engle and T. Bollerslev. Modelling the persistence of condicional variances. *Econometric Reviews*, 5:1–50, 1986.

[85] R. Engle and K. Kroner. Multivariate simultaneous generalized ARCH. *Econometric Theory*, 11:122–150, 1995.

[86] R. F. Engle. Autoregressive conditional heteroskedasticity with estimates of the variance of united kingdom inflations. *Econometrica*, 50:987–1007, 1982.

[87] R. F. Engle and S. Kozicki. Testing for common features. *Journal of Business and Economic Statistics*, 11(4):369–395, 1993.

[88] P. Evans, B. Kim, and K. Oh. Capital mobility in saving and investment: a time-varying coefficients approach. *Journal of International Money and Finance*, 27:806–815, 2008.

[89] F. Fabozzi and J. Francis. Beta as a random coefficient. *Journal of Financial and Quantitative Analysis*, 13:101–116, 1978.

[90] L. Fahrmeir and G. Tutz. *Multivariate Statistical Modelling Based on Generalized Linear Models*. Springer, New York, 2nd edition, 2001.

[91] J. Fan, J. Jiang, C. Zhang, and Z. Zhou. Time-dependent difussion models for term structure dynamics and the stock price volatility. *Statistica Sinica*, 13:965–992, 2003.

[92] J. Fan and W. Zhang. Statistical estimation in varying coefficient models. *Annals of Statistics*, 27, 5:1491–1518, 1999.

[93] J. Fan and W. Zhang. Statistical methods with varying coefficient models. *Statistics and Its Interface*, 1:179–195, 2008.

[94] W. Favoreel, B. De Moor, and P. Van Overschee. Subspace state space system identification for industrial processes. *Journal of Process Control*, 10:149–155, 2000.

[95] R. Fernandez. A methodological note on the estimation of time series. *The Review of Economics and Statistics*, 63:471–476, 1981.

[96] R. Fisher. On the mathematical foundations of theoretical statistics. *Philosophical Transactions of the Royal Society of London. Series A*, 222:309–368, 1921.

[97] R. Flores and G. Serrano. A generalized least squares estimation method for VARMA models. *Statistics*, 36(4):303–316, 2002.

[98] D. Y. T. Fong. *State Space Models and Filtering Methods in Longitudinal Studies*. PhD thesis, University of Waterloo, 1997.

[99] C. Francq and J. M. Zakoian. Estimating weak GARCH representations. *Econometric Theory*, 16:692–728, 2000.

[100] P. Franses. *Periodicity and Stochastic Trends in Economic Time Series*. Oxford University Press, 1996.

[101] P. Franses and R. Paap. Model selection in periodic autoregressions. *Oxford Bulletin of Economics and Statistics*, 56, 4:421–439, 1994.

[102] A. Garcia-Hiernaux. *Identificación de Modelos para Series Temporales mediante Métodos de Subespacios*. PhD thesis, Universidad Complutense de Madrid, Febrero 2005.

[103] A. Garcia-Hiernaux. Forecasting linear dynamical systems using subspace methods. *Journal of Time Series Analysis*, 32(5):462–468, 2011.

[104] A. Garcia-Hiernaux, M. Jerez, and J. Casals. Fast estimation methods for time series models in state-space form. *Journal of Statistical Computation and Simulation*, 79(2):121–134, 2009.

[105] A. Garcia-Hiernaux, M. Jerez, and J. Casals. Unit roots and cointegration modeling through a family of flexible information criteria. *Journal of Statistical Computation and Simulation*, 80(2):173–189, 2010.

[106] A. Garcia-Hiernaux, M. Jerez, and J. Casals. Estimating the system order by subspace methods. *Computational Statistics*, 27(3):411–425, 2012.

[107] E. Ghysels and D. Osborn. *The Econometric Analysis of Seasonal Time Series*. Cambridge University Press, 2001.

[108] K. Glover and J. C. Willems. Parameterizations of linear dynamical systems: Canonical forms and identifiability. *IEEE Transactions on Automatic Control*, 19, 6:640–646, 1974.

[109] G. Golub and C. Van Loan. *Matrix Computations*. John Hopkins University Press, Baltimore, 1996.

[110] V. Gomez and A. Maravall. Estimation prediction and interpolation for nonstationary series with the Kalman filter. *Journal of the American Statistical Association*, 89:611–624, 1994.

[111] V. Gomez., A. Maravall, and D. Pena. Missing observations in ARIMA models: Skipping approach versus additive outlier approach. *Journal of Econometrics*, 88, 2:341–363, 1998.

[112] R. L. Goodrich and P. E. Caines. Linear system identification from nonstationary cross-sectional data. *IEEE Transactions on Automatic Control*, 24:403—411, 1979.

[113] C. Granger and P. Siklos. Systematic sampling, temporal aggregation, seasonal adjustment and cointegration: Theory and evidence. *Journal of Econometrics*, 66:357–369, 1995.

[114] P. Hackl. Demand for international telecommunication: time-varying price elasticity. *Journal of Econometrics*, 70:243–260, 1996.

[115] H. Hamilton. *Time Series Analysis*. Princeton University Press, 1993.

[116] E. J. Hannan and M. Deistler. *The Statistical Theory of Linear Systems*. John Wiley, New York, 1988.

[117] E. J. Hannan and B. G. Quinn. The determination of the order of an autoregression. *Journal of the Royal Statistical Association, Series B*, 41:713–723, 1979.

[118] A. Harvey. *Forecasting, structural time series models and the* Kalman *Filter*. Cambridge University Press, 1989.

[119] A. Harvey, E. Ruiz, and N. Shephard. Multivariate stochastic variance models. *Review of Economic Studies*, 61, 2:247–264, 1994.

[120] A. Harvey and T. Trimbur. Trend estimation and the hodrick-prescott filter. *Journal of the Japan Statistical Society*, 38:41–49, 2008.

[121] D. A. Harville. *Matrix algebra from a statistician's perspective*. Springer, New York, 1997.

[122] T. Hastie and R. Tibshirani. Varying-coefficient models. *Journal of the Royal Statistical Society. Series B (Methodological)*, 55, 4:757–796, 1993.

[123] G. Hewer. An iterative technique for the computation of the steady state gains for the discrete optimal regulator. *IEEE Transaction on Automatic Control*, AC-16, 4:382–384, 1971.

[124] S. Hillmer and G. Tiao. Likelihood function of stationary multiple autoregressive moving average models. *Journal of the American Statistical Association*, 74:652–660, 1979.

[125] S. Hillmer and G. Tiao. An ARIMA-model-based approach to seasonal adjustment. *Journal of the American Statistical Association*, 77:63–70, 1982.

[126] R. Hodrick and E. Prescott. Post-war U.S. business cycles: An empirical investigation. *Journal of Money, Credit and Banking*, 29:1–16, 1997.

[127] C. Hsiao. *Analysis of Panel Data*. Cambridge University Press, New York, 2014.

[128] C. Ionescu, V. Oara and M. Weiss. General matrix pencil techniques for the solution of algebraic Riccati equations: A unified approach. *IEEE Transactions on Automatic Control*, 42, 8:1085–1097, 1997.

[129] M. Jamshidian and R. I. Jennrich. Standard errors for EM estimation. *Journal of the Royal Statistical Society. Series B, Statistical Methodology*, 62, 2:257–270, 2000.

[130] A. Jazwinsky. *Stochastic Processes and Filtering Theory*. Academic Press, New York, 1970.

[131] G. Jenkins and A. Alavi. Somes aspects of modelling and forecasting multivariate time series. *Journal of Time Series Analysis*, 2(1):1–47, 1981.

[132] R. Jones. Maximum likelihood fitting of ARMA models to time series with missing observations. *Technometrics*, 22, 3:389–395, 1980.

[133] R. H. Jones. *Longitudinal Data with Serial Correlation: A State-Space Approach*. Chapman & Hall/CRC, Boca Raton, 1993.

[134] R. Kalman. A new approach to linear filtering and prediction problems. *Transactions of the ASME-Journal of Basic Engineering (Series D)*, 82:35–45, 1960.

[135] G. Kauermann and G. Tutz. On model diagnostics using varying coefficient models. *Biometrika*, 86, 1:119–128, 1999.

[136] R. Kohn and C. Ansley. Signal extraction for finite nonstationary time series. *Biometrika*, 74, 2:411–421, 1987.

[137] R. Kohn and C. F. Ansley. Estimation, prediction and interpolation for ARIMA models with missing data. *Journal of the American Statistical Association*, 81:751–761, 1986.

[138] R. Kohn and C. F. Ansley. A fast algorithm for signal extraction, influence and cross-validation in state space models. *Biometrika*, 76, 1:65–79, 1989.

[139] S. Koopman. Exact initial Kalman filtering and smoothing for nonstationary time series models. *Journal of the American Statistical Association*, 92:1630–1638, 1997.

[140] S. Koopman and N. Shephard. Exact score for time series models in state space form. *Biometrika*, 79, 4:823–826, 1992.

[141] S. Koopman, N. Shephard, and J. Doornik. *SsfPack 3.0: Statistical Algorithms for Models in State Space Form.* Timberlake Consultants Press, London, 2008.

[142] S. J. Koopman and J. Durbin. Filtering and smoothing of state vector for diffuse state-space models. *Journal of Time Series Analysis*, 24, 1:85–98, 2003.

[143] S. G. Koreisha and T. H. Pukkila. Fast linear estimation methods for vector autoregressive moving average models. *Journal of Time Series Analysis*, 10:325–339, 1989.

[144] S. G. Koreisha and T. H. Pukkila. A generalized least-squares approach for estimation of autoregressive moving-average models. *Journal of Time Series Analysis*, 11:139–151, 1990.

[145] S. Y. Kung. A new identification and model reduction algorithm via singular value decompositions. In *Proceedings of the 12th Asilomar Conference on Circuits, Systems and Computers, Pacific Grove*, pages 705–714, 1978.

[146] N. M. Laird and J. H. Ware. Random-effects models for longitudinal data. *Biometrics*, 38:963—974, 1982.

[147] W. E. Larimore. Canonical variate analysis in identification, filtering and adaptive control. *Proceedings of the 29th Conference on Decision and Control, Hawaii*, pages 596–604, 1990.

[148] W. K. Li. *Diagnostic Checks in Time Series*. Chapman and Hall/CRC, Florida, 2004.

[149] R. Litterman. A random walk, markov model for the distribution of time series. *Journal of Business and Economic Statistics*, 1:169–173, 1983.

[150] T. D. Little. *Longitudinal Structural Equation Modeling*. Guilford Press, New York, 2013.

[151] G. M. Ljung and G. E. P. Box. On a measure of lack of fit in time series models. *Biometrika*, 65:297–303, 1978.

[152] L. Ljung. *System Identification, Theory for the User*. PTR Prentice Hall, New Jersey, 1999.

[153] L. Ljung and P. Caines. Asymptotic normality of prediction error estimators for approximate system models. *Stochastic*, 3:29–46, 1979.

[154] R. Lucas. Econometric policy evaluation: A critique. *Carnegie-Rochester Conference Series on Public Policy*, 1, 1:19–46, 1976.

[155] H. Lutkepohl. *Forecasting Aggregated Vector* ARMA *Processes*. Springer-Verlag, Berlin, 1987.

[156] H. Lutkepohl. *Introduction to Multiple Time Series Analysis*. Springer Verlag, Berlin, 1993.

[157] H. Lutkepohl. *Handbook of Matrices*. Wiley, Chichester, 1996.

[158] H. Lutkepohl. *New Introduction to Multiple Time Series Analysis*. Springer-Verlag, Berlin, 2005.

[159] H. Lutkepohl and D. S. Poskitt. Specification of echelon form VARMA models. *Journal of Business and Economic Statistics*, 14(1):69–79, 1996.

[160] S. G. Makridakis, S. C. Wheelwright, and V. E. McGee. *Forecasting: Methods and Applications*. John Wiley & Sons, 1983.

[161] M. Marcellino. Some consequences of temporal aggregation in empirical analysis. *Journal of Business and Economic Statistics*, 1:129–136, 1999.

[162] D. Margaritis. A time-varying model of rational learning. *Economics Letters*, 33:309–314, 1990.

[163] R. K. Mehra. Identification and estimation of the error-invariables model (EVM) in structural form. In R. Web, editor, *Stochastic Systems: Modelling Identification and Optimization*, volume 1. North-Holland, Amsterdam, 1976.

[164] G. Mélard, R. Roy, and A. Saidi. Exact maximum likelihood estimation of structured or unit root multivariate time series models. *Computational Statistics and Data Analysis*, 50,11:2958–2986, 2006.

[165] A. Naranjo, A. A. Trindade, and G. Casella. Extending the state-space model to accommodate missing values in responses and covariates. *Journal of the American Statistical Association*, 108, 501:202–216, 2013.

[166] M. Nerlove, D. M. Grether, and J. L. Carvalho. *Analysis of Economic Time Series: A Synthesis*. Academic Press, New York, 1995.

[167] D. Nicholls and B. Quinn. *Random Coefficient Autoregressive Models: An Introduction*. John Wiley & Sons, 1982.

[168] A. Novales and R. Flores. Forecasting with periodic models a comparison with time invariant coefficient models. *International Journal of Forecasting*, 13, 3:393–405, 1997.

[169] L. Nunes. Nowcasting quarterly GDP growth in a monthly coincident indicator model. *Journal of Forecasting*, 24:575–592, 2005.

[170] T. M. O'Donovan. *Short Term Forecasting*. John Wiley & Sons, 1983.

[171] D. Osborn and J. Smith. The performance of periodic autoregressive models in forecasting seasonal u.k. consumption. *Journal of Business & Economic Statistics*, 7, 1:117–127, 1989.

[172] J. H. L. Oud and R. A. R. Jansen. Continuous time state space modeling of panel data by means of SEM. *Psychometrika*, 65:199–215, 2000.

[173] A. Pagan. Some identification and estimation results for regression models with stochastically varying coefficients. *Journal of Econometrics*, 13, 3:341–363, 1980.

[174] R. Paige, A. Trindade, and R. Wickramasinghe. Extensions of saddlepoint-based bootstrap inference. *Annals of the Institute of Statistical Mathematics*, 66(5):961–981, 2014.

[175] K. Palda. *The Measurement of Cumulative Advertising Effects*. Prentice-Hall, Englewood Cliffs (N.J.), 1964.

[176] T. Pappas, A. Laub, and N. Sandell. On the numerical solution of the algebraic Riccati equation. *IEEE Transactions on Automatic Control*, AC-25, 4:631–641, 1980.

[177] B. U. Park, E. Mammen, Y. K. Lee, and E. Lee. Varying coefficient regression models: A review and new developments. *International Statistical Review*, 83, 1:36–64, 2014.

[178] S. Park and G. Zhao. An estimation of U.S. gasoline demand: A smooth time-varying cointegration approach. *Energy Economics*, 32, 1:110–120, 2010.

[179] M. M. Pelagatti. State space methods in Ox/SsfPack. *Journal of Statistical Software*, 41, 3:1–25, 2011.

[180] K. Peternell. *Identification of Linear Dynamical Systems by Subspace and Realization-Based Algorithms*. PhD thesis, TU Wien, 1995.

[181] P. Petkov, N. Christov, and M. Konstantinov. *Computational Methods for Linear Control Systems*. Prentice-Hall, Englewood Cliffs, New Jersey, 1991.

[182] G. Petris and S. Petrone. State space models in R. *Journal of Statistical Software*, 41:1–25, 2011.

[183] R. Pierse and A. Snell. Temporal aggregation and the power of tests for a unit root. *Journal of Econometrics*, 65:333–345, 1995.

[184] J. Pinheiro and D. Bates. *Mixed-Effects Models in S and S-PLUS*. Springer, New York, 2000.

[185] S. J. Qin. An overview of subspace identification. *Computers and Chemical Engineering*, 30:1502–1513, 2006.

[186] R Core Team. *R: A Language and Environment for Statistical Computing*. R Foundation for Statistical Computing, Vienna, Austria, 2014.

[187] C. G. Reinsel. *Elements of Multivariate Time Series Analysis*. Springer-Verlag, New York, 2nd edition, 1997.

[188] B. Rosenberg. *Varying Parameter Estimation*. PhD thesis, Harvard University, 1968.

[189] B. Rosenberg. The analysis of a cross section of time series by stochastically convergent parameter regression. *Annals of Economic and Social Measurement*, 2:399–428, 1973.

[190] H. H. Rosenbrock. *State Space and Multivariable Theory*. John Wiley, New York, 1970.

[191] G. Schwarz. Estimating the dimension of a model. *Annals of Statistics*, 6:461–464, 1978.

[192] F. Schweppe. Evaluation of likelihood functions for gaussian signals. *IEEE Transactions on Information Theory*, IT-11, 1:61–70, 1965.

[193] R. Selukar. State space modeling using sas. *Journal of Statistical Software*, 41, 12:1–13, 2011.

[194] R. H. Shumway and D. S. Stoffer. An approach to time series smoothing and forecasting using the EM algorithm. *Journal of Time Series Analysis*, 3:253–264, 1982.

[195] R. H. Shumway and D. S. Stoffer. *Time Series Analysis and Its Applications: With R Examples*. Springer, New York, 2nd edition, 2006.

[196] R. H. Shumway and D. S. Stoffer. *Time Series Analysis and Its Applications: With R Examples*. Springer, New York, 3rd edition, 2011.

[197] J. Sorelius. *Subspace-Based Parameter Estimation Problems in Signal Processing*. PhD thesis, Uppsala University, 1999.

[198] H. Spliid. A fast estimation method for the vector autoregressive moving average model with exogenous variables. *Journal of the American Statistical Association*, 78(384):843–849, 1983.

[199] J. Stock and M. Watson. Why has U.S. inflation become harder to forecast? *Journal of Money, Credit and Banking*, 39:3–33, 2007.

[200] D. S. Stoffer. *Estimation of Parameters in a Linear Dynamic System with Missing Observations*. PhD thesis, University of California, Davis, 1982.

[201] D. S. Stoffer and K. Wall. Bootstrapping state space models: Gaussian maximum likelihood estimation and the Kalman filter. *Journal of the American Statistical Association*, 86:1024–1033, 1991.

[202] P. Stoica, Y. Selen, and Jian Li. On information criteria and the generalized likelihood ratio test of model order selection. *Signal Processing Letters, IEEE*, 11(10):794–797, 2004.

[203] P. Stoica and T. Soderstrom. On nonsingular information matrices and local identifiability. *International Journal of Control*, 36:323–329, 1982.

[204] D. Stram and W. Wei. Temporal aggregation in the ARIMA process. *Journal of Time Series Analysis*, 7:279–292, 1986.

[205] P. Swamy and G. Tavlas. Random coefficient models: Theory and applications. *Journal of Economic Surveys*, 9:165–196, 1995.

[206] J. Terceiro. *Estimation of Dynamics Econometric Models with Errors in Variables*. Springer-Verlag, Berlin, 1990.

[207] G. Tiao. Asymptotic behavior of time series aggregates. *Biometrika*, 59:521–531, 1972.

[208] G. Tiao. *Vector* ARMA *Models*. John Wiley & Sons, 2001.

[209] G. C. Tiao and R. S. Tsay. Model specification in multivariate time series. *Journal of the Royal Statistical Society, Series B*, 51(2):157–213, 1989.

[210] H. Tong. *Nonlinear Time Series: A Dynamical System Approach*. Oxford University Press, Oxford, 1990.

[211] E. M. M. Toscano and V. A. Reisen. The use of canonical correlation analysis to identify the order of multivariate ARMA models: Simulation and application. *Journal of Forecasting*, 19:441–455, 2000.

[212] R. S. Tsay. Identifying multivariate time series models. *Journal of Time Series Analysis*, 10(4):357–372, 1989.

[213] F. Tusell. Kalman filtering in r. *Journal of Statistical Software*, 39:1–27, 2011.

[214] P. Van Dooren. A generalized eigenvalue approach for solving Riccati equations. *SIAM Journal of Scientific Statistic Computation*, 2:121–135, 1981.

[215] P. Van Overschee and B. De Moor. N4SID: Subspace algorithms for the identification of combined deterministic-stochastic systems. *Automatica*, 30(1):75–93, 1994.

[216] P. Van Overschee and B. De Moor. A unifying theorem for three subspace system identification algorithms. *Automatica*, 31(12):1853–1864, 1995.

[217] M. Verhaegen. Identification of the deterministic part of MIMO state space models given in innovations form from input-output data. *Automatica*, 30(1):61–74, 1994.

[218] A. Wald. Note on the consistency of the maximum likelihood estimate. *The Annals of Mathematical Statistics*, 20, 4:595–601, 1949.

[219] M. Watson and R. Engle. Alternative algorithms for the estimation of dynamic factor, MIMIC and varying coefficient regression models. *Journal of Econometrics*, 23, 3:385–400, 1983.

[220] W. Wei. *Some Consequences of Temporal Aggregation in Seasonal Time Series Models. In Seasonal Analysis of Economic Time Series*. Bureau of the Census, Washington D.C., 1978.

[221] W. Wei. *Time Series Analysis*. Addison Wesley, 2006.

[222] W. Wei and D. Stram. Disaggregation of time series models. *Journal of the Royal Statistical Society, Series B*, 52:453–467, 1990.

[223] M. West. Time series decomposition and analysis in a study of oxygen isotope records. *Biometrika*, 84:489–494, 1997.

[224] M. West and P. Harrison. *Bayesian Forecasting and Dynamic Models*. Springer-Verlag, New York, 1989.

[225] M. West and P. Harrison. *Bayesian Forecasting and Dynamic Models*. Springer-Verlag, New York, 2nd edition, 1997.

[226] H. White. Maximum likelihood estimation of misspecified models. *Econometrica*, 50, 1:1–25, 1982.

[227] H. O. A. Wold. *A Study in the Analysis of Stationary Time Series*. Almqvist and Wiksell, Uppsala, 1964.

[228] C. M. Woodside. Estimation of the order of linear systems. *Automatica*, 7:727–733, 1971.

Author Index

Ahn, S., 60
Akaike, H., 35, 47, 48, 89, 91
Alavi, A., 63, 121, 159
Alexander, G., 66
Amemiya, T., 154
Anderson, B., xxiii, 3, 12, 30, 35, 36, 47, 48, 76, 216
Ansley, C., 1, 34, 36, 47–49, 99, 235
Aoki, M., 89
Arun, K., 87

Baiocchi, G., xxiii
Barnard, R., 218
Bartels, R., 200
Bauer, D., 85, 89, 90, 99
Bell, .W, 48
Bengtsson, T., 91
Benson, P., 66
Bentarzi, M., 72
Bera, A., 66
Berger, R., 3
Bitmead, R., 207
Bollen, K., 169
Bollerslev, T., 74, 76, 228, 234
Boot, J., 165
Boswijk, P., 71
Box, G., xxiii, 60, 62, 95, 100, 104, 108, 121, 125, 143, 234
Brockwell, P., 218
Bujosa, M., 89, 135
Burman, J., 108, 114
Burr, I., 100

Cai, Z., 66
Caines, P., 169, 213, 214, 217–219, 238, 239
Carvalho, J., 141
Casals, J., 238
Casals, J., 1, 3, 13, 20, 24, 34, 37, 51, 52, 56, 70, 85, 90–92, 96, 99, 112, 132, 134, 150–153, 198, 203, 205, 236, 238–240
Casella, G., 1, 3, 34, 175, 181
Cavanaugh, J., 91
Chan, S., 204, 206
Chang, H., 66
Chatterjee, S., 183, 185
Chen, M., 126
Chow, G., 154, 165
Chow, S., 169
Christov, N., 11, 23, 48, 201, 206, 207
Chu-Chun-Lin, S., 13, 34, 47, 50, 52, 53, 57, 73, 112
Cleveland, W., 60, 108
Commandeur, J., xxiv
Cooley, T., 66
Cooper, D., 89
Cox, D., 95, 234
Cramer, H., 212
Crowder, M., 214

Datta, A., 194, 195
Davidson, R., 212, 213
Davis, R., 99, 126, 218
De Jong, P., 13, 34, 37, 47–50, 52–54, 57, 58, 73, 111, 112, 236
De Moor, B., 20, 56, 84–88
De Souza, C., 13, 39, 40, 204–207
Deistler, M., xxiii, 131, 154, 217
Demetry, J., 35
Dempster, A., 175
Dennis, J., 46, 238, 239
Denton, F., 165
Devlin, S., 60, 108

Di Fonzo, T., 148
Dickey, D., 27, 234
Dickinson, B., 135
Diggle, P., 169
Distaso, W., xxiii
Dolan, C., 169
Doornik, J., xxiii
Dufour, J., 92
Dunsmuir, W., 126, 218
Durbin, J., xxiv, 31, 48

Engle, R., 1, 74, 76, 120, 121, 228, 234
Evans, P., 66

Fabozzi, F., 66
Fahrmeir, L., 169
Fan, J., 66
Favoreel, W., 85
Feibes, W., 165
Fernandez, R., 165
Fisher, R., 212
Flores, R., 68, 92
Fong, D., 169
Francis, J., 66
Francq, C., 92
Franses, P., 68, 71
Fuller, W., 27, 234

Garcia–Ferrer, A., 89, 135
Garcia–Hiernaux, A., 3, 85, 90–92, 95, 99, 132, 134, 239, 240
Gevers, M., 13, 39, 40, 204–207
Ghysels, E., 68
Glover, K., 138
Golub, G., 23, 48, 195, 197
Gomez, V., 48, 59
Goodrich, R., 169
Goodwin, G., 13, 39, 40, 204–207
Granger, C., 153
Grether, D., 141

Hackl, P., 66
Hallin, M., 72
Hamaker, E., 169
Hamilton, H., xxiii
Handcock, M., 183, 185
Hannan, E., xxiii, 89, 91, 131, 154, 217
Harrison, P., xxiii
Harvey, A., xxiii, xxiv, 1, 6, 7, 14, 34, 35, 44, 49, 108, 117, 135, 141, 154, 224, 236
Harville, D., 215
Hastie, T., 66
Hewer, G., 204, 205
Higgins, M., 66
Hillmer, S., 48, 60, 108, 141
Ho, M., 169
Hodrick, R., 14, 44, 117, 141, 224
Hsiao, C., 169
Hsing, Y., 66

Ionescu, C., 14, 204, 206

Jamshidian, M., 213
Jansen, R., 169
Jazwinsky, A., xxiii, 35
Jenkins, G., xxiii, 63, 95, 100, 121, 125, 159, 234
Jennrich, R., 213
Jerez, M., 1, 3, 13, 20, 24, 37, 51, 52, 70, 85, 90–92, 95, 99, 112, 132, 134, 150–153, 236, 238–240
Jian, L., 91
Jones, R., 31, 169, 175

Kailath, T., 135
Kalman, R., 34
Kauermann, G., 66
Kim, B., 66
Kohn, R., 1, 34, 36, 47–49, 235
Konstantinov, M., 11, 23, 48, 201, 206, 207
Koopman, S., xxiii, xxiv, 48, 49, 58
Koreisha, S., 92
Kozicki, S., 120, 121
Kroner, K., 76
Kung, S., 87

Laird, N., 171, 175

Larimore, W., 87, 88
Laub, A., 204, 206
Lee, S., 66
Li, W., 135
Lin, A., 154, 165
Lisman, J., 165
Litterman, R., 165
Little, T., 169
Ljung, L., 3, 62, 86, 90, 104, 121, 143, 218, 238, 239
Lucas, R., 66
Lutkepohl, H., xxiii, 69, 78, 92, 131, 150, 154, 214–216

MacKinnon, J., 212, 213
Makridakis, S., 100
Maravall, A., 48, 59
Marcellino, M., 153, 154
Margaritis, D., 66
McGee, V., 100
Mehra, R., 179
Melard, G., 131
Moore, J., 216
Moore, J., xxiii, 3, 12, 30, 35, 36, 47, 48, 76
Morf, M., 135

Naranjo, A., 1, 34, 175, 181
Nelson, D., 74, 228, 234
Nerlove, M., 141
Newbold, P., 99
Nicholls, D., 66
Novales, A., 68
Nunes, L., 31, 158

O'Donovan, T., 100
Oara, V., 14, 204, 206
Oh, K., 66
Osborn, D., 68
Oud, J., 169

Paap, R., 71
Pagan, A., 1, 214
Paige, R., 99
Palda, K., 123
Pappas, T., 204, 206
Park, B., 66
Park, S., 66
Peña, D., 59
Pelagatti, M., xxiii
Pelletier, D., 92
Penzer, J., 236
Peternell, K., 89
Petkov, P., 11, 23, 48, 201, 206, 207
Petris, G., xxiii
Petrone, S., xxiii
Pierse, R., 153
Poskitt, D., 92, 131
Prescott, E., 14, 44, 66, 117, 141, 224
Pukkila, T., 92

Qin, S. J., 3
Quenneville, B., 31
Quinn, B., 66, 89, 91

Reinsel, G., 60, 95, 100, 125, 130, 234
Reisen, V., 89
Rosenberg, B., 36, 66
Rosenbrock, H., 10, 21, 109, 116
Roy, R., 131
Rubin, D., 175
Ruiz, E., xxiv

Saidi, A., 131
Sandell, N., 204, 206
Schnabel, R., 46, 238, 239
Schwarz, G., 89, 91
Schweppe, F., 47
Selukar, R., xxiii
Serrano, G., 92
Shephard, N., xxiii, xxiv
Shumway, R., 169, 175, 246
Siklos, P., 153
Simonoff, J., 183, 185
Sin, K., 204, 206
Smith, J., 68
Snell, A., 153
Soderstrom, T., 89
Song, L., 99
Sorelius, J., 89

Sotoca, S., 1, 13, 20, 24, 34, 37, 51, 52, 56, 70, 112, 150–153, 236, 238, 239
Spliid, H., 92
Stewart, G., 200
Stock, J., 66
Stoffer, D., 169, 175, 177, 217, 246
Stoica, P., 89, 91
Stram, D., 151–154
Swamy, P., 66

Tavlas, G., 66
Terceiro, J., 1, 3, 15, 24, 30, 34, 46, 179, 238, 239
Tiao, G., 60, 61, 89, 108, 141, 157
Tibshirani, R., 66
Tong, H., 42, 140
Toscano, E., 89
Trimbur, T., 14, 44, 141, 224
Trindade, A. A., 1, 34, 99, 175, 181, 218
Tsay, R., 89, 90
Tusell, F., xxiii
Tutz, G., 66, 169

Van Dooren, P., 204, 206
Van Loan, C., 23, 48, 195, 197
Van Overschee, P., 20, 56, 84–88
Verhaegen, M., 87

Wagner, M., 99
Wald, A., 212
Wall, K., 217
Ware, J., 171
Watson, M., 1, 66
Wei, W., xxiii, 98, 100, 150–154, 234
Weiss, M., 14, 204, 206
West, M., xxiii, 114
Wheelwright, S., 100
White, H., 238, 239
Wickramasinghe, R., 99, 218
Willem, J., 138
Wold, H., 100
Wood, E., 89
Woodside, C., 89
Wu, R., 154

Yan, S., 91
Yao, Q., 66
Young, P., 89, 135

Zakoian, J., 92
Zhang, W., 66
Zhao, G., 66

Subject Index

Aggregation, 145–163
algorithm, 154, 158, 160
constraints, xxiv, 1, 28, 30, 148, 150, 154
of model, xxiv, 2, 150–154, 158
of time series, 2–4, 30, 31, 34, 146–150, 152, 153
partial, 148–150, 154
Airline
model, 227, 240
Airline passengers
Airline dataset, *see* Datasets
Asymptotic properties, 4, 12, 46, 90, 168, 169, 174, 211–222
efficiency, 213
information matrix, 212–216, 222
normality, 213, 216, 219
unbiasedness, 213
Autocorrelation function, 64, 74, 121, 135, 143, 162, 233
Autoregression(s), 2, 17, 23, 24, 27, 28, 41, 43, 64, 67–69, 79, 89, 118, 121, 139, 140, 152, 154, 157, 159, 226–227, 229, 232, 235
Autoregressive Conditional Heteroscedastic (ARCH)
ARCH, 74
GARCH, 3, 65, 74, 79, 80, 92, 207–208, 228–230, 234, 236, 238, 239
IGARCH, 76
Autoregressive Distributed lag (ADL), 165
Autoregressive Integrated Moving Average (ARIMA)
ARIMA, xxiii, xxiv, 23, 25, 27, 29, 57, 79, 117, 118, 121, 140–142, 152, 182, 227–228, 233–235, 241
ARIMAX, 1, 21, 135, 136, 143
ARMA, 16, 139, 140, 152, 154, 173, 182, 183, 218
ARMAX, 185, 186, 188

Backshift operator, 15, 226
Bootstrap, 217

Canonical
correlation, 89–91, 197–198, 203
representation, 11, 92, 130–132, 135, 138, 153
Variate Analysis (CVA), 87, 90
Capital Asset Pricing Model (CAPM), 76, 77
Cointegration, 7, 52, 63, 64
matrix, 99
rank, 99
Consistency
of estimates, 74, 84–86, 91, 140, 212, 240
of models, 4, 146, 150, 154, 158–160
Controllability, 9–11, 21, 48, 55, 86, 109–111, 219
matrix, 86, 88, 113
Coprime, 130
Cramer-Rao lower bound, 213, 218
Cross-correlation, 64, 121, 125, 235
Cross-section(al), 4, 166–189, 211, 218–222
Cycle
economic, 44

seasonal, 3, 7, 27, 59, 68, 115, 146, 226, 227, 231, 232, 236
stationary, 2, 3, 7, 23, 24, 43, 59, 107–109, 114–116, 121–123, 140

Datasets
Airline passengers, 57–60, 95–98
Beluga whales, 183–189
Deutsch Mark and Japanese Yen returns, 79–80
GE shares and SP500, 76–77
Housing starts and sales, 60–64
Lydia Pinkham sales and advertising, 123–127
Mare ovarian follicles, 182–183
Spanish VAI and quarterly IND, 158–163
Spanish wheat prices, 120–123
US quarterly GDP, 44, 140–142
US short-term interest rates, 99–104
West German consumption, 78–79
Wolf sunspots, 42–43, 139–140
Decomposition (of a matrix), *see* Factorization
Decomposition (of a model), 2, 19, 117, 236
deterministic-stochastic, 2, 19, 20, 25, 31, 108, 114, 126
of model, 236
of prediction error, 214, 235
structural, 6, 7, 14, 24, 34, 49, 59, 107, 109, 110, 114, 116, 121, 123, 154, 157
Detectability, 11, 13, 39, 40, 152, 153, 205–207
Diagonalization (of a matrix), *see also* Decomposition (of a matrix)
block-diagonal form, 23, 31
block-diagonalization, 2, 23, 24, 52, 108, 110, 114–118, 121, 201–202
Diffuse, 50
component, 49, 51
initialization, 36, 48, 111
Kalman Filter, 49, 58
likelihood, 49–51
states, 50, 51
Disaggregation, 145–163
of model, xxv, 236
of time series, 4, 152, 153, 158–160, 162

EM Algorithm, 46, 175, 176, 186
Endogeneization, 2, 6, 24, 26, 31, 179–181, 230
Exogenous variables, 3, 6, 7, 12, 15, 25, 35, 40, 45, 49, 54, 55, 73, 74, 108, 113, 167–170, 172–174, 179–181, 185, 214, 218, 227, 230, 237

Factorization (of a matrix)
Cholesky, 73, 199
Jordan, 23, 52, 151, 152
QR, 193–194
Schur, 23, 48, 194–196
Singular Value Decomposition (SVD), 83, 87, 88, 90, 197
Filter, 3, 76, 120
diffuse, 49, 58, *see* Diffuse filter
Hodrick-Prescott, 14, 44, 117, 141, 142
information, 36, 48
Kalman, 3, 14, 34–39, 41, 42, 47–49, 52, 54, 56, 67, 72, 73, 75, 76, 93, 94, 167, 169–171, 175, 176, 179, 186, 188, 189, 208–210
Forecasting, 168, 169, 223, 230, 236–238, 241
filtering, 2–4, 33, 34, 36, 40, 41, 49, 52, 162

interpolation, xxiv, 4, 145, 146, 163
out-of-sample, xxiv, 4, 14, 34, 35, 59, 91, 108, 145, 146, 150, 157, 158, 160, 162, 163, 183–189
retropolation, 163
smoothing, 2–4, 33, 34, 36–38, 40–42, 77, 146, 153, 160

Gaussian, xxiv, 35, 54, 74, 162
errors/innovations, 47, 117, 203, 214, 216, 218, 219
likelihood, 3, 43, 45–47, 50, 65, 72, 74, 92, 93, 126, 143, 146, 169, 217, 219, 235, 238
process, 174, 216
Generalized Least Squares (GLS), 3, 92, 111, 112, 217
Gradient vector, 175, 238, 239, 241

Hessenberg, *see also* Factorization (of a matrix)
form of matrix, 195–196
Schur algorithm/procedure, 196–197, 200
Hessian matrix, 175, 212–216, 219–221, 240
Heteroscedasticity, 90, *see also* ARCH
conditional, 79, 228–229
of errors, 65, 66, 74, 237
Householder reflection matrix, 194, 196

Identifiability, 97, 105, 214, 219
global, 112, 129, 130, 135, 138, 211
local, 4, 11, 50, 138, 211, 213
overidentifiability, 138, 140, 142
underidentifiability, 138
Immemorial time assumption/hypothesis, 11–13, 48, 53, 55–57
Information criterion, 89, 90
AIC, 89, 91, 97, 98, 104
HQ, 89, 91, 97, 98, 104
SBC, 89, 91, 97, 98, 104
Information matrix, 3, 89, 104, 238, 241, *see also* Asymptotic properties
Innovations, 14, 84, 86, 92, 93, 111, *see* Residuals, *see also* Gaussian
Irregular component, 2, 3, 7, 107–109, 114, 116, 118, 123, 160, *see* Residuals

Kalman Filter, 3, 14, 34–39, 41, 42, 47–49, 52, 54, 56, 67, 72, 73, 75, 76, *see* Filter

Least Squares (LS), 3, 42, 84, 86, 88, 89, 92, 172, 186, 202–203, 217
Likelihood, xxiv, 3, 45–80, 91, 94, 140, 142, *see also* Gaussian
diffuse, 49–51, 58, *see* Diffuse likelihood
Gaussian, xxiii, 3, 4, 14, 34–36, 43, 45–50, 52, 54, 56, 57, 59, 65, 67, 68, 72–75, 77, 79, 146, 157, *see* Gaussian likelihood
maximum, 92, 95, 97–99, 104, 141
minimally conditioned, 49, 51, 52, 58, 73
ratio, 140
Longitudinal, 4, *see* Cross-section(al)
Luenberger Canonical Form (LFC), 113, 114, 133, 134
Lyapunov equation(s), 198–199, 201, 205–206

MEM (multiple error model), 5, 12–14, 19, 24, 29, 30, 33–35, 42, 45, 50, 54, 67, 84, 94, 116, 117, 120, 153, 168, 209, 214, 224
Minimality, 10, 21, 70, 71, 86, 108–110, 116, 132, 133, 146, 152, 157, 159, 205, 219

Missingness
in inputs/covariates, 31, 178–182, 230
in outputs/responses, 176–182, 238
in values, xxiv, 1–4, 28–31, 34, 52, 59, 60, 120, 121, 123, 125, 135, 158, 162, 168, 169, 176, 183–189, 235–237, 239

Nesting (models), 230–233
in errors, 230, 232–233
in inputs, *see* Endogeneization
NIDC (*n* identification criterion), 91, 97, 98, 104
Nonstationary, 9, 12, 36, 37, 48, 52, 53, 55–59, 73, 150, 157, 159, *see* Stationary
Normal, 35, 47, 54, 74, 162, *see* Gaussian

Observability, 4, 9–11, 21, 30, 51, 70, 146, 150–153, 219
indices, 130, 134
matrix, 85–88, 132–134
Observation errors, xxiv, 1, 2, 28, 29, 31

Panel, 4, 31, *see* Longitudinal
Periodic models, 3, 47, 65, 68–71, 78, 79, 239
Portmanteau statistics, 62, 64, 159
Prediction, 4, 150, 163, *see also* Forecasting
accuracy, 157, 158, 183
error decomposition, 46, 47, *see* Decomposition (of a model)
errors, 47, 171, *see also* Innovations
horizon, 183, 237
minimum prediction error, 170–171, 217, 219
of missing data, *see* Forecasting
Projection
operator, 86, 170, 208
orthogonal, 93, 203
Propagation, 39, 57, 72, 75
of Kalman Filter, 35, 47, 54, 72, 93, 175, 207
of missing values, 59
of system equations, 8, 13, 34, 36, 52, 94, 111, 113, 114, 147, 205
PVARMAX, 65, 68, 69, 71, *see* VARMAX

Random walk, 8, 14, 67
Reduced rank least squares, 86, 202–203
Regression, 3, 7, 8, 42, 65, 66, 76, 86, 91, 92, 136, 154, 182, 186, 202, 229–230
Residuals, 60, 84, 88, 90, 95–97, 104, 125, 157, 162, 174, 217, 235, 237, *see also* Innovations
Riccati equation, 13, 14, 35, 39, 40, 52, 72, 75, 76, 203–208
convergence of, 40, 204, 205
Robust
model fitting criterion, 91, 217–218, 242
model information matrix, 238, 239
Roots (of model polynomial), 9, 13, 23, 27, 40, 41, 117, 125, 130, 140, 231
complex, 24, 43
unit, 41, 45, 50, 62, 76, 95, 112, 153, 154, 157, 159, 205, 207, 229

Seasonal, 2, 4, 7, 27, 44, 57, 59, 61–63, 68, 78, 94–98, 107–109, 114–118, 124, 127, 141, 146, 147, 150–154, 157, 160, *see* Cycle
SEM (single error model), 5, 12–15, 21–23, 25, 26, 33, 35, 39–41, 52, 69, 72, 74, 84, 88–90, 94, 107–110, 116–118,

120, 121, 132–134, 146, 147, 153, 205, 224, 228, 231

Signal, 151
- extraction, xxiv, 3, 31, 38, 107, 109, 116, 160, 204, 223, 233, 235–238, 240
- to noise ratio, 44
- to-noise, 141

Singular Value Decomposition (SVD), *see* Factorization (of a matrix)

SISO (single-input single-output), 16, 20, 25

Smoothing, 2, 3, 235–237
- fixed-interval, 3, 4, 33, 34, 36–38, 40–44, 77, 146, 152, 153, 160, 236
- Kalman, *see* Filter

Stability, 242
- of model, 9, 11, 12, 47, 48, 50, 52, 55, 199
- stabilizability, 11, 204–206

Staircase algorithm, 10, 21, 109, 110, 116

Stationary, 9, 11, 12, 23, 35, 37, 50–55, 57–59, 67–69, 71–73, 79, 86, 99, 112, 121, 123, 150, 154, 157, 162, 173, *see* Nonstationary

Steady-state, 9, 14, 35, 204

Structural Time Series Model (STSM), xxiv, 1, 3, 6, 117, 118, 120, 154, 226–228, 237

Sunspots dataset, *see* Datasets

SVC (singular values criterion), 90, 97, 98, 104

SVD, *see* Factorization (of a matrix)

Transfer function, xxiv, 1, 2, 16, 20, 25, 59, 109, 114, 125, 154, 157, 226–228, 230

Trend, 2, 3, 7, 14, 44, 59, 107–109, 114–118, 121–123, 137, 141, 160, 185, 216, 236, 237

Unidentifiable, *see* Identifiability

VARMAX, xxiv, 1, 2, 15, 26, 64, 68, 69, 75, 80, 92, 129–135, 138, 139, 141, 153, 154, 157, 169, 223, 226–230
- form for state space model, xxv, 1, 3, 4, 11, 14, 21
- Periodic (PVARMAX), 65, 68–71, 73
- VAR, 17, 79, 157
- VARMA, 22, 26, 71, 92, 99, 104, 131, 154, 235, 241
- VARX, 71, 72
- VMA, 157, 160

Volatility, 74

White noise, 6, 15, 26, 28, 29, 43, 56, 62, 64, 67, 86, 94, 117, 123, 139, 140, 157, 183, 216, 229

Printed in the United States
by Baker & Taylor Publisher Services